DEVELOPMENTAL BIOLOGY
RESEARCH IN SPACE

ADVANCES IN SPACE BIOLOGY AND MEDICINE

Volume 9

Series Editor

Augusto Cogoli

Space Biology Group
ETH Technopark
Zurich, Switzerland

Developmental Biology Research in Space

Editor

Hans-Jürg Marthy

CNRS, Observatoire Océanologique
Banyuls sur mer, France

2003
ELSEVIER
Amsterdam – Boston – London – New York – Oxford – Paris – San Diego
San Francisco – Singapore – Sydney – Tokyo

ELSEVIER SCIENCE B.V.
Sara Burgerhartstraat 25
P.O. Box 211, 1000 AE Amsterdam, The Netherlands

First edition 2003

Library of Congress Cataloging in Publication Data
A catalog record from the Library of Congress has been applied for.

British Library Cataloguing in Publication Data
A catalogue record from the British Library has been applied for.

ISBN: 0-444-51353-1
ISSN: 1569-2574

♾ The paper used in this publication meets the requirements of ANSI/NISO Z39.48-1992 (Permanence of Paper).
Printed in The Netherlands.

Contents

Foreword of the Series Editor to Volume 9

I am writing my short remarks to this volume just a few days after Volume 8, dedicated to cell biology and biotechnology, has been published. I knew that Hendrik van Leusen and his team at Elsevier were doing their best for the success of the series, nevertheless I was pleased and surprised to see how well the work was done. After an interruption that lasted three years, mainly due to the transition from JAI Press – the publisher of the first seven volumes of ASBM – to Elsevier, the publication of the volumes of the series will proceed on a yearly base.

I was very happy when, one year ago, Hans-Jürg Marthy accepted with enthusiasm my invitation to be the editor of volume 9. I know Jürg since many years and it is always a great pleasure to run with him and other colleagues the ERASMUS and SOCRATES courses on Space Biology and Life in Space at the prestigious Oceanographic Institute in Banyuls sur Mer in South France where Jürg is working. I invited him to edit volume 9 because he is an internationally recognized expert in developmental biology as well as a veteran of space flight experimentation. Indeed he was able to collect excellent articles from nine world leaders in their respective fields of research and from one of the persons who mostly contributed as scientific coordinator at ESA to the success of our experiments in space.

I am sure that the content of this volume will be very useful and highly informative not only for the colleagues working in space and gravitational biology but also for all those interested in developmental biology in general.

Augusto Cogoli
Zürich, November 2002

Developmental Biology Research in Space: Introductory Remarks of the Volume Editor

Hans-Jürg Marthy*

CNRS, Observatoire Océanologique, Banyuls sur mer, France

This book on Developmental Biology Research in Space is specifically dedicated to Dr. med Heinz Oser, former chief scientist in Life Sciences of the Microgravity Office at the European Space Agency ESA. Dr. Oser was deeply convinced of the prime importance of developmental biology studies in altered gravity conditions and strongly supported the editor of this volume and the scientific community active in this research field.

The research field termed "Developmental Biology" comprises quite a number of scientific disciplines, each of them with many interdependances and thus the term may have different meanings. In a broad sense, research in Developmental Biology certainly means research activities which address overall biological aspects of embryonic development (behavioral, anatomical/morphological, physiological studies in developing and growing organisms). It also encompasses descriptive and experimental investigations at the cellular and subcellular levels (cell proliferation, cell differentiation, cell–cell recognition, cell interactions, cell communications and cell migration in embryonic tissues, apoptosis etc). This "classic" conception of Developmental Biology becomes increasingly overlapped, if not replaced, by a "molecular oriented Developmental Biology", which focuses on genes, gene expression, mechanisms of cell cycle regulation, inter- and intracellular signal transduction cascades, etc. With respect to fundamental scientific interests, each orientation is, on principle, equally important; in practice, however, recent research programmes set priorities clearly towards molecular analyses of oo- and embryogenetic events. Developmental Biology research "in Space" will logically evolve in a similar way.

In this volume an attempt is made to give an overview on experiments performed over the past 17 years, on the ground and in space, at different gravity levels and dealing with various developmental aspects in plant and

* E-mail: marthy@obs-banyuls.fr

animal models. Such an overview appears particularly in time since continued research on the International Space Station (ISS) becomes imminent.

Mary E. Musgrave, University of Connecticut, and her collaborator Anxiu Kuang review plant reproductive development during spaceflight, in well studied *Arabidopsis, Brassica* and *Triticum.* Whereas androecium and gynoecium development occur normally in microgravity in that functional propagules are produced, qualitative changes in anther and pistil development are reported as well as in storage reserve deposition during seed development. Consequences of diminished seed quality when plants are grown in the absence of gravity will evidently detract from the utility of plant-based life support systems.

Claudia Ricci and her collaborator Chiara Boschetti of the University of Milan present a not (yet) well known animal system, from which new insights are expected as to the role of gravity in developmental processes: the rotifers. The simplicity of the system (these multicellular animals are built up by about 1000 cells and can easily enter dormancy) may allow to disentangle complex problems much easier than in organisms built up by millions of cells. In view of the experiments to be performed in a near future on the ISS, a number of ground-based experiments under increased and decreased gravity conditions have been performed on the model rotifer *Macrotrachela quadricornifera.* Eggs experiencing up to $20 \times g$ do not show permanent changes in the embryo's morphology.

Roberto Marco, Autonomous University of Madrid, is a pioneer in space biology and as such he is especially well placed to remember some important experiments performed since 1961 onwards. However, his contribution is not limited to a simple description of experiments (including those done by his own with the *Drosophila* system), but extends largely beyond by "looking into the future". He presents and discusses in depth technical advances in view of the performance of long-term multi-generation experiments on the ISS for which the development of new hardware is a prerequisite. Theoretical considerations as to a future "evo-devo-biology" in Space, including even an outlook towards problems of colonization of other planets are additional stimuli within this chapter.

Akemi Izumi-Kurotani and Masato Kiyomoto, our Japanese colleagues, present a series of experiments with newts (*Cynops phyrrogaster*), termed "Astronewt", in which they have been involved. They claim that for a thorough knowledge of the role of gravity in morphogenesis, investigations on gravitational effects are necessary on single embryonic cells and therefore suggest a new experimental system with isolated micromeres of sea urchin embryos. The study is in progress.

Christian Dournon, from the French University of Nancy, reviews various experiments performed with newts and, in particular, with the salamander model *Pleurodeles waltl.* The advantages of urodeles for space flight experiments are quite convincing. For instance: mated females keep spermatozoa in special glands for many months and thus, eggs can be laid in the absence of males; the rate of development is slow and their maintenance is possible in a moisturized environment, etc. Despite the fact that development results by normal larvae, early development in microgravity is characterized by some abnormalities such as disturbed cortical cytoplasmic movements, decrease of cell adhesion and/or loss of cells. It is also reported that space flights have clear effects on organs in regeneration.

Eberhard R. Horn, gravitational physiologist at the German University of Ulm, reports on the development of gravity related reflex behavior in lower vertebrates (*Xenopus laevis*, *Oreochromis mossambicus*) and insects (*Acheta domisticus*, *Gryllus bimaculatus*). In fact, gravity related behavior and the underlying neuronal networks appear as most suitable model systems to study basic effects of altered gravitational input on the development of neuronal systems. Based on experiments using spaceflights, centrifugation, lesion and loading or deloading, all species studied so far reveal a significant susceptibility to modifications of the gravity sensory input during development. Nevertheless, since clear arguments for the existence of a critical period in the development of the sense of gravity is still missing, long-term experiments on the ISS will be particularly informative.

Ralf H. Anken, University of Stuttgart-Hohenheim in Germany, contributes with a review on "neurophysiology of developing fish at altered gravity". Special emphasis focuses on effects of altered gravitational conditions on the developing brain and vestibular system, comprising investigations on behaviour and plastic reactivities of the brain and inner ear. Clues and insights into the possible basic causes of space motion sickness-phenomena (a kinetosis) are further provided. Short- and long-termed altered gravity have different effects. Whereas the former (one day) can induce transitional aberrant behavior, in the latter case (several days) feedback mechanisms between the central nervous system and the vestibular organs are activated for the regain of normal behavior.

Kenichi Ijiri, University of Tokyo, is known by the scientific community for his experiments in space with the medaka fish (*Oryzias latipes*) as well as his public contributions due to his video film "The first Vertebrate mating in Space". In this book, he presents life-cycle experiments to be performed in the near future aboard the ISS. Various ground-based studies using different strains of mutant fish are progressing. It is conceivable indeed, that fish are the most likely candidates among Vertebrates to live their whole life cycle aboard the station and suitable housing facilities and other hardware are at an advanced conceptual level or already under construction.

April E. Ronca works at the NASA Ames Research Center. She reviews the current state of our knowledge of spaceflight effects on developing mammals (rodents). A particular focus is on the fetal and early postnatal life, the developmental periods for which the greatest amounts of spaceflight data have been amassed. During these early formative phases, the so-called "maternal-offspring system" (coordinated interaction between mother and young) is of primary importance. There is good reason to assume that such biologically meaningful interactions between mothers and offspring are changed in weightlessness. The findings support the view that the environment within which mammals develop is of paramount importance in interpreting spaceflight effects.

Enno Brinckmann from ESA gives, last but not least, an overview on new facilities and instruments for developmental biology research in Space. He presents *Biopack* for the Space Shuttle and two new instruments for the ISS, the *Biolab,* which will be integrated into the European "Columbus" Laboratory and the *EMCS* (European Modular Cultivation System) to be housed in the US Lab "Destiny". The highly automated Biolab and EMCS facilities will provide a dedicated atmosphere (controlled CO_2, O_2, relative humidity and trace gas removal) as well as power and data lines. Design concepts are also developed for experiments with small eggs, cells and tissues, small aquatic animals, insects and plants.

I would like to thank all the authors, who made contributions to this volume 9 of the ASBM series, particularly dedicated to Developmental Biology. The cooperation was a real pleasure for me. My thanks also to the Editor of the series, Dr. Augusto Cogoli, who invited and encouraged me to edit this book. The French space agency CNES is acknowledged for its year-long support of my experimental work. Finally, I wish to thank the Elsevier Team for the careful publishing work.

Complementary reference

Marthy, H.-J. (2002) Developmental biology of animal models under varied gravity conditions: a review. Vie Milieu, 52 (4), 149–166.

Developmental Biology Research in Space
H.-J. Marthy (editor)

Plant Reproductive Development during Spaceflight

Mary E. Musgrave[1,*] and Anxiu Kuang[2]

[1]*Department of Plant Science, University of Connecticut, 1376 Storrs Road, Unit 4067, Storrs, CT 06269, USA*
[2]*Department of Biology, University of Texas Pan American, Edinburg, TX 78539, USA*

Abstract

Reproductive development in microgravity has now been studied in a variety of plants; *Arabidopsis*, *Brassica*, and *Triticum* have been especially well studied. Earlier indications that gravity might be required for some stage of reproductive development have now been refuted. Nevertheless, the spaceflight environment presents many unique challenges that have often compromised the ability of plants to reproduce. These include limitations in hardware design to compensate for the unique environmental characteristics of microgravity, especially absence of convective air movement. Pollen development has been shown to be sensitive to high concentrations of ethylene prevailing on various orbital platforms. Barring these gross environmental problems, androecium and gynoecium development occur normally in microgravity, in that functional propagules are produced. Nonetheless, qualitative changes in anther and pistil development have been shown, and significant qualitative changes occur in storage reserve deposition during seed development. Apart from the intrinsic biological importance of these results, consequences of diminished seed quality when plants are grown in the absence of gravity will detract from the utility of plant-based life support systems. By understanding gravity's role in determining the microenvironments that prevail during reproductive development, countermeasures to these obstacles can be found, while at the same time providing basic knowledge that will have broader agricultural significance.

Introduction

Although plants have been included in the biological payloads of orbital platforms for decades, we are only beginning to understand how the absence of

* E-mail: mary.musgrave@uconn.edu

gravity impacts long-duration growth and reproductive development. Because of the well-known tropic responses of plant organs to gravity, the utility of the microgravity environment for studying signal transduction in plants has been appreciated (Antonsen and Johnsson, 1998; Kiss *et al.*, 2000; Perbal and Drissecole, 1994; Volkmann *et al.*, 1986). However, as the Soviets tried to use plants to supplement the psychological environment of space craft, as well as give variety and fresh food to astronauts engaged in long-duration missions, it became clear that the absence of gravity posed problems for plant reproductive development (Nechitailo and Mashinsky, 1993).

This review will discuss the obstacles to plant reproductive development in microgravity. Historically, these have included hardware design issues, since early plant hardware design did not provide an adequate environment to support plant growth, and plants died at the transition from vegetative to reproductive development (Halstead and Dutcher, 1984, 1987; Nechitailo and Mashinsky, 1993). In more recent studies, better hardware design has permitted an examination of developmental obstacles to reproduction imposed by microgravity. Reproduction is a complex developmental process. Not only must the vegetative portion of the plant be sufficiently vigorous to support this energetically demanding process, but all of the components of the process must function well. Megasporogenesis, microsporogenesis, pollination, pollen tube growth, fertilization, embryo development, and seed maturation are all complex events. This review will show that while none of these processes is absolutely dependent on gravity for its completion, environmental constraints imposed by microgravity have been implicated in qualitative changes in these processes.

Hardware for reproduction studies

More than 10 different types of plant hardware have been used in the microgravity environment to study some aspect of plant reproductive development (Table 1). On Kosmos 1129, a simple Plexiglas beaker containing moist soil had been taken to orbit containing flowering *Arabidopsis* plants, which subsequently produced seeds while in microgravity (albeit with only 55% fertility; Parfenov and Abramova, 1981). On Salyut 6, two open plant growth chambers, Oasis, and Malachite, were used to grow garden pea and *Epidendrum* orchid, respectively (Nechitailo and Mashinsky, 1993). In the case of peas growing in the Oasis module, plants were started from seeds in microgravity, but did not develop into flowering plants. Similarly, orchids taken to Salyut in flower in the Malachite hardware produced no additional flowers while on orbit.

Building on the success of work with the small plant *Arabidopsis*, Ukrainian scientists tried to grow it in a hermetically sealed hardware called Svetoblok (Fig. 1a). Designed to clip under a light on the Salyut 6 station, Svetoblok provided a low light environment for *Arabidopsis* plants growing on an agar medium. Plants were developmentally delayed, but eventually flowered although the androecium and gynoecium aborted (Kordyum *et al.*, 1983).

Table 1

Experiments on plant reproduction during spaceflight have utilized a variety of plant materials and growth chambers. The results of these experiments have been reviewed elsewhere (Halstead and Dutcher, 1984, 1987; Musgrave *et al.*, 1997; Nechitailo and Mashinsky, 1993)

Starting material	Chamber	Ventilation	Flowers	Seeds	Reference
Plants or seedlings					
Arabidopsis	(Kosmos 1129)[a]	Open	+	+	Parfenov and Abramova, 1981
	PGU[c] (STS-54)	Closed	+	–	Kuang *et al.*, 1995
	PGU (STS-51)	" + CO_2	+	–	Kuang *et al.*, 1996a
	PGU (STS-68)	active	+	+	Kuang *et al.*, 1996b
Brassica	PGF[b] (STS-87)	active	+	+	Kuang *et al.*, 2000a
	BPS[c] (ISS)	active	+	+	Morrow, pers. comm.
Epidendrum	Malachite (Salyut 6)	open	–[d]	–	Nechitailo & Mashinsky, 1993
Seeds					
Arabidopsis	Svetoblok (Salyut 6)	closed	+	–	Kordyum *et al.*, 1983
	Phyton (Salyut 7)	passive	+	+	Merkys and Laurinavicius, 1983
	Advanced Astroculture[d]	active	+	+	Stankovic *et al.*, 2001
Pisum	Oasis (Salyut 6)	open	–	–	Nechitailo and Mashinsky, 1993
Brassica	Svet (Mir)	open w/fan	+	+	Musgrave *et al.*, 2000
Triticum	Svetoblok M (Mir)	passive	+	–	Mashinsky *et al.*, 1994
	Svet (Mir)	open w/fan	+	–	Strickland *et al.*, 1997
	Svet (Mir)	open w/fan	+	+	Bingham *et al.*, 1999

[a]The hardware in this case was an open beaker containing moist soil.
[b]Plant Growth Facility.
[c]Biomass Production System.
[d]Commercial hardware used on the International Space Station.

Fig. 1. Evolution in plant hardware design has made plant reproduction during spaceflight a reliable occurrence. (A) The first sophisticated hardware used to grow *Arabidopsis* in space was Svetoblok (on Salyut 6), a small, hermetically sealed plant growth chamber that could be clipped on to a light in the cabin. The metal cylinder would snap over the chamber for protection during transport back to Earth. Low light and lack of gas flow within the chamber were probably responsible for the reproductive failure reported (Kordyum *et al.*, 1983), as absence of convective air movement in microgravity makes such unventilated chambers problematic for long-duration growth of plants. (B) Mature *Arabidopsis* plants were grown in the Advanced Astroculture™ unit on the ISS in 2001. The entire root tray is visible. Photo courtesy of Bratislav Stankovic, Wisconsin Center for Space Automation and Robotics (WCSAR). (C) *Brassica* and wheat were grown in the Biomass Production System on the ISS in 2002. The BPS is the precursor for the Plant Research Unit, a habitat designed for use in the planned Centrifuge Accommodation Module of ISS. (D) Images acquired during flight were relayed to the ground in real time, making daily monitoring of reproduction by *Brassica* in the BPS possible. Photos C and D courtesy of Robert Morrow, Orbital Technologies, Inc., Madison, WI.

The first successful plant life cycle in microgravity occurred on Salyut 7 in a small, ventilated plant growth chamber called Phyton. *Arabidopsis* seeds sown in microgravity on an agar surface by a "seed shooter" grew into plants that eventually flowered and produced seeds themselves (Merkys and Laurinavicius, 1983).

Attempts to design hardware to support more vigorous plant growth resulted in a variety of units. Bulgarian-made Svet provided a well-ventilated environment with a fan that drew cabin air through the unit. Subsequent upgrades resulted in brighter lights, a root module with regulated water injection, and a gas exchange module that permitted measurement of rootzone moisture and canopy carbon dioxide gas exchange. Using Svet,

Bingham *et al.* (1996) showed that a persistent problem in plant hardware design is the tendency to overwater in microgravity, since a dose of water on the ground resides lower in the root matrix than it does in microgravity. Using moisture sensors that reported actual water status in the root matrix volume, they were able to grow plants full term in Svet. Wheat (Bingham *et al.*, 1999) and *Brassica* (Musgrave *et al.*, 2000) would be grown from seed to maturity in this unit prior to the deorbiting of the Mir space station in 2000.

Units with more modestly sized head spaces were developed for use in the shuttle mid-deck. Of these, the Plant Growth Unit and Plant Growth Facility both supported studies on reproductive development in *Arabidopsis* (Kuang *et al.*, 1995, 1996a,b) and *Brassica* (Kuang *et al.*, 2000a), respectively. Individual small chambers (5 or 6) within the units each supported six pre-grown plants through flowering and early reproductive development over periods of up to 16 days on orbit.

Two plant growth hardware units originally designed for the shuttle are currently in use for long-duration plant growth studies on the International Space Station. The Advanced Astroculture Unit (Fig. 1b) grew *Arabidopsis* through to seed set on ISS in 2001 (Stankovic *et al.*, 2001). In 2002 the Biomass Production System (Fig. 1c) supported growth by the larger plant, *Brassica* (Fig. 1d), through seed production. The BPS is the precursor design for the Plant Research Unit hardware that will be housed in the Centrifuge Accommodation Module on ISS at assembly complete (Morrow *et al.*, 2001).

Transition from vegetative to flowering stage

As chronicled in the previous section, researchers interested in plant reproductive development in microgravity were initially unable to confirm any specific role for gravity in plant reproduction because of general problems in the growth of plants in space. The plant growth over extended periods that is necessary for reproduction to occur was not possible, and plants frequently died in the transition from vegetative to reproductive stage (Table 1). Even with experiments with the small plant *Arabidopsis*, which has among the lowest light and nutrient requirements of any higher plant, delays in transition to the reproductive stage were the norm in microgravity.

We conducted a series of experiments to determine if the problem with plant reproduction in space was just a consequence of overall poor plant growth in microgravity, or if there was a specific effect of microgravity on the reproductive apparatus per se. Using *Arabidopsis* plants that had been grown to the pre-flowering stage, we quantified the development of flowers during three experiments in microgravity: 6 (Kuang *et al.*, 1995), 10 (Kuang *et al.*, 1996a) and 11 (Kuang *et al.*, 1996b) days in duration on the space shuttle. Table 2 summarizes details of the three experiments, and the floral production in each on a per-plant basis. Unlike previous experiments in which flowering was delayed, these experiments, using plants at the pre-flowering stage at the

Table 2

Flower production in microgravity by *Arabidopsis* plants launched at the pre-flowering stage, compared with corresponding ground controls. Results of three experiments, of different durations, in the Plant Growth Unit

		Flowers per plant		
Flight #	Duration (days)	Flight	Ground	Configuration
STS-54	6	16.1	16.1	Closed chambers
STS-51	10	44.1	39.8	Closed + CO_2
STS-68	11	79.5	77.5	With AES[a]

[a]Air Exchange System, provided a slow exchange (90 ml/min) of chamber air with filtered air from the crew cabin.

time of introduction to the microgravity environment, yielded flower production at the same magnitude and schedule as in the ground control (Table 2 and Fig. 2).

Despite the equivalent rates of floral initiation in these three experiments, success of the transition to reproductive development was highly dependent upon other factors listed in Table 2 and Figure 3. In the experiment on STS-54, both androecium and gynoecium aborted at an early stage (Kuang *et al.*, 1995; and see pertinent sections below for details). Analysis of foliar material showed a significantly lower amount of carbohydrate in the spaceflight material (Musgrave *et al.*, 1998). In the next experiment on STS-51, supplementation of the closed headspace of the Plant Growth Chamber with high carbon dioxide (8000 ppm) raised the foliar carbohydrates, and early development of both androecium and gynoecium proceeded normally, however without pollination due to high humidity in the closed chamber (Kuang *et al.*, 1996a). In the third experiment on STS-68, a flow through ventilation system was used to maintain a supply of carbon dioxide to the foliage, resulting in higher concentrations of foliar starch. Again, androecium and gynoecium development proceeded normally, and the flow through system reduced humidity sufficiently to allow anther dehiscence, pollination, and subsequent embryo development (Kuang *et al.*, 1996b).

The important role played by the chamber environment in promoting successful transition from vegetative to reproductive development is illustrated in Figure 3. Note that the closed chambers in both the experiments on STS-54 and STS-51 resulted in significantly lower soluble carbohydrates in the spaceflight material, while the flow-through system in use during the experiment on STS-68 allowed comparable amounts of soluble carbohydrates to accumulate in the foliage in flight and ground control. Starch concentration in the foliage followed a similar pattern. Although the closed systems supported reproductive development in 1-*g*, lack of convective mixing in microgravity significantly slows the resupply of metabolic gases to the plant.

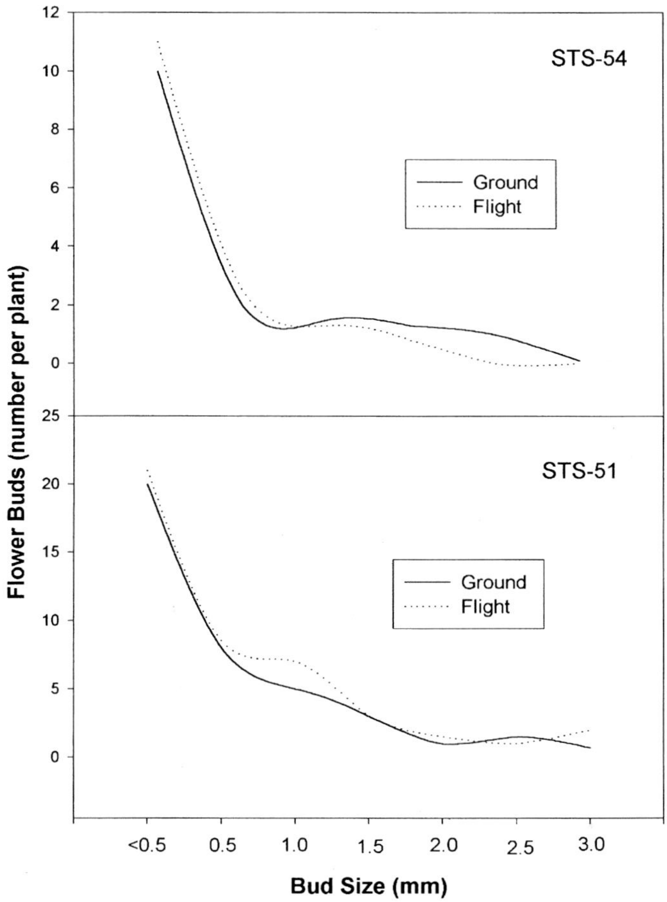

Fig. 2. Floral development timing was the same in spaceflight and ground control *Arabidopsis* plants as judged by the post-flight census of bud sizes in experiments on STS-54 (A) and STS-51 (B), respectively. Approximately twice as many buds were present after the 10-day exposure to microgravity (on STS-51) as following the 6-day exposure (on STS-54). Plants were at the pre-flowering stage at the time of launch.

Table 1 highlights the important role played by ventilation in fostering successful plant reproduction in microgravity (Musgrave *et al.*, 1997). Clearly, hardware providing an open or flow-through system is needed in order to permit successful transition from the vegetative to reproductive stage and subsequent seed development. As described in this section, carbohydrate status of the plant in transition from the vegetative to reproductive stage must be sufficient to support the energy demands. This is readily accomplished with an open system if the watering demands can be met but has been problematic in closed chambers in microgravity. Recently, Porterfield *et al.* (2000) and Kitaya *et al.* (2000) demonstrated the important role played by gravity in maintaining the small-scale gas transport necessary to sustain metabolic processes.

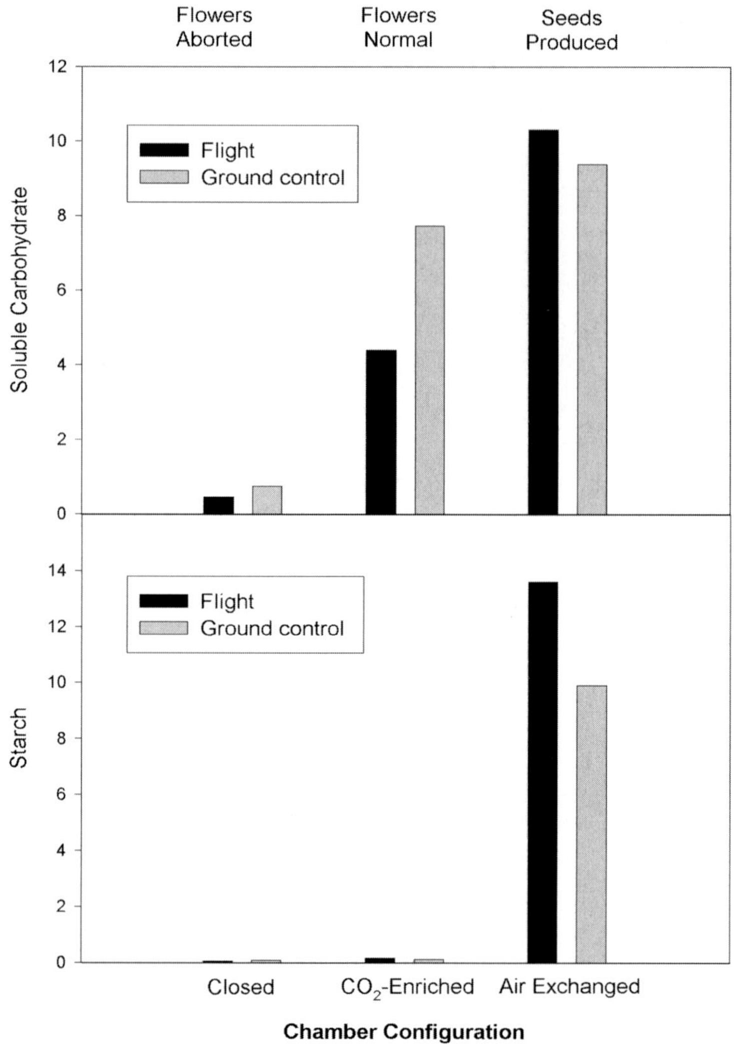

Fig. 3. A series of three experiments with *Arabidopsis* in the Plant Growth Unit on STS-54, STS-51 and STS-68 established a link between carbohydrate status of the foliage and successful transition to the reproductive stage in the spaceflight environment. Refer to Table 2 and the text for additional details. Based on data from Musgrave *et al.* (1998).

Anther development and pollen quality

In the series of experiments described above, many assays were employed post-flight to assess the success of every aspect of the reproductive process. Anther development aborted at an early stage in *Arabidopsis* during 6 days of spaceflight exposure on STS-54 (Fig. 4), providing very similar material to that

Fig. 4. Scanning electron micrographs of flowers of ground control (A) and spaceflight flowers formed during STS-54 (B). The pistil is shrunken and the anthers are collapsed due to aborted gynoecium and androecium development at an early stage. Bars = 100 μm. From Kuang *et al.* (1995), reproduced with permission of the American Journal of Botany.

observed by Kordyum *et al.* (1983) in *Arabidopsis* grown in Svetoblok. A pollen viability test, assessed by fluorescein diacetate staining (Heslop-Harrison *et al.*, 1984), indicated that there was no viable pollen produced in the spaceflight material. Young microspores were deformed and empty. At a late stage, the cytoplasm of the pollen contracted and became disorganized, even though the pollen wall developed and the exine appeared normal. The tapetum degenerated at an early stage (Kuang *et al.*, 1995).

Supplementation of the chamber headspace with carbon dioxide (on STS-51; Kuang *et al.*, 1996a), improved the plant carbon budget and allowed normal pollen development to occur in microgravity in *Arabidopsis*. Viability in this material was high, as assessed by fluorescein diacetate staining. In these closed chambers, high humidity prevailed. Pollen, although normally developed and fully functional, was evidently not released from the anthers, however, and a filmy material was observed in the tapetum. In the STS-68 experiment in which flow-through ventilation was provided, all phases of pollen development occurred normally, including release of the pollen from the dehiscent anthers (Kuang *et al.*, 1996b).

Normal pollen development during spaceflight was also initially elusive in wheat (Strickland *et al.*, 1997). Despite numerous attempts to grow SuperDwarf wheat through a complete life cycle in the Svet greenhouse on Mir, plants either failed to produce heads or else produced empty heads. Gas samples on the Mir space station revealed that high concentrations of ethylene (1.1–1.7 μmol/mol) prevailed, and a ground-based experiment confirmed that

SuperDwarf pollen development ceased prior to anthesis and anthers did not dehisce when grown under such high ethylene concentrations (Levinskikh *et al.*, 2000; Campbell *et al.*, 2001). Another cultivar, Apogee, was less sensitive to ethylene, and when grown on the Mir station produced wheat seed through several generations (Bingham *et al.*, 1999). Ethylene accumulated on Mir as a result of contributions from biogenic sources while at the same time lack of removal by the environmental control system.

Just as wheat cultivars vary in their sensitivity to ethylene, species are differentially affected. *Brassica rapa* grown on the Mir station and in a high-fidelity post-flight ground control produced functional pollen despite high ethylene concentrations. Anther structure from the spaceflight flowers showed that numerous large starch grains were present in the cells of the anther wall and filaments, while starch grains were rarely observed in those cells of the ground control plants. There was no difference in the organization of the anther wall between space and ground control materials. However, the inner layer of anther wall, tapetum, was swollen and the tapetal plasmodium intruded into the anther chambers in the spaceflight tissues (Kuang and Musgrave, 2001). Despite the storage of starch grains in the anther wall and the intrusion of tapetal plasmodium, most pollen grains appeared normal. Viability of *Brassica rapa* pollen produced by plants grown from seed in the Plant Growth Facility on STS-87 was 93%, as assessed by fluorescein diacetate staining, and not significantly different from the ground control (Kuang *et al.*, 2000a).

Pistil development

Ovules that degenerated during development in microgravity have been described in *Arabidopsis* by several authors (Kordyum *et al.*, 1983; Kuang *et al.*, 1995). As a result of insufficient reserves, gynoecium development aborted at an early stage in the STS-54 experiment (Fig. 4). Ovules stopped growing and the integuments and nucellus collapsed and degenerated. The megasporocytes appeared abnormal and rarely underwent meiosis. Apparently they enlarged, or occasionally produced a dyad or tetrad, to assume the form of a female gametophyte with the single nucleus located in an egglike cell that lacks a cell wall. Synergids, polar nuclei, and antipodals were not observed. All tissues, whether in the ovary or in the anther, stained poorly with TBO. This suggests that cells contain aberrant vegetative cytoplasm and lack reserve substances, suggesting a lack of an energy supply during flower development (Kuang *et al.*, 1995). Upon supplementation of the chamber headspace with carbon dioxide, the female gametophyte developed normally, although unusually large starch grains were present in the central cell (Kuang *et al.*, 1996a). Because pollination did not occur in this closed chamber configuration due to high humidity, empty embryo sacs were observed in the older material. During the experiment on STS-68, air flow through the chambers resulted in normal pistil development (Fig. 5a; Kuang *et al.*, 1996b). Additional assays were employed to assess the

Fig. 5. Pistil development and pollen transfer were normal in *Arabidopsis* when plants were grown in ventilated Plant Growth Unit chambers on STS-68. (A) Morphology of flower in anthesis, formed in microgravity, was indistinguishable from that of the ground controls. (B) The stigmatic surface gave a positive reaction to an esterase test, indicating that this component was functioning properly despite development in microgravity. (C) Pollen was mechanically deposited on the stigmatic papillae in microgravity, as indicated by this scanning electron micrograph. (D) Normal growth of pollen tubes through the style in microgravity was detected by the fluorescence of callose when stained with aniline blue. For additional details, refer to Musgrave *et al.* (1997).

functional status of the gynoecium. Esterases, important components of the stigma surface proteins, were localized cytochemically, based on the hydrolysis of the substrate α-naphthyl acetate, to form a reddish insoluble complex upon treatment with fast blue B (Shivanna and Rangaswamy, 1992). The esterase reaction obtained in the flight material (Fig. 5b) was indistinguishable from that of the ground control, indicating full functionality of this component of the stigma following development in microgravity.

Pollination

Arabidopsis is generally self-pollinating. As the stamens elongate, the anthers dehisce and discharge pollen on the stigmatic papillae as they grow alongside the pistil. None of the pistils examined immediately post-flight on STS-51 or fixed for scanning electron microscopy had pollen grains visible on the stigmatic papillae, leading us to conclude that pollination had failed in this material. Following use of an Air Exchange System on STS-68, scanning electron micrographs detected the remains of germinated pollen grains on the stigmatic papillae (Fig. 5c), indicating that pollination had occurred in microgravity.

In an effort to identify any obstacles to reproductive development by *Arabidopsis* in microgravity, we had set up an experiment to score pollen tube growth through the stigmas of spaceflight material from STS-68. Germination of pollen on the stigma surface results in the formation of tubes that carry sperm cells to the ovules. These tubes contain a high concentration of callose, which fluoresces when stained with aniline blue (Preuss *et al.*, 1993). Using this test on fresh material immediately post-flight, we saw that pollen tubes had grown down through the style (Fig. 5d).

Other types of flowers require manual pollination (Fig. 6a). This was the case in another model plant used for spaceflight studies, *Brassica rapa*. For laboratory work, pollen is routinely collected using a bee stick, the thorax of a dead bee glued to a toothpick (Fig. 6b). Because pollen is a small, charged particle, there was some concern that manual collection and transfer of pollen during spaceflight could be compromised due to the microgravity environment. A comparison of the number of pollen grains collected in microgravity with the number collected in 1-*g* revealed no statistically significant difference (Garber and Musgrave, 1997). Similarly, numbers of pollen grains transferred from the bee stick in microgravity were quantified using actual pistils as well as artificial stigmas made of moist filter paper squares, and were found to be comparable to the results in 1-*g* (Tabor and Musgrave, 1997; Smith and Musgrave, 1997).

Early events following pollination were also tracked in *Brassica* by employing inflight fixation of flowers 2, 5, and 24 h following manual pollination on STS-87. Steric interactions between the pollen grains and the stigmatic papillae were not different in those flowers fixed in microgravity and in 1-*g* (Fig. 6c—24 h following pollination). *Brassica* plants pollinated during microgravity exposure on the KC-135 produced equal numbers of seeds per siliques (Guillory and Musgrave, 1997). Another indication of equivalent pollen transfer and function in microgravity and in the ground control is the rate of successful seed formation within individual *Brassica* siliques on STS-87. Spaceflight and ground control plants did not differ significantly with regard to number of seeds per silique (Kuang *et al.*, 2000a).

In the normally wind-pollinated species, wheat, there was concern that pollen might be lost upward in Mir without landing on receptive stigmas, due to the draw of the Svet hardware fan. Astronauts were requested to manually disturb the canopy during anthesis to ensure adequate dispersal of pollen to the stigmatic papillae.

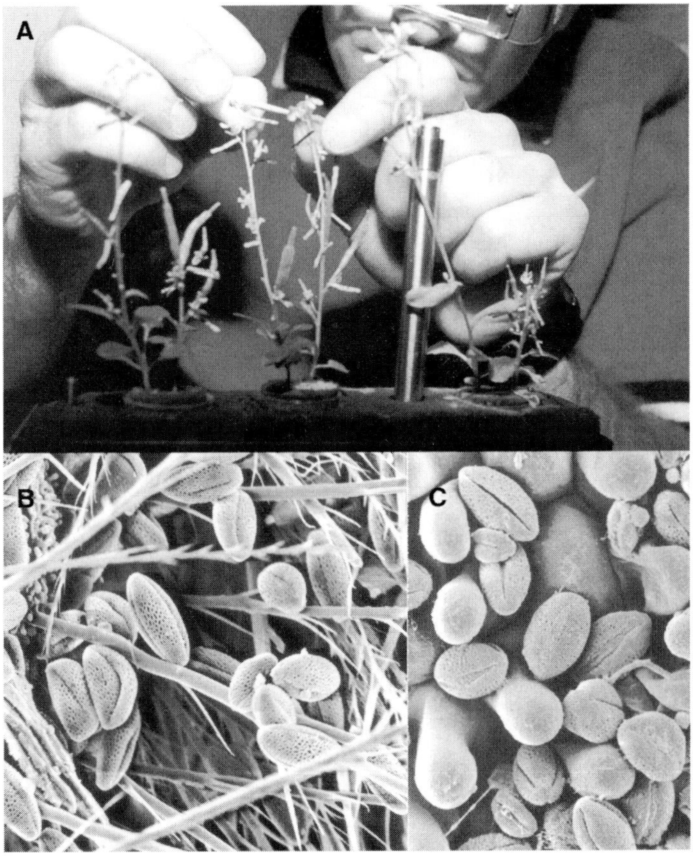

Fig. 6. *Brassica rapa* plants require manual pollination, permitting the generation of cohorts of siliques of known ages for purposes of a detailed embryo development study. (A) Payload specialist Leonid Kadenyuk pollinated *Brassica* plants grown in the Plant Growth Facility on STS-87. Siliques that had developed from earlier pollinations can be seen on the plants. (B) Pollen was collected and transferred using a bee stick. This scanning electron micrograph shows how the pollen grains were collected by the hairs on the bee thorax. No difference was found in the rate of pollen collection and transfer in 1-*g* and microgravity. (C) A scanning electron micrograph of the stigmatic surface of a *Brassica* flower, 24 h following pollination in microgravity, showing the spatial relationship between pollen grains and stigmatic papillae. The flowers were fixed in space and processed post-flight for microscopy.

Early embryogenesis

Pollen transfer and subsequent fertilization were successful in *Arabidopsis thaliana* plants grown in flow-through chambers in the Plant Growth Unit on STS-68. All developmental milestones up to the curled cotyledon stage of embryo development were observed in both the spaceflight and ground control material, consistent with the maximum possible age of embryos from the first

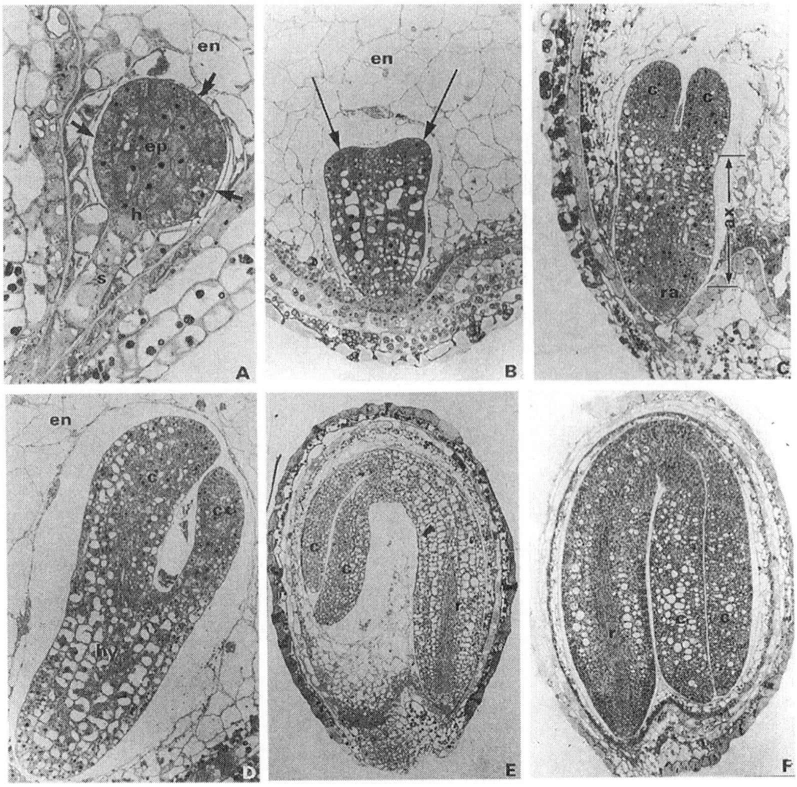

Fig. 7. Representative embryos at different developmental stages in *Arabidopsis* plants grown under spaceflight conditions on STS-68. (A) Late globular or transition stage. en, endosperm; ep, embryo proper; s, suspensor; h, hypophysis cell; arrowheads, protoderm. (B) Cotyledon initiation stage. arrows, cotyledon primordia; en, endosperm. (C) An early linear cotyledon stage. ax, hypocotyl axis; c, developing cotyledon; ra, radicle apical meristem. (D) Late linear cotyledon stage. c, cotyledon; en, endosperm; hy, hypocotyl. (E) Curled cotyledon stage. c, cotyledon; r, radicle. (F) mature embryo. c, cotyledon; r, radicle. Figure reproduced with permission, from Kuang *et al.* (1996b), © Annals of Botany.

day of flower opening. No differences were seen between the two treatments (Fig. 7, from Kuang *et al.*, 1996a).

The oldest seeds from this material contained completely developed embryos and seed coats. Radicle, hypocotyl, mersistematic apical tissues and differentiated cotyledons were observed. Protoderm, procambium, and primary ground tissue had differentiated. Protein and starch were deposited in the embryos during tissue differentiation, and at this stage, the aleurone layer contained a large quantity of storage protein and starch grains. Seed coat development from integuments of the ovule was progressing normally, with a gradual change in cell composition and cell material deposition. Seed quality could not be evaluated functionally because the seeds could not reach full maturity during the 11-day mission.

Proportions of pistils elongating following pollination and reaching a certain size class were not different in the spaceflight and ground control (Musgrave *et al.*, 1997). Nevertheless, the self-pollinating character of *Arabidopsis* left the true age of the embryos observed in question. For this reason, a second cycle of spaceflight experiments was mounted with a closely related plant, *Brassica rapa*, which requires mechanical transfer of pollen from another plant in order to achieve fertilization. In this experimental design, individual flowers were tagged with narrow, color-coded tape at the time of pollination, so that when plants were returned to Earth after the 16-day mission, ages of individual fruits could be unequivocally determined (Fig. 6a).

Embryos aged from 8 to 15 days after pollination. The embryos developed along the same schedule in spaceflight and ground control. By the torpedo stage, 3 days after pollination, the protoderm and rootcap were formed. At the curled cotyledon stage, 9 days after pollination, cotyledons and radicle were well developed, shoot apical meristem protruded between cotyledons. By 15 days after pollination, storage reserve deposition was already well underway in the cotyledons (Kuang *et al.*, 2000b).

Despite the similar morphological appearance and developmental schedule of these spaceflight embryos, a functional test using an embryo rescue technique showed that the spaceflight embryos were significantly delayed in producing plants (Kuang *et al.*, 2000a). Although the plants produced eventually had the same characteristics as in the ground control, the spaceflight embryos produced plants that lagged 12 days behind the ground controls. Subsequent electron microscopic analysis of the storage reserves of the 15-day old embryos, discussed below in the section on seed maturation, would reveal that protein, lipid, and starch storage was occurring along a different timetable in the spaceflight material than in the ground control (Kuang *et al.*, 2000b).

Somatic embryos in the spaceflight environment

The effect of the spaceflight environment on young developing embryos has also been studied extensively without the potential complication of microgravity effects on maternal tissues by growing somatic embryos in tissue culture. Using the spaceflight environment to study the effect of gravity unloading at the level of the plant nucleus, Krikorian's lab group found that low mitotic index and chromosomal anomalies were frequent but not invariable consequences of growth in the unique environment of orbital spacecraft (Krikorian and O'Connor, 1984; Krikorian and Levine, 1991; Krikorian 1996a). After a series of experiments with somatic embryos, they concluded that indirect effects on the microenvironment were responsible for many of the deleterious effects of spaceflight (Krikorian, 1998).

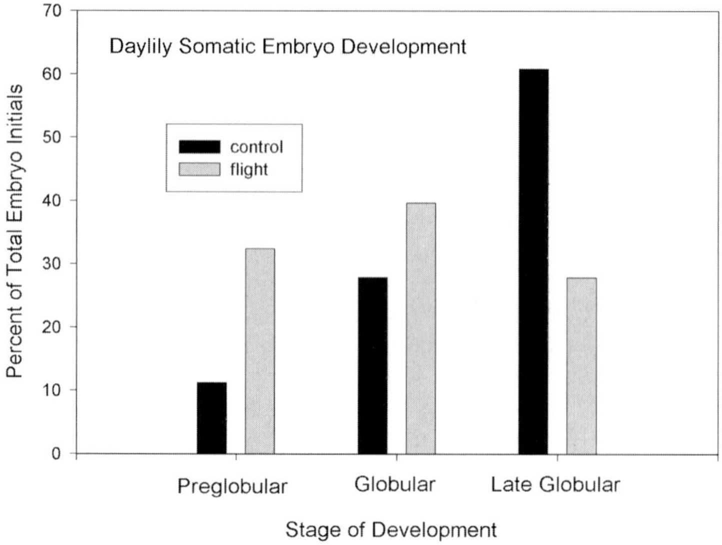

Fig. 8. Comparison of degree of embryonic development achieved by embryo initials of daylily following 7 days in space (or ground control). Based on data of Krikorian (1996b).

Krikorian (1996b) found that development of daylily somatic embryos was delayed in microgravity compared to ground-based controls (Fig. 8). Somatic embryos did develop in space but the number of units that progressed to later stages of embryo development was diminished in flight samples. Flight samples also appeared to be hyperhydrated (Krikorian, 1996b).

Orbital spacecraft maintain a microgravity environment (10^{-6} to 10^{-4} g) where convective currents and buoyancy are lacking. Surface tension dominates (Krikorian and Levine, 1991; Todd, 1989). Hypoxic responses of tissues in microgravity have been well-documented (Stout *et al.*, 2001; Paul *et al.*, 2001; Porterfield *et al.*, 1997; Porterfield and Musgrave, 1998; Porterfield *et al.*, 2000), and effects on surface temperature and photosynthesis have also been reported (Kitaya *et al.*, 2000). In fact, many of the differences noted between clinostat-grown and spaceflight material (Sievers and Hejnowicz, 1992; Moore, 1990) have been attributed to physical differences in the spaceflight environment. Krikorian's group found that the way the biological component of the system reacts to these changes in the microenvironment is modulated by several factors. In the case of embryoids, the younger they are in their developmental progressions, the more sensitive they are (Krikorian, 1996b); the more advanced developmentally, the less damage.

Fruit ripening and seed maturation

The *Brassica* silique has been the most extensively studied fruit ripening in microgravity. Using matched-aged siliques from STS-87, we found that silique

Table 3

Comparison between characteristics of the mature *Brassica rapa* seeds produced in space on Mir and the corresponding ground control seeds

Characteristics	Seed source	
	Spaceflight	Ground control
Seed weight (mg)	1.3[a]	1.7
Cell number/cotyledon × 10^2	10.3[a]	60.0
Cell size (μm^2)	303	435
Protein bodies		
Number per cell	3.4	2.8
Size (μm^2)	23.9[a]	42.4
Size ratio (pb area/cell size)	26.1	33.2
Starch grains per cell	2.7	0

[a]Significantly different from ground control, $P < 0.05$; (modified from Kuang *et al.*, 2000b).

development was initially comparable in microgravity and 1-*g* (Fig. 6a). At later stages of maturity, *Brassica* silique ripening in microgravity on Mir consistently occurred from the tip of the silique to the point of attachment on the plant, compared to the uniform ripening pattern seen in 1-*g* (Musgrave *et al.*, 2000; Kuang *et al.*, 2000b). This suggests that a gradient develops inside the silique in microgravity due to the absence of buoyancy-driven convection (Kuang *et al.*, 2000b).

Seeds of *Brassica rapa* produced on the Mir space station were smaller and weighed less than ground control seeds (Table 3). Although cell size in the cotyledons was not significantly lower in the spaceflight material than in the ground control, the cotyledons contained substantially fewer cells. Storage reserves were significantly altered as well, with smaller protein bodies and the retention of starch at maturity in the spaceflight cotyledons (Kuang *et al.*, 2000b).

The spaceflight-induced differences in storage reserves were already manifest in the cotyledon cells of developing *Brassica* seeds that were 16 days post-pollination, as was found in our experiment on STS-87. At this stage, starch grains are still evident in the ground control material, but many more are present in the spaceflight cotyledon cells (Kuang and Musgrave, 2000; Kuang *et al.*, 2000b). Protein bodies are already well-developed in the ground control material, while lipid bodies and starch are the major storage reserves in the spaceflight tissue (Fig. 9).

To summarize, the results of these two spaceflight experiments studying seed development in *Brassica* indicate that development in microgravity causes major shifts in the storage reserve composition. *Brassica* seeds are important agriculturally because they are a rich source of oil and protein (Musgrave, 2000). As such, *Brassica* is grown worldwide for both edible and industrial oils,

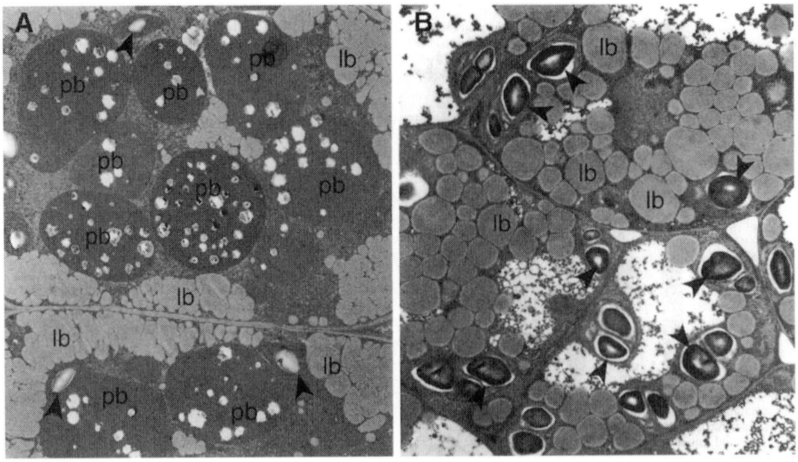

Fig. 9. Electron microscope pictures showing the ultrastructure of cotyledon cells of immature *Brassica rapa* seeds 15 days after pollination from the STS-87 shuttle experiment. Note that although the ground control (A) and spaceflight seeds (B) are the same age, the internal structure of the cells is different. In the ground control seeds (A), large protein bodies (pb) have formed, and lipid bodies (lb) and starch grains (arrowheads) are present. In the space-produced seeds (B), no protein bodies were observed, but lipid bodies (lb) and numerous large starch grains (arrowheads) are major storage reserves. Figure reproduced with permission, from Kuang *et al.* (2000b), © Annals of Botany.

which are extracted from the seeds by pressing. The remaining material is so high in protein that it has commercial value in animal feed production (Robbelen and Thies, 1980). The normal developmental process by which the starch that is stored initially in young seeds is converted into the lipid and protein storage reserves is delayed or disrupted in microgravity. This finding has significance for: (1) the vigor of plants arising from seeds produced in microgravity; (2) the nutritional quality of the seeds produced in microgravity; and hence (3) the feasibility of using plants in a microgravity setting to provide biological life support functions for humans on long-duration missions (Frick *et al.*, 1994; Henninger, 2001; NRC, 1997). A further point of basic significance of the finding is a demonstration of gravity's role in shaping the physical environment that defines our biological reality.

Conclusions

Environmental factors play a large role in determining the success of plant reproductive development. Studies on plant development in microgravity initially disallowed investigation of the reproductive stage because the physical challenges of the microgravity environment were not being met by early hardware designs. Active gaseous exchange around the aerial portions of the plant and control of substrate moisture levels have been key to supporting plant growth sufficiently to provide for reproductive development. At the same time, limited long-duration spaceflight opportunities have required that these processes be studied piecemeal—only with the Mir station was sufficient time available for the study for a complete life cycle. The International Space Station may provide similar opportunities.

Early reproductive development in microgravity has been especially susceptible to the environment and failed historically due to inadequate ventilation (closed growth chambers in microgravity) or high concentrations of ethylene in the spacecraft environment. Late stages of reproductive development, fruit ripening and seed maturation, are adversely affected in *Brassica* by the spaceflight environment. We suspect that these problems are caused by abnormal gaseous microenvironments prevailing inside the fruit due to lack of convective air movement in microgravity. Oxygen produced at the interior silique wall surface through photosynthesis must traverse the lumen before it becomes available to the seed (Porterfield *et al.*, 1999; Kuang *et al.*, 1998); likewise the high carbon dioxide concentration prevailing inside the silique becomes a source for photosynthetic carbon dioxide fixation. Without buoyancy-driven convection to speed this gas exchange, these processes become limited by the rate of diffusion.

Matching the metabolic needs of the developing seed with protection from desiccation and herbivory has resulted in fruit architectures that we view as narrowly balanced trade-offs between ventilation and protection (Musgrave and Blasiak, 2001). The role of buoyancy-driven convection in maintaining this tight balance is just beginning to be recognized, and has implications for a number of agronomically important problems in seed development and maturation. As more varied plant species are grown in microgravity, it is likely that additional manifestations of this phenomenon will be found to affect their reproductive processes.

Acknowledgements

Supported by NASA grants NAG-100139, NAG2-1020 and NAG2-1375.

References

Antonsen, F. and Johnsson, A. (1998) Effects of microgravity on the growth of *Lepidium* roots. J. Gravita. Physiol. 5, 13–21.

Bingham, G.E., Jones, S.B., Podolsky, I. and Yendler, B.S. (1996) Porous substrate water relations observed during the Greenhouse-2 flight experiment. Soc. Auto. Eng. Technical Paper #961547.

Bingham, G.E., Sytchev, V.N., Levinskikh, M.A. and Podolsky, I.G. (1999) Final plant experiments on Mir provide second generation wheat and seeds. Gravita. Space Biol. Bull. 13, 48.

Campbell, W.F., Salisbury, F.B., Bugbee, B., Klassen, S., Naegle, E., Strickland, D.T., Bingham, G.E., Levinskikh, M., Iljina, G.M., Veselova, T.D., Sytchev, V.N., Podolsky, I., McManus, W.R., Bubenheim, D.L., Stieber, J. and Jahns, G. (2001) Comparative floral development of Mir-grown and ethylene-treated, earth-grown Super Dwarf wheat. J. Plant Physiol. 158, 1051–1060.

Frick, J., Nielsen, S.S. and Mitchell, C.A. (1994) Yield and seed oil content response of dwarf, rapid-cycling *Brassica* to nitrogen treatments, planting density, and carbon dioxide enrichment. J. Am. Soc. Hort. Sci. 119, 1137–1143.

Garber, M. and Musgrave, M.E. (1997) Determination of microgravity effects on pollen collection. Gravita. Space Biol. Bull. 11, 70.

Guillory, N.G. and Musgrave, M.E. (1997) Microgravity effects on successful fertilization in *Brassica rapa*. Gravita. Space Biol. Bull. 11, 38.

Halstead, T. W. and Dutcher, F.R. (1984) Status and prospects. Ann. Bot. 54(Suppl. 3), 3–18.

Halstead, T.W. and Dutcher, F.R. (1987) Plants in space. Annu. Rev. Plant Physiol. 38, 317–345.

Henninger, D.L. (2001) Advanced life support for a human mission to Mars. Gravita. Space Biol. Bull. 15(1), 80.

Heslop-Harrison, J., Heslop-Harrison, Y. and Shivanna, K.R. (1984) The evaluation of pollen quality and a further appraisal of the fluorochromatic (FCR) test procedure. Theore. Appl. Gen. 67, 367–375.

Kiss, J.Z., Brinckmann, E. and Brillouet, C. (2000) Development and growth of several strains of *Arabidopsis* seedlings in microgravity. Int. J. Plant Sci. 161, 55–62.

Kitaya, Y., Kawai, M., Tsuruyama, J., Takahashi, H., Tani, A., Goto, E., Saito, T. and Kiyota, M. (2000) The effect of gravity on surface temperatures and net photosynthetic rates of plant leaves, 33rd COSPAR Scientific Assembly, Warsaw, Poland, 16–23 July 2000.

Kordyum, E.L., Sytnik, K.M. and Chernyaeva, I.I. (1983) Peculiarities of genital organ formation in *Arabidopsis thaliana* (L.) Heynh. under spaceflight conditions. Adv. Space Res. 3, 247–250.

Krikorian, A.D. (1996a) Space stress and genome shock in developing plant cells. Physiol. Plant 98, 901–908.

Krikorian, A.D. (1996b) Embryogenic somatic cell cultures of daylily (*Hemerocallis*): a system to probe spaceflight-associated mitotic disturbances. pp. 491–555. *In* Suge, H. and Takahashi, H. (eds.) *Plants in Space Biology*. Tohoku University Press, Sendai, Japan.

Krikorian, A.D. (1998) Plants and somatic embryos in space: what have we learned? Gravita. Space Biol. Bull. 11, 5–14.

Krikorian, A.D. and Levine, H.G. (1991) Development and growth in space. pp. 491–555. *In* Steward, F.C. (ed.) *Plant Physiology: Growth and Development*. New York, Academic Press.

Krikorian, A.D. and O'Connor, S.A. (1984) Experiments on plants grown in space. Karyological observations. Ann. Bot. 54(Suppl. 3), 49–63.

Kuang, A., Crispi, M.L. and Musgrave, M.E. (1998) Control of seed development in *Arabidopsis thaliana* by atmospheric oxygen. Plant Cell Environ. 21, 71–78.

Kuang, A. and Musgrave, M.E. (2000) Changes in cotyledon cell ultrastructure during *Brassica rapa* seed development in microgravity. Gravita. Space Biol. Bull. 14, 34.

Kuang, A. and Musgrave, M.E. (2001) Anther structure and pollen development in *Brassica* plants grown during spaceflight. Gravita. Space Biol. Bull. 15(1), 48.

Kuang, A., Musgrave, M.E., Matthews, S.W., Cummins, D.B. and Tucker, S.C. (1995) Pollen and ovule development in *Arabidopsis thaliana* under spaceflight conditions. Amer. J. Bot. 82, 585–595.

Kuang, A., Musgrave, M.E. and Matthews, S.W. (1996a) Modification of reproductive development in *Arabidopsis thaliana* under spaceflight conditions. Planta 198, 588–594.

Kuang, A., Xiao, Y. and Musgrave, M.E. (1996b) Cytochemical localization of reserves during seed development in *Arabidopsis thaliana* under spaceflight conditions. Ann. Bot. 78, 343–351.

Kuang, A., Popova, A., Xiao, Y. and Musgrave, M.E. (2000a) Pollination and embryo development in *Brassica rapa* L. in microgravity. Int. J. Plant Sci. 161, 203–211.

Kuang, A., Xiao, Y., McClure, G. and Musgrave, M.E. (2000b) Influence of microgravity on ultrastructure and storage reserves in seeds of *Brassica rapa* L.. Ann. Bot. 85, 851–859.

Levinskikh, M.A., Sychev, V.N., Derendyaeva, T.A., Signalova, O.B., Salisbury, F.B., Campbell, W.F., Bingham, G.E., Bubenheim, D.L. and Jahns, G. (2000) Analysis of the spaceflight effects on growth and development of super dwarf wheat grown on the space station Mir. J. Plant Physiol. 156, 522–529.

Mashinsky, A., Ivanova, I., Derendyaeva, T., Nechitailo, G. and Salisbury, F. (1994) 'From seed-to-seed' experiment with wheat plants under space-flight conditions. Adv. Space Res. 14, 13–19.

Merkys, A.I. and Laurinavicius, R.S. (1983) Complete cycle of individual development of *Arabidopsis thaliana* (L.) Heynh. plants on board the Salyut-7 orbital station. Dokladi Akademii Nauk SSSR 271, 509–512.

Moore, R. (1990) Comparative effectiveness of a clinostat and a slow-turning lateral vessel at mimicking the ultrastructural effects of microgravity in plant cells. Ann. Bot. 66, 541–549.

Morrow, R.C., Frank, J.G., Stolp, K. M., Guetshow, S.M. and Crabb, T.M. (2001) Science and technology overview of the Biomass Production System mission verification test. Gravita. Space Biol. Bull. 15(1), 66.

Musgrave, M.E. (2000) Realizing the potential of rapid-cycling *Brassica* as a model system for use in plant biology research. J. Plant Growth Regul. 19, 314–325.

Musgrave, M.E. and Blasiak, J. (2001) Internal atmosphere dynamics in peppers: a model for spaceflight. Gravita. Space Biol. Bull. 15(1), 49.

Musgrave, M.E., Kuang, A. and Matthews, S.W. (1997) Plant reproduction during spaceflight: importance of the gaseous environment. Planta 203, Suppl. 177–184.

Musgrave, M.E., Kuang, A., Brown, C. S. and Matthews, S.W. (1998) Changes in *Arabidopsis* leaf ultrastructure, chlorophyll, and carbohydrate content during spaceflight depend on ventilation. Ann. Bot. 81, 503–512.

Musgrave, M.E., Kuang, A., Xiao, Y., Stout, S.C., Bingham, G.E., Briarty, L.G., Levinskikh, M.A., Sychev, V.N. and Podolski, I. G. (2000) Gravity-independence of seed-to-seed cycling in *Brassica rapa*. Planta 210, 400–406.

NRC (National Research Council) (1997) Advanced technology for human support in space. Committee on advanced technology for human support in space. Aeronautics and Space Engineering Board, Washington, DC, National Academy Press.

Nechitailo, G.S. and Mashinsky, A.L. (1993) Space biology: Studies at orbital station. pp. 293–330. Moscow, Mir Publishers.

Parfenov, G.P. and Abramova, V.N. (1981) Blossoming and maturation of *Arabidopsis* seed: Experiment on Biosatellite Kosmos 1129. Dokladi Akademii Nauk SSR 256, 254.

Paul, A.L., Daugherty, C.J., Bihn, E.A., Chapman, D.K., Norwood, K.L. and Ferl, R.J. (2001) Transgene expression patterns indicate that spaceflight affects stress signal perception and transduction in *Arabidopsis*. Plant Physiol. 126, 613–621.

Perbal, G. and Drissecole, D. (1994) Sensitivity to gravistimulus of lentil seedling roots grown in space during the IML-1 mission of Spacelab. Physiol. Plant 90, 313–318.

Porterfield, D.M., Matthews, S.W., Daugherty, C.J. and Musgrave, M.E. (1997) Spaceflight exposure effects on transcription, activity and localization of alcohol dehydrogenase in the roots of *Arabidopsis thaliana*. Plant Physiol. 113, 685–693.

Porterfield, D.M. and Musgrave, M.E. (1998) The tropic response of plant roots to oxygen: oxytropism in *Pisum sativum* L. Planta 206, 1–6.

Porterfield, D.M., Kuang, A., Smith, P.J.S., Crispi, M.L. and Musgrave, M.E. (1999) Oxygen-depleted zones inside reproductive structures of *Brassicaceae*: implications for oxygen control of seed development. Can. J. Bot. 77(10), 1439–1446.

Porterfield, D.M., Monje, O., Stutte, G.W. and Musgrave, M.E. (2000) Rootzone hypoxic responses result from inhibition of gravity dependent oxygen transport in microgravity. Gravita. Space Biol. Bull. 14, 50.

Preuss, D., Lemieux, B., Yen, G. and Davis, R.W. (1993) A conditional sterile mutation eliminates surface components from *Arabidopsis* pollen and disrupts cell signaling during Fertilization. Gen. Devel. 7, 974–985.

Robbelen, G. and Thies, W. (1980) Biosynthesis of seed oil and breeding for improved oil quality of rapeseed. pp. 253–284. *In* Tsunoda, S., Hinata, K. and Gomez-Campo, C. (eds.) *Brassica Crops and Wild Allies, Biology and Breeding*. Japan Scientific Societies Press, Tokyo.

Shivanna, K.R. and Rangaswamy, N.S. (1992) Pollen biology: a laboratory manual. pp. 45–47. Springer, New York.

Sievers, A. and Hejnowicz, Z. (1992) How well does the clinostat mimic the effect of microgravity on plant cells and organs? ASGSB Bull. 5, 69–75.
Smith, C. and Musgrave, M.E. (1997) Pollination of Brassica rapa under microgravity conditions on the KC-135. Gravita. Space Biol. Bull. 11, 70.
Stankovic, B., Link, B. and Zhou, W. (2001) *Arabidopsis* rocket science. Gravita. Space Biol. Bull. 15(1), 54.
Stout, S.C., Porterfield, D.M., Briarty, L.G., Kuang, A. and Musgrave, M.E. (2001) Evidence of root zone hypoxia in *Brassica rapa* L. grown in microgravity. Int. J. Plant Sci. 162, 249–255.
Strickland, D.T., Campbell, W.F., Salisbury, F.B. and Bingham, G.E. (1997) Morphological assessment of reproductive structures of wheat grown on Mir. Gravita Space Biol. Bull. 11, 14.
Tabor, M. and Musgrave, M.E. (1997) Microgravity effects on collection and transfer of pollen in *Brassica rapa*. Gravita. Space Biol. Bull. 11, 38.
Todd, P. (1989) Gravity-dependent phenomena at the scale of the single cell. ASGSB Bull. 2, 95–113.
Volkmann, D., Behrens, H.M. and Sievers, A. (1986) Development and gravity sensing of cress roots under microgravity. Naturwiss. 73, 438–441.

Developmental Biology Research in Space
H.-J. Marthy (editor)

Bdelloid Rotifers as Model System to Study Developmental Biology in Space

Claudia Ricci* and Chiara Boschetti

Department of Biology, State University of Milan, Milan, Italy

Abstract

Bdelloid rotifers are suitable model systems for space experiments. Due to their developmental pattern they appear adequate to investigate the role of the cytoskeleton during oogenesis and during early developmental stages, and to reflect the effects of disturbances in the spatial arrangement of cytoskeletal components. The effect of weightlessness on the developmental pattern of a bdelloid rotifer will be studied in the International Space Station: in preparation for it we are performing ground-based experiments on the development of rotifer embryos under either increased or decreased gravity. The model studied is *Macrotrachela quadricornifera*, a species of rotifers belonging to the Bdelloidea class. Samples exposed to gravity disturbance were analyzed for morphology and fitness-related parameters.

Rotifers were exposed over several days to altered gravity conditions and the morphology of eggs laid during this period were investigated using a confocal laser microscope. A subset of eggs was allowed to hatch to determine newborn developmental time and age at maturity. High (up to 20*g*) gravity was obtained in a slow centrifuge suitable for animal cultivation over several days. To produce low (simulated 0.0001*g*) gravity a Random Positioning Machine equipped with a 'rotifer bioreactor' was used. Under all conditions the rotifer retained normal life-history traits, and did not show permanent changes in embryo morphology, regardless to the stresses to which it was exposed. Only some modification of the shape of early embryos, experiencing 20*g*, has been noted, but later developmental stages appeared unaffected, and normal juveniles hatched. Whether this result indicates any capacity to repair damage during embryogenesis of these Spiralia experiencing 20*g* is an open question. The significance of the result as well as the use of instruments to simulate gravity perturbations are discussed.

* E-mail: claudia.ricci@unimi.it

Introduction

The development of a multicellular animal from a single cell, the egg, follows a rigid sequence of events during which the egg divides to give rise to thousands (often millions) of cells, that will form almost all structures of the animal. The complete development typically goes through three different stages: cleavage, gastrulation and organogenesis. Cleavage consists of a number of mitotic divisions occurring on smaller and smaller cells, because blastomeres (cells originated during cleavage divisions) do not grow between subsequent mitoses. Gastrulation consists of the rearrangement of the cells and the formation of the so-called germ layers, establishing the basic body plan of the animal. Organogenesis consists of the differentiation of specialized cells and gives rise to the typical animal features. All of these stages are under genetical control, but are also sensitive to environmental conditions.

The orientation of first cleavage divisions is critical for establishing the body axis of the future animal. The distribution of cytoskeletal elements and yolk granules in the egg cytoplasm is important in determining the spindle orientation and consequently the future body pattern of the embryo. Thus if the organization of the cytoskeleton is disturbed by any perturbation, altered embryos might be produced. A possible way to perturb cytoskeletal components is to modify the gravity force, assuming that the assembly of long polymeric structures such as the filaments of actin or tubulin can be affected. The effects of gravity on the embryo cytoskeleton has been studied by following the developmental pattern of several animal taxa. Embryos or unsegmented eggs were exposed to altered gravity and their subsequent development was studied (see Moore *et al.*, 1996, and issues of J. Grav. Physiol.). While gravity can be easily increased experimentally by placing the developing embryo in a fine-tuned slow centrifuge (commonly called hyperfuge), the only condition where absence or decrease of gravity can be realized is in space. The results obtained so far from embryos which have undergone development under approximately 0*g* conditions are often contrasting and unclear. There are several possible reasons for these differing results. Firstly, experiments in space are difficult to replicate, due to the extremely high costs and restricted number of flight opportunities. Further variation can be ascribed to the characteristics of the animals used as experimental models. Most of the taxa used were Echinoderms and Vertebrates (e.g., Marthy, 1997; Ubbels *et al.*, 1995; Ubbels, 1997). These animals are named RADIALIA, because the mitotic furrows during cleavage are either parallel with or at right angles to the major axis of the embryo. The radial pattern of cleavage is associated with a ‘plastic’ development pattern, which implies that the fate of the blastomeres is not pre-determined and that damage occurring to them during development at one stage can commonly be rectified at a later stage. If defects can be repaired later during development the effects of perturbations during embryo maturation could be reduced or neutralized.

In contrast, if the fate of the blastomeres is fixed very early during development, defects occurring at any time during embryogenesis could be long-lasting and could produce consequences on the future animal. This sort of development, where each embryonic cell is committed early to a given fate on the developing body, is called mosaic development. Several animals have mosaic development, and since the disposition of blastomeres during cleavage resembles a spiral pattern, the animals possessing these characteristics are named SPIRALIA. Several taxa of invertebrates belong to the Spiralia, including the annelids, the nematodes and the rotifers.

In this chapter we present the results of tests carried out in preparation of a long-term experiment that will take place on International Space Station (ISS). The experiment consists of culturing several generations of one species of rotifers and one of nematodes under space conditions to study the effect of weightlessness on animals with spiral cleavage. Here we present the results of ground-based experiments on rotifer embryo development under either increased or decreased gravity.

Rotifers

Macrotrachela quadricornifera, used as model system, belongs to the class Bdelloidea of the phylum Rotifera. Like all rotifers, its development follows a modified spiral cleavage pattern. Bdelloid rotifers have small body size (100–600 μm), fast development (1–5 days), obligatory parthenogenetic reproduction (only females are present in populations), short life cycle (about 30 days), eutely (their cell number is fixed at birth and does not change during life) and the capacity to survive harsh conditions by entering dormancy (Fig. 1).

Dormancy is a reversible process that allows animals to escape uninhabitable conditions and to resume active life when conditions again become suitable (Keilin, 1959). The most common form of dormancy in the bdelloid rotifers is represented by anhydrobiosis, induced by the dessication of the environment at any time during the animal's life time. Resistance to dessication is a natural feature of most bdelloids, able to inhabit ephemeral habitats such as the thin water film surrounding soil particles or moss leaves. Anhydrobiosis may extend the animal's life span from months to years, but has costs in terms of mortality when animals are re-hydrated (Ricci *et al.*, 1987; Ricci, 2001). Anhydrobiotic rotifers are said to tolerate conditions that cannot be successfully tolerated by the active forms (Gilbert, 1974). They were found to resist extreme temperature and pressure (Rahm, 1923; Becquerel, 1950) and to be unaffected by space flight conditions, such as microgravity and vibrations (Ricci and Caprioli, 1998). Anhydrobiosis can be experimentally induced by subtracting water medium from rotifer cultures, providing a useful system to manipulate the animals under safe conditions. This is an appropriate method to preserve organisms during stressful conditions maintaining them in a viable state capable of resuming normal life when convenient (Ricci, 1996).

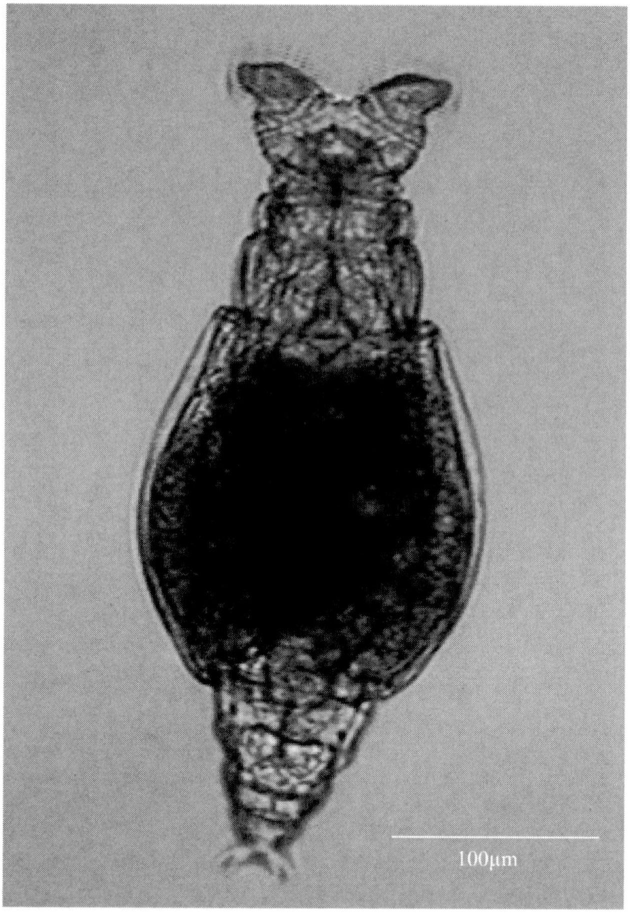

Fig. 1. *Macrotrachela quadricornifera*, Rotifera Bdelloidea.

Outline of ISS experiment

M. quadricornifera is easily cultivated under laboratory conditions, and at 22 °C it lives for about 30 days and lays approximately 30 eggs (Ricci, 1991, 1995). The specimens that will be flown to the ISS will be transferred by the Shuttle in dry conditions and will be activated by adding culture medium (deionized water plus food) when convenient. The first batch of experimental animals, when recovered, will be cultivated by renewing the culture medium and will reproduce. These animals will form the P (=parental) generation. At prefixed times, their eggs will be removed and transferred to a second culture chamber to start the second population that will constitute the F1 generation. The same protocol will be followed for subsequent populations in order to perform a multigenerational experiment, in which each step represents one generation. After egg transfer, the populations will be desiccated and maintained in a dormant state until return to earth. Using this procedure, alive dormant rotifers belonging to different generations will return to the laboratory, where they will be re-hydrated and cultivated or processed for further studies. While in space, the rotifers will experience hypogravity and this condition is expected to affect the cytoskeletal organization of the embryo and hence its role in establishing the division furrow of the cells.

In addition to the synthesis of cytoskeletal components occurring during embryo development, another potential source of damage to the cytoplasmic organelles of the rotifers must be considered, since the cytoplasm of the maturing egg is entirely synthesized by the mother's glands, called vitellaria (sing. vitellarium) and transferred to the egg (Amsellem and Ricci, 1982). Mistakes in synthesis and assembly of cytoskeletal elements could occur both during oogenesis and during early developmental stages. A "mosaic" development pattern should make repair of embryo damage almost impossible. In addition, the eutely of rotifers does exclude repair mechanisms involving cell reproduction during the life time of the animal. Thus damage during development is expected to be maintained by the adult.

Bdelloid rotifers provide a suitable system to study the role of the cytoskeleton during oogenesis and early developmental stages, reflecting the effects of disturbances in the spatial arrangement of cytoskeletal components. However, our knowledge of the pattern of embryo development of rotifers and of bdelloids in particular is very poor.

Eggs and embryos

Most rotifers lay unsegmented eggs, with only a minority of species keeping the developing egg inside the body till birth. From the egg a miniature rotifer hatches, with the time necessary to complete embryogenesis depending upon the species and temperature. A newly laid egg of *M. quadricornifera* takes about 4 days to hatch at 22 °C.

Studies on rotifer embryogenesis have mostly used species from the Monogononta class of rotifers. Independent studies have used different species, and it is doubtful if a general model of development is applicable to all species (e.g., Tessin, 1886; Tannreuther, 1920; Natchwey, 1925; de Beauchamp, 1956). Data on development of bdelloid rotifers are even scarcer with the only detailed study being published at the end of XIX century (Zelinka, 1891).

Because of the parthenogenetic reproduction, the eggs of bdelloids do not undergo meiosis but only equational divisions (= mitosis), with the extrusion of two polar bodies (Hsu, 1956a,b). The egg matures inside the mother's gonad, where, due to rotifer's eutely, the definitive number of oocyte nuclei are already present at birth (Pagani *et al.*, 1993). The ovary, called the germarium, is embedded into a syncytial nurse gland, called vitellarium, and both are surrounded by a follicular layer. The three structures, germarium, vitellarium

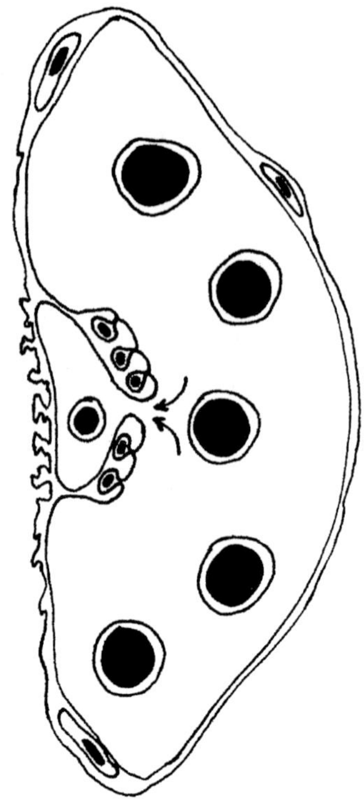

Fig. 2. Germovitellarium (gonad and nurse gland) of a bdelloid rotifer. Vitellarium is syncytial, and provides the cytoplasm to the maturing egg. The arrows indicate cytoplasm flowing from the gland to the egg through a cytoplasmic bridge (redrawn after Amsellem and Ricci, 1982).

and follicular layer, are paired in the bdelloids, leading to the name of Digononta (=two gonads), and are located at both sides of the gut. When an oocyte is going to be laid, a cytoplasmic bridge opens temporarily between the oocyte and the vitellarium and the cytoplasm and its organelles are transferred from vitellarium to growing oocyte (Amsellem and Ricci, 1982) (Fig. 2). When mature, the egg is surrounded by a transparent resistant shell that is produced by the mother along its oviduct. The oviparous bdelloids lay unsegmented eggs that are commonly kept close to the mother's foot for a while, and later attached to the substrate via a sticky secretion produced by mother's pedal glands.

Rotifer eggs undergo holoblastic (=total), unequal segmentation of a modified spiral type. Cleavage stages of *Macrotrachela quadricornifera* are illustrated in Figure 3. The egg is laid at the one-cell stage, and before undergoing the first cleavage division, extrudes one polar body. Cleavage consists of a series of unsynchronized divisions of blastomeres, one of which remains larger throughout cleavage. After the fourth division, a typical 16-cell embryo, composed of four rows of four cells each, is formed. This peculiar stage is

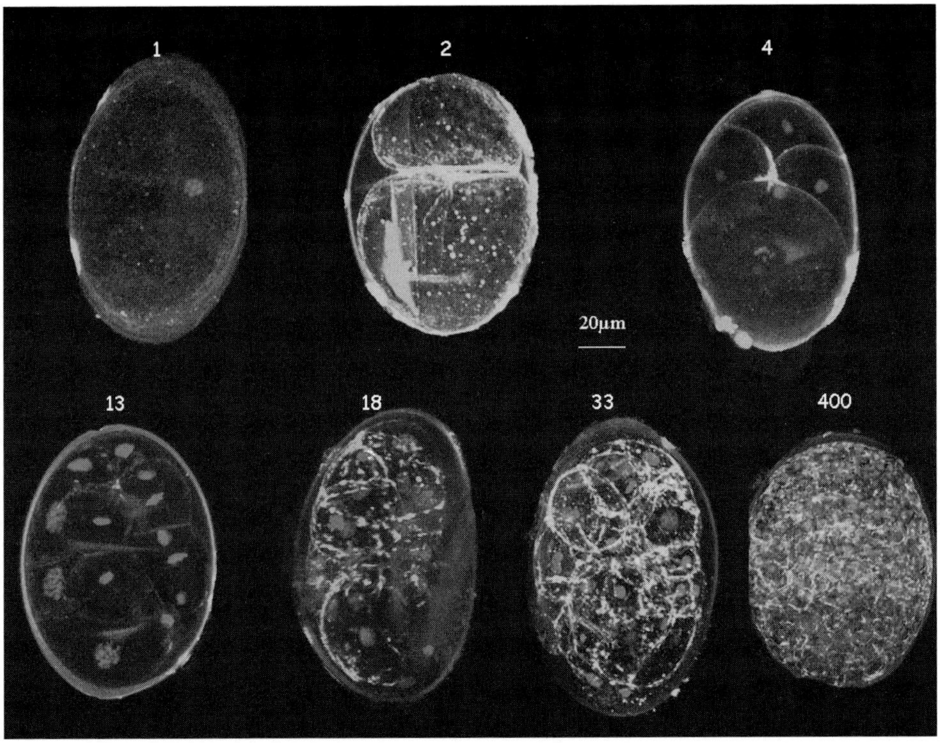

Fig. 3. Embryos from 1 cell to 400 cells (cleavage + gastrulation) of *Macrotrachela quadricornifera*, stained with phalloidin and DAPI and observed at a confocal laser scanning microscope. Numbers indicate the number of cells for each stage.

common to all rotifers (see Gilbert, 1989) and more or less marks the start of gastrulation. Rotifer gastrulation consists of epibolic movements of smaller blastomeres (from the animal pole) that, probably with minor differences among species, grow over and surround the large cell derived by the large blastomere (belonging to D quadrant), pushing it inward, up to the opposite extremity of the egg according to Pray (1965). Gastrulation is a long-lasting process during which blastomeres move along the embryo axis and keep dividing, till the definitive number typical for each species is reached.

Following gastrulation, the stages of organogenesis occur. Cells differentiate into the definitive types and organize into tissues and organs. Although most cells do not undergo further mitoses, organogenesis appears the longest phase of the entire development of rotifers. The first structure of the *M. quadricornifera* embryo to be identified is the mastax, a peculiar masticatory structure made of muscles and skeletal elements located in the pharynx; this is distinguishable in 2–3 day-old embryos, corresponding to about one half of the entire developmental timespan of the species. Many tissues of adult rotifers are syncytial; cell membranes break down at different times during development in the different structures.

Experimental studies of the cell-lineage have only been performed for *Asplanchna girodi*, a monogonont rotifer species (Lechner, 1966), while amost nothing is known about bdelloids so far.

Ground-based experiments

Since the cytoskeleton formation of bdelloid rotifers can be affected both during egg formation inside the mother gonad and during development, we decided to use two approaches to test the effects of gravity force on the cytoskeleton. Whenever possible, both (A) unsegmented eggs and (B) reproductive animals were exposed to altered conditions of gravity and the resulting eggs were studied. (A) Synchronized eggs were collected before cleavage divisions started, and exposed to the experimental condition till hatching. (B) Adult rotifers were exposed to altered gravity conditions till egg production and the eggs were then removed and allowed to develop under normal conditions. Parallel controls were kept at 1*g*, and represented the reference result for comparisons. Developing eggs were either processed for morphology (cytoskeleton) investigation at prefixed times, or hatched to establish newborn life-history traits, i.e. developmental time and age at maturity.

To visualize cytoskeletal elements, actin and tubulin of embryos at different stages of development were stained using immunofluorescence techniques and the embryos were studied using a Leica TCSNT confocal microscope equipped with Argon-Krypton laser. Actin microfilaments were stained with phalloidin conjugated with rodamin, and microtubules were stained with monoclonal antibody anti α-tubulin and antibody anti-mouse IgG conjugated with fluorescein. To determine the embryonic stage the DNA of blastomeres was stained with DAPI.

We decided to test the effects of both increased and decreased gravity on embryo formation under the assumption that the two conditions could produce similar, but opposite effects.

Hypergravity

Increased gravity (hypergravity) was performed using a centrifuge, constructed to simulate a gravity force up to 20*g* for a long period of time (several days) and suitable for culturing small organisms (Fig. 4).

The overall size is 800 × 650 × 300 mm. The working plate is an aluminium disk, 620 mm in diameter, with a number of sample holders for sealed round glass beakers of about 2–3 ml capacity. Maximum sample weight compatible with the centrifuge is 100 grams, but our samples were well below this limit. A rubber belt secured the beakers to sample holders. The disk is operated by a 24 volt motor. An optical encoder provides the signal to detect speed and the

Fig. 4. Hyperfuge for the rotifers, that allows cultivation of aquatic organisms and substitution of food and medium without interrupting the experiment. One semitrasparent covers is removed to show the rotating plate with sample holders.

motor is driven by a constant voltage power supply. The value of gravity can be set continuously from 0.0 to 20.0*g* at 600 mm from the center of the plate. The accuracy and stability of the control of the acceleration is $\pm 0.2g$. Two LCD displays show both speed and acceleration with a resolution of 1 rpm and 0.1*g*. The required power is less than 100 W. Semi transparent covers and safety micro switches avoid risks of contact with the rotating plate.

Medium and food can be supplied manually with a syringe or automatically with a pump, built into the instrument. Each sample beaker is connected to the dispenser through tubes and needles; excess of liquid and waste are collected in a container inside the centrifuge chassis.

Several problems in using this instrument came from the biocompatibility of a number of parts that directly or indirectly were in contact with the animals. Surprisingly, materials known to be biocompatible in general were lethal to the rotifers. After a series of preliminary tests, the rotifers, fed once a day and exposed to several *g*, maintained their life-history traits, in terms of fertility and survival as high as the 1-*g*-controls. Experiments were run setting the gravity force at 3, 5, 7, 15 and 20*g*. As no remarkable effect was found at the lower values of *g*, we present results on rotifers exposed to 20*g*.

Unsegmented eggs developed into embryos under 20*g* conditions were somewhat flattened with a crescent-like space evident between blastomeres and the shell. This effect can be ascribed to exposure to hypergravity, but subsequent development took place normally as shown by the late embryos that developed at 20*g* (Fig. 5). No difference was observed in the duration of

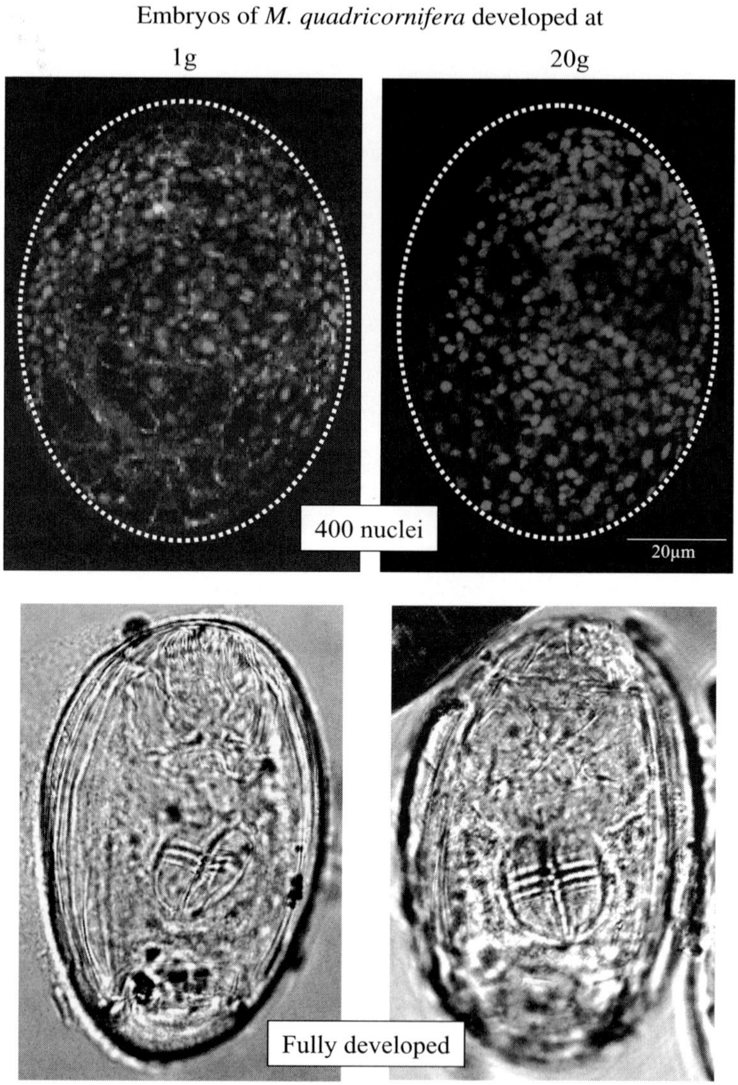

Fig. 5. *Macrotrachela quadricornifera* embryos. Upper panel: embryos at 400 cells, stained with phalloidin and DAPI, developed under 20*g* (right) and 1*g* (left) (at confocal laser scanning microscope). Lower panel: fully developed embryos maintained under 20*g* (right) and 1*g* (left) (at phase contrast illumination).

development and age at first reproduction of the rotifers hatched from these eggs. Eggs matured under 20*g* and developed at 1*g* did not have any shape modification, as expected, and gave rise at the due time to normal juveniles, that reproduced normally. Thus rotifers that either developed under or were produced by mothers exposed to 20*g* gave similar results and both conditions

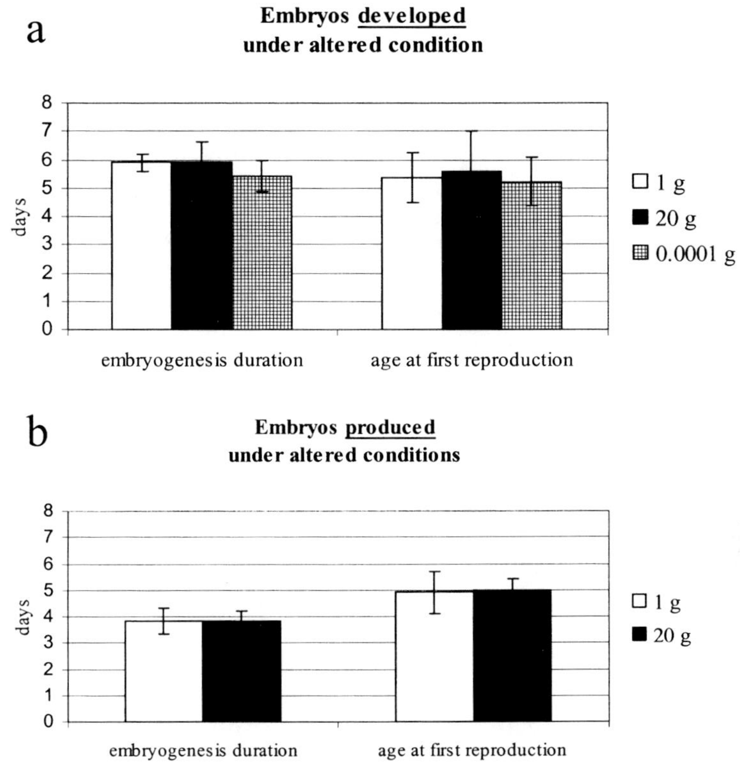

Fig. 6. (a) Life-history traits (embryogenesis duration and age at first reproduction) of rotifers hatched from eggs developed at 1*g* (open hystogram – control), 20*g* (solid hystogram – hypergravity) and ~0.0001*g* (shaded – hypogravity). (b) Life-history traits (embryogenesis duration and age at first reproduction) of rotifers hatched from eggs produced by mothers exposed to 1*g* (open hystogram – control) and 20*g* (solid hystogram – hypergravity). Bars represent standard deviation.

did not differ from the controls (Fig. 6 a, b). Whether the shape modification of early embryos exposed to 20*g* has to be considered as true damage is unclear. Certainly the deformation observed is temporary and it is not visible in later embryos. This could indicate that the correct shape and position inside the shell can be reestablished during organogenesis, suggesting that minor damages can be repaired by the rotifers during development. This seems to contrast with the assumption that spiral embryos are incapable of repair. In any case, the rotifer life-history traits were unaffected by gravity experienced either during development (Fig. 6a) or during egg formation (Fig. 6b).

Hypogravity

On Earth, it is not possible to get rid of gravity, and it is not possible to obtain true weightless condition to test. Real lack of gravity is limited to space conditions. Nevertheless, reduced gravity can be simulated by specific instruments, such as clinostats. Experiments on the development of rotifer eggs were run using a desktop Random Positioning Machine (RPM) (Dutch Space). It consists of two perpendicular cardanic frames, the inner one equipped with an experiment platform, that are moved according to a random trajectory generated by a computer software. Hypogravity is simulated by continuous random change of orientation of the sample relative to the gravity vector and can reach forces as low as 0.0001*g* (Huijser, 2000). The setting we used were 'random mode scenario' at 100 $°s^{-1}$.

The experiment platform of RPM has been equipped with a 'bioreactor' designed for the rotifers, but suitable for culturing most aquatic microinvertebrates. The animals are accommodated into a closed container, connected to a reservoir with medium and food. The automatic supply is controlled by a timer and operated by a small pump.

Not all microinvertebrates survive easily under simulated hypogravity conditions. Preliminary tests run on different bdelloid species showed that animals unable to swim cannot survive RPM conditions. The study species, *M. quadricornifera*, can live for several days in RPM conditions and feeds and reproduces successfully. We tested the effects of simulated gravity only on unsegmented eggs.

Synchronized eggs of the rotifer were kept in the RPM till hatching, and the newborns were isolated, cultured under normal conditions and their age at maturity was recorded. Again, both duration of embryogenesis and age at first reproduction of the rotifer embryos exposed to RPM were very similar to controls (Fig. 6a). We did not study the arrangement of the cytoskeleton of such embryos yet, but strongly doubt that any modification takes place, since the life-table parameters remain constant. However it is too early to conclude whether these results can be extended to true microgravity, such as that occurring in space, and whether the rotifer cytoskeleton is really insensitive to gravity forces.

Ground facilities like RPM provide a good simulation of the weightless condition of space, but are rather far from generating true hypogravity. The condition of weightlessness performed by the RPM results from the sum of vectors oriented in every direction. In other words, organisms are 'shaken', that is they are maintained in suspension by changing the position of the experiment platform continuously, but the gravity vector does still exist, although changing direction and preventing objects to settle. We suspect that animals that are endowed with georeceptors (statocysts, statoliths, etc.) and integrate the information though a nervous system, cannot easily be fooled by the shaker, because they can continuously perceive the direction of gravity vector. The overall effect could be simply a sensorial disorder. In fact, bottom-dweller rotifers, unable to swim, do not resist RPM treatment. This observation could indicate that the effect of agitating the rotifers produces equilibrium disorders that can be compensated by animals able to use the three dimensions of space, but cannot be overcome by animals that move in a two-dimension world. Absence of gravity should produce different stimuli to georeceptors, or no stimulus at all, and this could cause other disturbances.

Experimentation on ground gives a useful contribution to our knowledge, but cannot replace space experiments.

Our results show that *Macrotrachela quadricornifera* seems to be unaffected by exposure to high (up to 20*g*) or low (as low as 0.0001*g*) gravity force. The rotifer development was constant regardless of the different treatments to which it was exposed. Only some modification of the blastomere position in early embryos experiencing 20*g* was visible, but the effect was transitory and did not produce apparent damage to the rotifers, judging from life traits and morphology of the animals. The absence of relevant permanent damage in response to alterations of the gravity indicated the incredible flexibility of the animals and their ability to adapt to different conditions. In particular, the lack of response to exposure to 20*g* was unexpected. This result might be explained by assuming that, despite common assumption, embryos with mosaic development do retain some capacity of repair. Genetic control of the development pattern of rotifers may make the developmental process insensitive to external disturbances. Alternatively, actin and tubulin may be insensitive to gravity forces, due to the small mass of the molecules and to the weak force that gravity could exert on them.

Conclusions

The rotifers proposed for the ISS experiment are appropriate as experimental model for space experiments for a number of reasons. (A) Complexity. They are metazoa (multicellular animals), have tissues and organs, have a complete gut, yet are very simple animals. Rotifers are made of about 1000 cells, while a human is made of several millions of cells. This simplification allows to disentangle complex problems through easier approaches. (B) Miniaturization. Rotifers are small in size; a relevant number of animals requires small room. (C) Distribution. Rotifers are easily found in nature, and are easily cultivated under controlled conditions. (D) Life span. Rotifers have short life cycles that can be studied in a reasonable time period. (E) Density. They reproduce parthenogenetically. This provides two major advantages: fast population growth because all animals contribute to a next generation, and reliability of their characteristics, because the genome is entirely transmitted to the progeny. (F) Resistance. Rotifers can easily enter dormancy, and under such conditions they become extremely resistant to a number of stresses. Furthermore, their dormancy represents a simple suspension of time during their life cycle and has mortality costs only.

The present study contributes to the study of the effects of gravity disturbances on suitable animals, but many points remain open. Whether altered gravity produces effects on rotifer embryos remains doubtful and our results do not provide any certain answer so far. Some of the data presented are preliminary: more tests on the RPM and more research on cytoskeletal molecules must be carried out. If hypergravity disorganises F-tubulin assembly and position is yet to be assessed, and if animals with mosaic development retain the capacity to repair damage requires further study.

Acknowledgements

Several colleagues, students and technicians contributed to these studies. We are deeply indebted to Umberto Fascio, Cristina Sotgia, Valeria Cenini, Nadia Santo, Manuela Caprioli and Nausicaa Zarantonello for their technical help and scientific advises. Thanks are due to Professor Alan Marchant, University of Nottinghan (UK) for reading an earlier draft of the paper and correcting the English form. Alberto Musetti designed and realized both hyperfuge and bioreactor for the rotifers. His contribution is invaluable and his collaboration is gratefully acknowledged. Financial support for the study came from ASI grants to C.R.

References

Amsellem, J. and Ricci, C. (1982) Fine structure of female genital apparatus of *Philodina* (Rotifera, Bdelloidea). Zoomorphology 100, 89–105.

Becquerel, P. (1950) La suspension de la vie au-dessous de 1/20° K absolu par démagnétisation adiabatique de l'alun de fer dans le vide le plus élevé. Compt. Rend. Acad. Sci. 231, 261–263.

de Beauchamp, P. (1956) Le développement de *Plœsoma hudsoni* (Imhof) et l'origine des feuillets chez les rotifères. Bull. Soc. Zool. France 81, 374–383.

Gilbert, J.J. (1974) Dormancy in rotifers. Trans. Am. Microsc. Soc. 93, 490–513.

Gilbert, J.J. (1989) Rotifera. In Adiyodi, K.G., Adiyodi, R.G. (eds.). Reproductive biology of invertebrates. Fertilization, Development, and Parental Care, Volume IV, Part A. Oxford & IBH Publishing Co. Pvt. Ltd.

Hsu, W.S. (1956a) Oogenesis in *Habrotrocha tridens* (Milne). Biol. Bull. 111, 364–374.

Hsu, W.S. (1956b) Oogenesis in the bdelloidea rotifer *Philodina roseola* Ehr. La Cellule 57, 283–296.

Huijser, R.H. (2000) Desktop RPM: new small size microgravity simulator for the bioscience laboratory, FS-MG-R00-017 v2 Fokker Space, 1–5.

Keilin, D. (1959) The problem of anabiosis or latent life: history and current concept. Proc. Roy. Soc. London 150, 149–191.

Lechner, M. (1966) Untersuchungen zur Embryonalentwicklung des Rädertieres *Asplanchna girodi* de Guerne. Roux' Archiv für Entwicklungsmechanik 157, 117–173.

Marthy, H.-J. (1997) Sea urchin eggs under microgravity conditions. ESA SP-1206, 81–91.

Moore, D., Bie, P. and Oser, H. (eds.) (1996) Biological and medical research in space. An Overview of Life Science Research in Microgravity. Springer-Verlag, Berlin, Heidelberg, New York.

Nachey, R. (1925) Untersuchungen über die Keimbahn, Organogenese und Anatomie von *Asplanchna priodonta* Gosse. Z. wiss. Zool. 126, 239–492.

Pagani, M., Ricci, C. and Redi, C.A. (1993) Oogenesis in *Macrotrachela quadricornifera*. I. Germarium nuclei, caryotype and DNA content. Hydrobiologia 255/256, 225–230.

Pray, F.S. (1965) Studies on the early development of the rotifer *Monostyla cornuta* Müller. Trans. Am. Microsc. Soc. 84, 210–216.

Rahm, P.G. (1923) Biologische und physiologische Beiträge zur Kenntnis der Moosfauna. Z. Allg. Physiol. 20, 1–34.

Ricci, C. (1991) Comparison of five strains of a parthenogenetic species *Macrotrachela quadricornifera* (Rotifera Bdelloidea). I. Life history traits. Hydrobiologia 211, 147–156.

Ricci, C. (1995) Growth pattern of four strains of a Bdelloid Rotifer species: egg size and number. Hydrobiologia 313/314, 157–163.

Ricci, C. (1996) Desiccation as a switch for some microinvertebrates, VIth Life Science Research in Space. ESA SP-390, 273–275.

Ricci, C. (2001) Dormancy patterns in rotifers. Hydrobiologia 446/447, 1–11.

Ricci, C. and Caprioli, M. (1998) Stress during dormancy: effect on recovery rates and life-history traits of anhydrobiotic animals. Aquatic Ecology 32, 353–359.

Ricci, C., Vaghi, L. and Manzini, M. L. (1987) Desiccation of rotifers (*Macrotrachela quadricornifera*): survival and reproduction. Ecology 68, 1488–1494.

Tannreuther, G. W. (1920) The development of *Asplanchna ebbesbornii* (Rotifer). J. Morphol. 33, 389–437.

Tessin, G. (1886) Über Eibildung und Entwicklung der Rotatorien. Z. wiss. Zool. 44, 273–302.

Ubbels, G. A. (1997) Fertilization and development of *Xenopus* eggs on Sounding Rockets and in a clinostat. ESA SP-1206, 25–36.

Ubbels, G.A., Reijnen, M., Meijerink, J., Narrawaj, J. (1995) The role of gravity in the establishment of the dorso-ventral axis in the amphibian embryo. ESA SP-1162, 175–185.

Zelinka, C. (1891) Studien über Räderthiere. III. Zur Entwicklungsgeschichte der Räderthiere nebst Bemerkungen über ihre Anatomie und Biologie. Z. wiss. Zool. 53, 1–159.

Drosophila melanogaster and the Future of 'Evo-Devo' Biology in Space. Challenges and Problems in the Path of an Eventual Colonization Project Outside the Earth

Roberto Marco[1,*], David Husson[1,2], Raul Herranz[1], Jesús Mateos[1] and F. Javier Medina[3]

[1]*Departamento de Bioquímica de la UAM e Instituto de Investigaciones Biomédicas 'Alberto Sols' UAM-CSIC, C/Arzobispo Morcillo, 4, Madrid 28029, Spain*
[2]*Centre de Biologie du Développement, UMR 5547 CNRS-Université Paul Sabatier, 118 route de Narbonne, 31062 Toulouse cedex, France*
[3]*Centro de Investigaciones Biológicas del CSIC, C/Velázquez 144, Madrid 28006, Spain*

Abstract

Space exploration, especially its future phase involving the International Space Station (ISS) makes possible the study of the effects on living systems of long-term expositions to such a strange environment. This phase is being initiated when Biological Sciences are crossing a no-return line into a new territory where the connection between phenotype and genotype may be finally made. We briefly review the paradoxical results obtained in Space experiments performed during the last third of the XX Century. They reveal that simple unicellular systems sense the absence of gravity changing their cytoskeletal organization and the signal transduction pathways, while animal development proceeds unaltered in these conditions, in spite of the fact that these processes are heavily involved in embryogenesis. Longer-term experiments possible in the ISS may solve this apparent contradiction. On the other hand, the current constraints on the scientific use of the ISS makes necessary the development of new hardware and the modification of current techniques to start taking advantage of this extraordinary technological facility. We discuss our advances in this direction using one of the current key biological model systems, *Drosophila melanogaster*. In addition, the future phase of Space exploration, possibly leading to the exploration and, may be, the colonization of another planet, will provide the means of performing interesting evolutionary

* E-mail: roberto.marco@uam.es

experiments, studying how the terrestrial biological systems will change in their long-term adaptation to new, very different environments. In this way, Biological Research in Space may contribute to the advancement of the new Biology, in particular to the branch known as "Evo-Devo". On the other hand, as much as the Space Adventure will continue involving human beings as the main actors in the play, long-term multi-generation experiments using a fast reproducing species, such as *Drosophila melanogaster*, capable of producing more than 300 generations in 15 years, the useful life foreseen for ISS, will be important. Among other useful information, they will help in detecting the possible changes that a biological species may undergo in such an environment, preventing the uncontrolled occurrence of irreversible deleterious effects with catastrophic consequences on the living beings participating in this endeavour.

Introduction

The major problem faced by the rapid biological advances in the years of the transition from the XX to the XXI Century is understanding how the phenotype is actually fabricated from the instructions stored in the genotype and how the environment and additional non-coded physicochemical forces and mechanisms are playing critical roles in the process (Newman and Comper, 1990).

Biological scientists are living in an exciting period. The incredible power of current experimental techniques makes possible to obtain information undreamed a few decades ago. In particular, the explosion of knowledge associated with the completion of the genome project for an increasing number of selected biological species is worth noting. For humans, *Drosophila*, *Cenorhabditis*, *Arabidopsis* and yeast, to name the more relevant eucaryotic organisms, we now have access to the complete genomic sequences, making possible to approach in a new way the problems in understanding how biological organisms function, adapt and evolve. It is true that there are errors in the sequences so far assembled by the industrial-like, mass production sequencing laboratories. It is true that parts of the genome, which present special difficulties and/or seem less valuable, have been left out of the sequencing effort. For example, heterochromatic DNA is not included in the *Drosophila* genomic sequence. Nevertheless, it is clear that these shortcomings will be eventually corrected in the future. A more formidable obstacle will be to make sense out of the huge amount of data, to identify the functional role of all the genes, old and new, revealed by this sequencing effort. Biological research is equipped now with very powerful research tools to approach this task and accomplish, at least in theory, but extremely costly in terms of manpower and economic research support. It is therefore critical to increase the resolution and to overcome the limitations of present day Biology, still, in our view, a descriptive, qualitative discipline. Biology has began to develop new methods to analyze and integrate the information on the behavior of large networks of elements that we increasingly recognize as being underneath the function of biological systems and their adaptations to their environments, life styles and survival rates.

An important area of Research that we believe could contribute to this new phase of investigation can be linked to the exploration and the mastery of Space, to the challenge brought about by the objective of sending and maintaining living systems outside the limits of our Biosphere (Moody and Golden, 2000).

Biological Space Research has been pursued during the last decades with increasingly sophisticated means. Nevertheless, so far only relatively short-term experiments have been possible. Even in the Russian MIR Station, biological experiments have been operated usually during a few weeks. Exception has to be taken of the few human beings that have been exposed there during longer periods of time, periods not so long in relative terms if we take into account the human life-span. Another exception could be found in the uncontrolled specimens, collateral to these experiments, such as the microbiological flora that colonized and evolved in the Station during its years of operation, but which has not been studied in this connection or this knowledge made available to the scientific community.

The next phase of the Space Exploration will allow the performance of much longer biological experiments in Space. For example, once the International Space Station is completed, controlled long-term experiments using key biological model systems will be for the first time possible. It is true that financial limitations are putting enormous constraints on this possibility. The crew number has been reduced for safety reasons to a number that makes the availability of crew time one of the more precious resources in the ISS. The original dream of the scientists involved in Space Research was that laboratories completely equivalent to those that we use in our Ground Institutes would be installed in the ISS modules. At least for the next few years, the scientific community interested in using the ISS will have to cope with these limitations. This means that experiments have to be run using mostly automatized Hardware and postponing the more work intensive analysis until samples are recovered on the Ground, something that will only occur at certain times, taking advantage of the periodic refurbishing flights to the ISS. Thus, it will be critical to advance in the improvement of methods involved in the preservation of samples, both living and fixed, as well as samples maintained in ways that will allow the recovery of critical molecules (DNA, RNA, proteins) to run specific genomic and proteomic experiments on them after these periods of time. The availability of microarray methodology is an exciting development linked to the Genome project, making possible the investigation of the overall response of the genome of key biological model systems as they experience and adapt to different situations.

In fact, in this Chapter we will discuss the objectives and the problems behind our purpose of establishing a permanent colony of *Drosophila* in Space to study how a highly organized multicellular organism will respond and adapt to this foreign environment.

Furthermore, the phenotypic and genotypic changes actually observed in these long-term experiments will provide useful information on the ways current earth-based biological systems could adapt to unusual environments, probably providing clues on how evolution could have proceeded in the past in our planet. Another complementary line of research is founded on the efforts to build and control self-sustaining life support systems that will probably be the initial scenarios in which the problems behind the project of the colonization outside our Planet will be identified and tested.

In addition, even if relatively little could be learned about evolution of new adaptations, the exposure during long periods of time and many generations of selected model systems will be a necessary step forward to establish the conditions for the future colonization of another planet, especially if human beings are going to participate in this undertaking. This precaution would help unveiling any unforeseen problems before they affect the persons participating in the enterprise.

The earlier evidence. The earliest phase (1961–2001)

What do we know about the exposure of living systems to the Space environment and to the possible adaptations to the Terrestrial environment that may be affected in long-term exposures to an alien environment? We will briefly summarize in the following paragraphs what we think we know, bringing into focus some paradoxes that seem to have emerged from this work. The dates indicated above cover the period between the first Space flight in which an organism was recovered alive (Yuri Gagarin possibly accompanied by a few fruitflies in his pockets) and the disappearance of the Russian MIR Station.

Effects on animal development

Although, Biological Research in Space has been limited by many constraints, during the past fifty years this type of experimentation has been steadily pursued by laboratories in USA, the former Soviet Union, Japan and Europe (see Claassen and Spooner, 1994, for a review). As a consequence, a certain body of knowledge on the response of animal organisms, tissues and cells to the Space Environment has been accumulating. In fact, the starting point for our argument is the following apparent paradox. While most cells and tissues appear to sense and change their behavior in response to Microgravity, whole animal organisms have developed normally in spite of the clear involvement in the developmental process of the reported affected cellular processes and components. The paradox is made more forceful by the realization that gravity is a weak physical force and should be hardly effective on tiny microscopic objects, such as cells in the absence of specialized organelles of high density like those existing in plant systems (Albrecht-Bühler, 1990; Mesland, 1992). In spite of this fact, a relatively large number of experiments indicate that the smaller, simpler systems made up by isolated cells and/or tissues feel the absence of the normal gravity vector. As a consequence, they change many of their basic intrinsic processes, such as cell proliferation, cytoskeletal organization, signal transduction pathways, particularly those involving G proteins and protein kinase C action, etc. (for reviews, Claassen and Spooner, 1994; Moore and Cogoli, 1996; Cogoli and Cogoli-Greuter, 1997). Examples of these responses are: diverse ciliated and flagellated protozoa (Hemmersbach *et al.*, 1999), lymphocytes, monocytes and macrophages (Cogoli *et al.*, 1993; Cogoli, 1993; Chapes *et al.*, 1994; Schmitt *et al.*, 1996; Cogoli-Greuter *et al.*, 1996; Lewis *et al.*, 1998, 2001), epidermoid carcinoma cells (de Groot *et al.*, 1990, 1991), connective tissue cell lines (Chapes *et al.*, 1994), bone cells (Bikle *et al.*, 1994; Van Loon *et al.*, 1995; Vico *et al.*, 1998, Burger, Klein-Nulend, 1998), muscle and nerve cells (Gruener and Hoeger, 1990; Reitstetter, 1994; Gruener *et al.*, 1994), Drosophila Schneider S-1 cells (Schatten *et al.*, 2001), etc. In conclusion, simple, isolated cell systems, in principle too small to be evident targets of the weak force of gravity, show in general a clear response when exposed to a Microgravity environment. They sense the absence of gravity and change important cellular responses such as signal transduction pathways and the cytoskeletal organization. Equivalent changes have been reported for experiments using simulated Microgravity (Walther *et al.*, 1998; Clejan *et al.*, 2001; Uva *et al.*, 2002; Grimm *et al.*, 2002)

On the other hand and in spite of the extensive involvement of these processes (Signal transduction pathways and CSK organization) in Development, the larger and more complex multicellular organisms seem more robust, showing completely normal morphologies and functions, even when their development has occurred totally in Microgravity. This result has been verified in a wide variety of model systems organisms (nematodes, *Drosophila*, sea urchins, fishes and amphibians, etc.) from low to highly evolved invertebrates to vertebrates (see Marco *et al.*, 1996 and Marthy 2002 for reviews). Occasional reports of developmental alterations have not been reproduced in posterior, better conducted and controlled experiments. On the other hand, in other experiments, transient modifications in early developmental events have been observed, modifications that eventually are rescued. There are hints that some developmental processes may be affected (for example, in amphibians), but the formidable regulatory capabilities shown by developmental systems, are able to correct and overcome in the long run any alterations that Microgravity conditions introduce in the normal course of animal development (Souza *et al.*, 1995; Husson, 1998; Duprat *et al.*, 1998). We have little understanding of the correcting processes. Therefore, complete normal development has been the usual outcome of these experiments. It has to be pointed out that development, i.e., the production of complex highly differentiated organisms, begins with a single huge cell, the embryonic zygote.

This make these results even more paradoxical since larger size should make the earlier embryonic cells much better targets for the weak force of gravity. Not many of these experiments have been able to study the cellular events themselves during development due to the problems involved in the successful handling of these complex organisms during the Space flights. In some cases, morphological details of some intermediate structures during development suggest abnormalities in cell proliferation during development, such as the finding of a thicker blastocoel roof during early *Xenopus laevis* development in Microgravity (Souza *et al.*, 1995, Ubbels *et al.*, personal communication). Similar results have been obtained by French investigators using a different amphibian species, *Pleurodeles waltl* (Husson, 1998). Different anomalies were detected in the early phases of development (cleavage and neurulation) but the majority of the embryos achieved a normal differentiation and morphogenesis. In general, the reported changes were fully compatible with normal development. This is not an unusual result in Developmental Biology where the robustness of the systems has been traditionally ascribed to the so-called regulatory behavior at certain stages of embryos and parts of embryos. In relation to this regulatory behavior, there is the question of the redundancy of many elements that build the regulatory networks and pathways controlling development (Malacinski, 1990; Li and Noll, 1994; Cooke *et al.*, 1997; Tautz, 2000). This redundancy is especially remarkable in certain processes and pathways such as those involved in cytoskeleton organization and signal transduction, precisely those more often reported to be affected by exposure of isolated cells to Microgravity.

In several cases, alterations in functional properties of the organisms linked to the strong reduction in gravity forces have been described. In plants, geotropic response is impaired and in mammals, loss of bone tissue and changes in functioning correlated with switches in the specific proteins isoforms expressed in these muscles have been reported (Kraemer *et al.*, 2000). In sea urchins, effects on the mineralization under exposure to Microgravity have been described, although it is not yet clear if they are direct or indirect (Marthy *et al.*, 1998). Effects on mitochondria and cilia in sea urchin embryos have also been reported (Schatten *et al.*, 1998).

Our group has been involved in the study of this problem since almost two decades ago using mainly as the experimental system, *Drosophila melanogaster*, a small and resilient arthropod, probably the higher animal organism better known from a developmental point of view. *Drosophila* has been one of the organisms more often used in the earlier history of Space Research, both in Russian and American Space Flights (for reviews, see Miquel, 1984; Miquel and Souza, 1991). These earlier experiments established the compatibility of this small organism with the constrains and stresses of Space Travel, showing that it was possible to send and recover many developed fruit flies that had underwent development under Microgravity conditions. The logical constraints in these earlier flights made difficult to apply strict quantitative analyses to the

Table 1

Results of IML-2 Development Experiment *Drosophila* larvae/flies developed from eggs laid by flies at the end of the third day of a Shuttle Flight of 18 days after the first embryo collection had been removed

Number of developed animals recovered	Flight containers		Simultaneous ground control		Trays*
	Static (0 *g*)	Centrifuge (1 *g*)	Static (1 *g*)	Centrifuge (1.4 *g*)	
Frozen	2155	1875	1822	2028	7
Alive	719	947	554	558	4
Total	2874	2823	2376	2586	11

*The number of trays recovered from the containers (two for each experimental condition) with 40 female and 10 male flies each.

recoveries and properties of these organisms, leaving open the possibility that developmental effects had remained undetected or were eventually rescued during the operation. Our initial experiments carried out in the middle 80's and early 90's (Marco *et al.*, 1986; Vernós *et al.*, 1989; Marco *et al.*, 1992) seemed to support some of these expectations. More recent experiments (see Marco *et al.*, 1996) carried out with better hardware allowing a better oxygenation of the samples have clarified these issues, showing that development of *Drosophila* in Microgravity is quantitatively and qualitatively normal, at least when one looks to the final product, the developed larva and adult fruit fly. Large quantities of larval and adult animals in the order of thousands were recovered from these experiments (Table 1). The detailed analysis of the results failed to verify the earlier found 50% reduction in the number of normal developed flies in Microgravity, pointing even more clearly to the paradox posed by the results obtained with simpler unicellular systems. Furthermore, the enlarging list of organisms that have been sent into Space and whose development in Microgravity has been studied, indicates the generality of this conclusion. Thus, the complex process of development integrating all cellular processes, such as cell proliferation and movements, cell communication and signal transduction, cellular, extracellular and tissular differentiation, appears to be normal in Space, while the same processes studied in many isolated cell systems seem to be altered.

Aging and behavior

One of the main challenges in current Biological Research is to understand the aging process in complex multicellular organisms (Partridge and Barton, 1993; Shigenaga *et al.*, 1994; Finch and Tanzi, 1997; Johnson *et al.*, 1999). The similarities of the physiological changes and adaptations to Microgravity endured by astronauts with the aging response in humans have been repeatedly emphasized (Wang, 1999; Burger and Klein-Nulend, 1999). Although, a species life span most probably is an evolutionary acquired trait (Rose, 1991), it is still unclear how much of it is genetically or epigenetically modifiable.

Insects, mostly composed by post-mitotic cells with relatively short life spans are a group of organisms where some features of the aging response can be optimally studied. The idea that an enhanced metabolism, the so-called rate-of-living theory, accelerates the rate of aging has been one of the explanations of the phenomenon, earlier advanced by scientists (Pearl, 1928). The idea was further refined (Miquel and Fleming, 1984, 1986) linking the aging phenomenon to the increase in mitochondrial function, consequence of the increase in metabolism (Sohal and Weindruch, 1996; Wallace, 1999). Environmental epigenetic treatments such as changes in the temperature of rearing are known to modify the aging response in *Drosophila* (Miquel *et al.*, 1976; Lints and Soliman, 1988; Rose, 1991). Measurement of the oxygen consumption at different temperatures provided evidence in agreement with the mitochondrial involvement in *Drosophila* aging (Miquel *et al.*, 1976). Using *Musca domestica*, the common housefly, several environmental treatments, including the level of physical activity, have been shown to modify the rate of aging (Agarwal and Sohal, 1994). Exposure to the Microgravity environment of *Drosophila* male imagoes was another epigenetic treatment influencing the aging process in *Drosophila*, as previously reported by our groups (Miquel and Philpott, 1978; Marco *et al.*, 1986). In these experiments, exposure to the space environment during one to 3 weeks as young imagoes resulted in a shortening of life-span after recovery of the animals from space, when compared to the ground control animals measured in parallel in the same conditions.

An idea that could unify these results was that exposure to Microgravity resulted in behavioral changes, such as an increased locomotor activity. The consequent enhanced mitochondrial respiration of the flies may accelerate aging, as has been shown for raised temperatures, hyperoxia or high physical activity on the ground (Miquel *et al.*, 1980; Agarwal and Sohal, 1994). Moreover, the idea that the free radicals produced in metabolism could be related to the aging process has received support from experiments carried out in *Drosophila*. An artificially induced increased genetic dosage of two enzymes, copper-zinc superoxide dismutase and catalase, involved in the removal of deleterious free radicals, oxidative products of mitochondrial metabolism, resulted in the slowing down of senescence (Orr and Sohal, 1994). The hypothesis that a progressive alteration in mitochondrial activity is central to the decrease in functional performance linked to the aging process in high organisms obtained direct experimental evidence from our laboratory. We showed that in *Drosophila* this decrease in mitochondrial function is linked not to a modification in the quantity and/or quality of the mitochondrial DNA but rather caused by a general and significant decrease in the levels of RNAs coding for mitochondrial components (Calleja *et al.*, 1993). These results have been later confirmed in other laboratories (Schwarze *et al.*, 1998a,b).

As already mentioned, earlier Space experiments had indicated that young male *Drosophila* flies exposed to microgravity showed an acceleration in aging. In a 14.5 days Space Shuttle Flight (IML-2), we sent 300 young male flies with

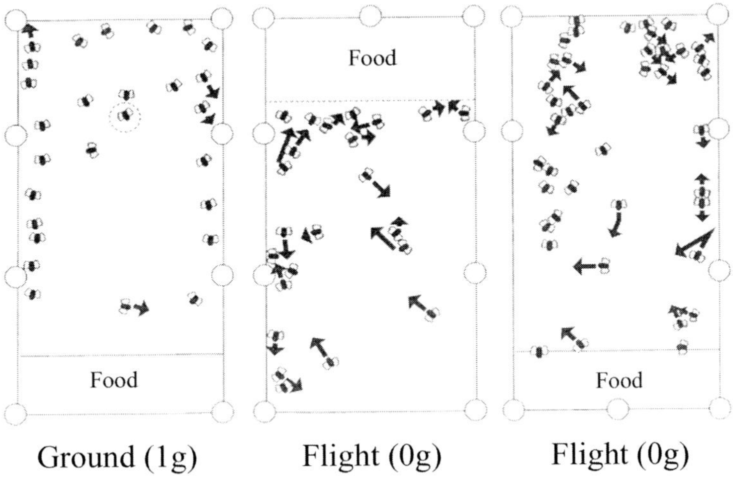

	Ground (1g)	Flight (0g)	Flight (0g)
Number of Moving Flies	3%	60%	40%
Average distance walked (mm) in 1/6 of a second	0.007± 0.002	0.318± 0.06	0.140± 0.03

Fig. 1. Results of *Drosophila* Motility video-recording in the IML-2 experiment. In the Figure, a reproduction of a snapshot (video image) from the experiment is presented including two images from two flight containers and one from the Ground experiment. The arrows in front of of the flies indicate which ones are moving during the short period of time following the recorded image. The size of the arrow measures the distance walked by the flies in 1/6 of a second afterwards. Summary numbers are given below. The fly enclosed in a dotted circle jumped during the recording of the image. While the flies in Microgravity do walk very actively, they rarely jump (this is the initial movement to initiate flying).

the purpose of confirming these findings. We could establish that changes in fly behavior were responsible of the effect in accordance with the proposal that alterations in mitochondrial metabolism may be involved in the aging response. Through repeatedly video recording, we found a very marked increase in the fruit-flies locomotor activity in Space (Fig. 1). The males showed accordingly an accelerated aging response upon recovery, both in terms of physiological vitality assays (mating and negative gravitaxis) and of life-span curves (Benguría *et al.*, 1996). Recently, an American team has been able to verify that the increase in motility is observable quite rapidly, even in short microgravity exposures, such as those achieved in Sounding rockets or in parabollic flights (Miller and Keller, 2000).

The involvement of mitochondrial metabolism is also suggested by the finding of a greater decrease in mitochondrial 16S ribosomal RNA in Microgravity exposed flies than in ground controls. On the other hand, a parallel 1 *g*-centrifuge control did not show such differences in the life-span curves when compared to flies exposed to a similar centrifugation on the Ground.

Drosophila females also increased their locomotor activity but did not show differential changes in the life-span curves, although they showed some initial signs of increased aging.

Another point is why the *Drosophila* imagoes increase their locomotor activity in Microgravity. Although, we do not know the reason, we suspect that this behavioral response of increased motility may be related to the immaturity of the young flies used in the experiments and the negative gravitaxis escape response shown by *Drosophila* on the Ground (Miquel *et al.*, 1972). Young imagoes may have a more immature behavioral response and therefore, they may overreact to the absence of a normal reference gravity vector, resulting in a continuous crawling in search for a 'higher' site in the container that obviously cannot be found in Microgravity. Earlier results indicate that mature older flies (two or more weeks old) exposed to Microgravity did not show the acceleration in the aging response that the young flies experience (Miquel and Philpott, 1978). More recent and unpublished work from our laboratories (the Bion-10 satellite flight) confirmed this difference for the more mature flies exposed to Microgravity (de Juan and Marco, unpublished). Interestingly, since older animals exposed to all the circumstances associated with a Space flight showed less acceleration in the aging response, nor such an increase in locomotor activity in space, strongly reinforces the conclusions of the IML-2 experiment and the connection between *Drosophila* metabolic rate and senescence. In this case, the fully mature older imagoes may use additional cues and, thus, avoid or show a less marked behavioral response. A testable prediction would be that *Drosophila* mutants affected in their negative gravitaxis response may show a change in the Microgravity response or an increase in the locomotor activity on the Ground, depending on which phenotype they show. *Drosophila* strains with altered response to the gravity vector on the Ground have been repeatedly selected in the past (Hirsch and Ksander, 1969; Dobzhansky *et al.*, 1972) and in our parallel Ground effort, we will reproduce and study the selection of similar gravitaxically altered strains. Finally, the long term cultivation of *Drosophila* flies in abnormal gravitational conditions and the consequent production of multiple generations will allow to test if there is a change in behavior as the culture become established and a corresponding mitigation of the effects on the life spans of the flies. Information on how insects do sense the gravity level is becoming available (Horn and Foller, 1998; Texada *et al.*, 2002) and the physiological basis of the altered response in *Drosophila* may be linked to changes in the sensing or in the responding mechanisms.

Our current goals in biological space research

As already described, there is evidence that multicellular organisms as well as simple unicellular organisms are able to sense the absence of gravity and modify their functional, cellular and molecular properties in ways that suggest the possibility of adaptations to the Space Environment. Obviously, a more

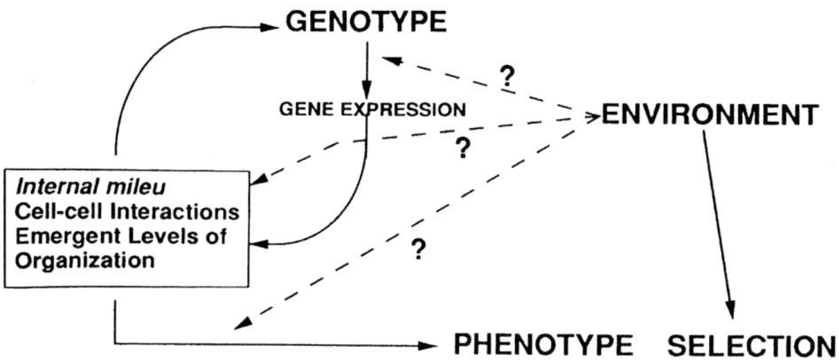

Fig. 2. Scheme of potential links between the environment and the genotype-phenotype connection.

thorough study of this possibility will have to wait until we have access to the capability of performing long-term experiments in the Space Environment, something feasible with the facilities that will become available once the International Space Station becomes fully operational. Therefore, it is worthwhile to concentrate now in the preparation of the experimental set-ups that will be able to take advantage of this unique research environment.

Our objective in this enterprise will be to use the possibilities associated to one of the current more popular research systems, *Drosophila melanogaster*, to gather new data that helps in solving one of the more important questions in Biology, how the instructions contained in the genotype are transformed into particular phenotypes under the influence of the environment (Fig. 2). Current existing biological systems are buffered against the alterations of the Environment. In fact, they tend to be relatively insensitive to big changes in environmental conditions, explaining why there has been a general failure in detecting developmental effects in Microgravity, as mentioned above. On the other hand, we can easily perceive that living systems are well adapted to the environmental conditions in which they actually live. In addition, we are also well aware that biological systems, such as *Drosophila*, easily respond to artificial selection protocols producing animals that have altered morphological or physiological features that are prompted by the selection protocols (Pigliucci, 2001). Furthermore the current explosion in basic knowledge in these systems makes for the first time feasible to try to clarify how the selection protocols work.

The current phase. Preparation and use of the ISS

Potential impact of the Space Exploration Program in the field of Biological Evolution

The current phase in the Space Exploration Program is the building and putting into operation of the International Space Station. It can be argued that the ISS is not the final goal of the Space Exploration Program itself. It may be just a milestone in a more ambitious long-term Project, the transformation of the environmental conditions and the colonization of other Planets and Celestial

Bodies by earth-originated living systems. In any case, if the presence of man in Space is going to be more than episodic, it is important to initiate the investigation of the long-terms effects of the exposure of living organisms to completely different environments outside the Earth. The use of appropriate model systems is mandatory, among other reasons, to prevent that man itself serves as the guinea pig for these long-term experiments. Thus, the establishment of more or less permanent and autonomous (self-sustaining) colonies of living beings in Space should be given priority in the Space Programs. In addition to testing and solving the practical problems behind this Enterprise, an additional question that we may ask is how the living systems will respond to this challenge? Although, many of the details are still unclear, very likely life has appeared and evolved on Earth by relative gradual mechanisms and processes known as Biological Evolution. Once biological systems of a certain complexity appeared, systems endowed of the main basic properties of cells (genetic information, catalytic activities and membranes), neo-Darwinian mechanisms of Evolution could start to operate and, eventually, the early sets of pre-biotic systems were replaced by more complex and autonomous organisms. The immense diversification of Living Systems currently found in our Planet has been achieved as a consequence of dialectical tensions operating on and in living systems, i.e., the maintenance of short-term 'exact' replication together with long-term variation, selection and adaptation to environmental conditions. In fact, the importance of getting a better understanding of Biological Adaptation, including the solution to the long standing controversy on the relative importance between neutral and selective forces in the shaping of this diversity is being increasingly discussed and studied (Goodwin, 1990; Rose and Lauder, 1996; Raff, 1996; Gerhart and Kirschner, 1997; Marco *et al.*, 1999; Arthur, 2000; Brock, 2000).

Although, the increase in our knowledge of the actual molecular, cell, organism and population related mechanisms involved in these processes is remarkable and continues increasing at an accelerated path, most questions remain unanswered. Some are quite difficult to study, in part because the process is quite slow, the evidence insufficient to discriminate among powerful theoretical alternatives and the results many times unexpected even under laboratory conditions (Rose and Lauder, 1996). It is not even clear what makes an environment radically new and different and how the response is influenced by random events.

One interesting consequence of performing relatively modest and feasible experiments using the ground and flight facilities that the new phase of the Space Exploration will be offering to the Scientific Community is that it will contribute to the answer of these questions in addition to the identification of any potential deleterious effect on living beings of the long-term exposure to the Space environment. This is an important goal of the group of experiments discussed in this Chapter. A shorter-term goal is the development of the hardware and techniques that will make possible the establishment of permanent colonies of certain Model Organisms in Space. Obviously, we intend to take advantage of all the current powerful techniques available to examine the modifications that the organisms will experience during this long-term exposure to this unusual environment. In our case, using *Drosophila*, we will monitor the phenotypic and genotypic changes that they will undergo as well as to identify the particular molecular and cellular mechanisms that will be modified in this process. This type of experimentation should follow equivalent ground experimentation in modified environments, using similar facilities and techniques as will be available later in the ISS. In these ground experiments, we will also monitor the overall changes that *Drosophila* may undergo when permanently exposed to modified gravity environments on the Ground or selected for different responses to the gravity vector.

The use of Ground Simulation Facilities is a key part of the present phase of Research. It is true that gravity is a pervasive force and it is impossible to really eliminate its action when standing in our Planet. Nevertheless, a growing arsenal of techniques and equipment is becoming available in ground research laboratories, particularly in special Centers supported by Space Agencies, such as the User Support Operation Centers that the European Space Agency is fostering in Europe. This approach allows at least three things: (1) to test the hardware and technical support of Space Experiments much before they are submitted to the real conditions in Sounding Rockets or in parabolic flights; (2) to test the feasibility of the actual experiments selected for Space Missions; and (3) and more importantly in a first approximation, to select from the enormous list of possible experiments, animal strains, gene markers, etc., those that more probably could give information in real Microgravity and Space conditions. Among the types of equipment currently available in these Centers the following ones seem the better adapted for the *Drosophila* experiments. (1) Compensating gravitation equipment such as clinostats, where sample orientation is regularly changed with respect to the normal gravity vector. Originally the equipment consisted of rotating clinostats (slow and fast) that could not easily accommodate large samples and pieces of hardware and observation equipment such as microscopes and/or video cameras. The recent availability of the 3D clinostat also called the Random Position Machine has solved many of these problems. (2) The Magnetic Levitation Machine is based on the magnetic susceptibility of the individual atoms that make up the molecules when subjected to magnetic fields of enough intensity (Berry and Geim, 1997). The size of the exposed organisms is limited in the available equipment, but it can easily accommodate *Drosophila* samples. On the other hand, it may not be used for long-term exposures. (3) Hypergravity delivering centrifuges in the range of a few g-level and capable of accommodating observation equipment are also available.

In this respect, it has to be pointed out that applying centrifugal forces, i.e., hypergravity conditions, to developmental systems was an often used experimental tool in the hands of certain classical developmental biologists, leading to important phenotypic effects on the development of certain biological species (Kalthoff, 1996). On the other hand, there is hardly any information on the effects of the exposure of *Drosophila* during development to Ground altered gravity simulation facilities.

Finally, we believe that the selection of the key model organisms that could be incorporated in experiments of long-term exposure to the Space Environment is very important. They could become part of the Life Support Systems that will allow the participation of Human beings in this new phase of Space Exploration. Furthermore and in relation to the problem of how evolutionary adaptation of living beings actually occurs, it is important to study the adaptation of these organisms to extraterrestrial environments. As stated by Edgerton and Roy in a relatively recent review (2000) 'it is not really a question of whether humans have the resolve to explore space. The question is which generation will be able to take advantage of this resolve'.

The reasons to select Drosophila for long-term experiments in Space

As already mentioned, *Drosophila* is a relatively small resilient organism, which has been used quite often in Space Research in the past. It is possible to grow relatively large quantities of flies in relatively small containers (but see below for some open questions). We are setting experimental conditions that will allow to obtain as many generations as desired (or as the storage capability and the occupancy of the International Space Station will make possible) to monitor the long-term effects of the exposure to the Space Environment of a population of a complex multicellular organism. Since the life cycle of *Drosophila* takes approximately 2 weeks at room temperature, this means that per year we will be able to grow more than twenty generations in Microgravity. Furthermore, since *Drosophila* is the higher organism for which we know more in terms of genetics and development (Powell, 1997), it will be possible to study what genetic changes do occur in this environment and how they can explain the phenotypic changes, biochemical, morphological, functional and behavioral that may be occurring during this long-term experiment.

For example, as already mentioned, the complete euchromatic sequence of *Drosophila* is now accessible, completed by a joint effort from private and public sources. The current impression is that the sequence is very useful, although it will be corrected in the future to become completely operative. Nevertheless, the availability of the sequence and the corresponding identification of the genes included in the sequence is providing tools for the examination of the gene expression profile in the fly when exposed to Microgravity and as it adapts to the Space environment. The usefulness of this approach is indicated in many publications, where the applications of microarrays to the detection of the changes in whole group of genes have been studied. In particular, the differential expression of the one-third of the genome complement during *Drosophila* development has began to be explored

(Arbeitman *et al.*, 2002). They show data generated at multiple time points throughout *Drosophila melanogaster* development, and identify new genes engaged in a broad spectrum of processes, including the patterning of the early embryo and senescence in adults. There are already publications that indicate the potentiality of this approach in detecting the overall changes of gene expression in Space (Hammond *et al.*, 1999; Lewis *et al.*, 1998, 2001) or in simulated Microgravity on the Ground (Khaoustov *et al.*, 2001). Many genes change expression when compared to the ground controls confirming what we have already mentioned, i.e. that isolated cells sense the absence of gravity and change their responses. What is still left to know are the functional consequences and the adaptations produced by these changes in response. During a long period of time, classical genetics have been merging with molecular genetics, but now this confluence is being extended to quantitative genetics. The recent advances in Genomics have defined the number of genes in a series of familiar organisms. Even though these numbers may change as information on Functional Genomics becomes available and as information on different strains in the same species is obtained, they put a limit on the number of ways that this information can be organized. The limit is extremely high and new methods of analyzing this information will emerge, they are indeed emerging already. Information on how these organisms are changing in different environments may provide critical inputs on how to test these methods and ideas. Space research may contribute to the solving of this challenge pushing forward how we understand biological processes.

So far we have defended the optimal nature of *Drosophila melanogaster* for this objective. Nevertheless, there may be also shortcomings in the decision of choosing this organism as any other organism that has become since many years ago the material of research in biological laboratories. The plasticity of biological systems may have produced a special type of organism. The standard laboratory wild type *Drosophila* is the product of a century of laboratory work (Kohler, 1994). They have become adapted to the relatively simple life style in them. They have been stripped out of their natural repository of variability. Selective inbreeding and additional manipulations could have suppressed much of the hidden genetic variability necessary for survival in the wild. Competition between different individuals derived from changing environmental factors has been minimized allowing the survival of less fit individuals. Therefore, are laboratory strains of *Drosophila melanogaster* the ones to be used in these experiments or should we turn to specimens caught directly from nature? Has not *Drosophila melanogaster* been following peasant populations originally from South Asia around the world? These are queries that can only hopefully be addressed by the types of experiments that the new Biology, based on Genomics and Proteomics, allows us to perform nowadays. The advantages of the huge information available on this organism makes difficult to escape from the conclusion that it should be one of the main actors of the new phase of Space exploration, but is it really the best one?

The experimental approach with Drosophila

(1) *In vivo* observation of *Drosophila* development using specific probes adapted to the detection of different developmental processes, namely, cell proliferation, cytoskeleton (CSK) and gene expression patterns including signal transduction pathways.

Embryonic development is usually very robust and apparently independent of most environmental conditions. As we have mentioned, this robustness may be responsible of the normal development in Microgravity of wild type embryos. One approach in ground control experiments is to find out whether transient developmental alterations caused by simulated Microgravity or Hypergravity treatments may be more readily seen or enhanced in appropriate hypomorphic mutants in one of the processes known to be altered in cells cultured in Microgravity, such as the cytoskeleton organization, the signal transduction pathways or the cell proliferation patterns.

Two types of observations are foreseen: The main one is the dynamic *in vivo* observations using appropriated markers. This will be complemented by the post-experimental analysis of fixed samples. In the ISS, it will be possible to monitor the *in vivo* behavior of the fluorescent markers in Microgravity using one of the ISS microscopes equipped with fluorescent illumination. Moreover, in the ground simulation experiments, this dynamic observation will be also possible and the experiments will involve the immediate observation of embryos exposed during different periods of time to the simulation conditions. In these experiments the particular expression patterns of certain developmental regulatory and selector genes that are under the control of cascades of similar genes expressed in the same cell or in near-by cells will be studied.

The current power of the molecular methodology has made possible to visualize *in vivo* the behavior and switching on/off of defined genes. The method employs chimeric DNA molecules where the relevant sequences of these marker genes are coupled to a so-called reporter gene. The more popular one is the green fluorescent protein (GFP) from the cnidarian *Aequorea victoria* which is capable of producing fluorescence without an exogenously added substrate (Amsterdam *et al.*, 1995). It can be linked into chimeric genes which have been visualized *in vivo* in *Drosophila* embryos, larvae and adults (Yeh *et al.*, 1995; Barthmaier and Fyrberg, 1995). This can be done directly, coupling the reporter gene to the actual promoter of interest or indirectly, for instance, through the use of a specific activator systems, for example, one derived from yeast, the Gal-4 system. Thus, the many Gal-4 enhancer trap lines currently available in *Drosophila* can be used to achieve this objective. In them, the Gal-4 activator is under the control of one of the many controlling sequences in *Drosophila* reproducing the patterns of the endogenous genes. This is achieved by crossing the flies which have these enhancer traps with transgenic flies in which the reporter gene is expressed under the transcriptional control of the upstream activating sequence recognized by Gal-4. In many cases, these flies are fully viable, providing embryos that will directly produce a fluorescent product marking a particular tissue or cell structure that can be monitored under a fluorescent microscope without further manipulation.

In this way, the main development of *Drosophila* can be followed by microscopic observation of live unstained embryos. As will be described below a whole set of fluorescent components, markers of different subcellular elements and processes are now available. Nuclei, cytoskeletal components, mitochondria, membrane receptors, etc. can be now expressed in defined tissues and/or moments of development, providing a simple method to observe dynamically how different embryonic processes actually proceed.

For example, even before gastrulation, the embryo is subdivided in zones: namely, (a) Mitotic domains (Foe, 1989); (b) the precursor zones of the segments, marked by a series of stripes of defined gene expression domains in the antero-posterior axis; and (c) several control genes are activated in the dorso-ventral axis, marking the subdivision into mesoderm, neuroectoderm and epidermis. Many of these control genes are transcription factors, but others are cell signaling and/or signal transduction pathways components. Morphogens are synthesized and secreted by certain cells to interact with neighboring cells endowed with appropriate receptors to induce the activation of a particular different state, usually by a chain of signal transduction pathways. These inducing phenomena due to the interaction of neighboring territories is one of the earlier recognized phenomena participating and shaping the complexity of embryonic development.

The actual number of components whose cell expression can be monitored is very large. Information on the transcription regulation and signal transduction cascades is becoming readily available in *Drosophila*. The main question is to decide which ones to try or to focus on. As mentioned before, looking into the changes in overall gene expression patterns of adult flies in Microgravity using microarray techniques may help in this decision. Since the processes so far found altered in Microgravity, real or simulated, are cellular related events, implicated in the mechanisms of proliferation, signal transduction and/or cytoskeletal organization, we are going initially to chose markers of events occurring before and after blastoderm cellularization. Before blastoderm cellularization the whole embryo is a huge single cell territory where nuclei are moving and dividing. After cellularization cells are much more smaller. Therefore, these different stages provide very different targets to the alteration in the gravitation vector. The experiments performed in the Ground Simulation Facilities should be critical to clarify which ones of the many possible markers may be more informative to be used in Space.

As a first approximation, three processes seem the best candidates for *in vivo* observation: (1) Markers of general cellular components, which can be used to follow cell division, movement and shaping during development (nuclei, cytoskeleton and cell membranes seem reasonable first choices). (2) Markers of signal transduction pathways. As a first possibility, we propose to use a particular well-studied process such as the embryonic dorsal closure at the end of germ band retraction since the involvement of several signal transduction pathways has been well established in this process (Jacinto *et al.*, 2002). If we turn to defined genes, three types of genes seem prime candidates for this Investigation: (A) The group of pair-rule genes, that produce seven stripes immediately after cellularization: *fushi-tarazu*, *even-skipped* or *hairy* among the better known ones. (B) The group of segment polarity genes, such as *engrailed*, *wingless*, *hedgehog* or *patched*, that produce fourteen stripes, one for each segment and where *hedgehog*, one of the better characterized intercellular signals is operative. (C) Tissue specific gene markers, such as those involved in mesoderm, ectoderm or nervous system specification. In addition, it would be interesting to monitor the cytoskeleton in the cells (Fig. 3). This could be attempted with cytoskeletal-GFP chimeric genes that express proteins capable of polymerizing into a particular cytoskeletal structure. The construction of such a hybrid with the *Drosophila* MER homolog (Moesin-like gene product) has been reported (Edwards *et al.*, 1997). In this way it would be possible to monitor the Microfilaments Cytoskeletal Network *in vivo*. The production of a hybrid with the Nonclaret disjunctional (Ncd) microtubule motor protein fused to the green fluorescent protein (GFP) is capable of monitoring the microtubules (Endow, 1999). The incorporation of a nuclear signalling sequence in the GFP product labels the nuclei, etc. The selection of the more appropiate markers and the developmental processes to be followed will be done during the ground work in gravity simulation facilities. During this phase of the work, a hypomorphic mutant affecting dorsal closure will also be used, mutant selected with the help of Dr Enrique Martín-Blanco (Centro de Investigación y Desarrollo, CSIC, Barcelona, Spain), one of the main investigators studying this process. An important point to make is that the GFP protein maintains its fluorescence in fixed material, what makes possible to combine *in vivo* on flight/simulation observations with post-experiment analysis after recovery.

As mentioned, these developmental experiments can be complemented in the future with similar experiments using selected *Drosophila* strains, such as strains selected on the Ground for a different response in the developmental process found to be modified in Ground Gravity Simulation conditions. The increasing availability in *Drosophila* of a whole range of mutations that may reproduce or affect the same developmental process is well known. In this way, the possible mechanism of the Gravity environment on Development may be further investigated.

(2) Aging and behavior in Gravity altered conditions. *Observation of Drosophila behaviour and aging in Ground Simulation Facilities*. The methodology to be used is straightforward. The life span of the groups of flies will be followed by counting the number of flies that remain alive after

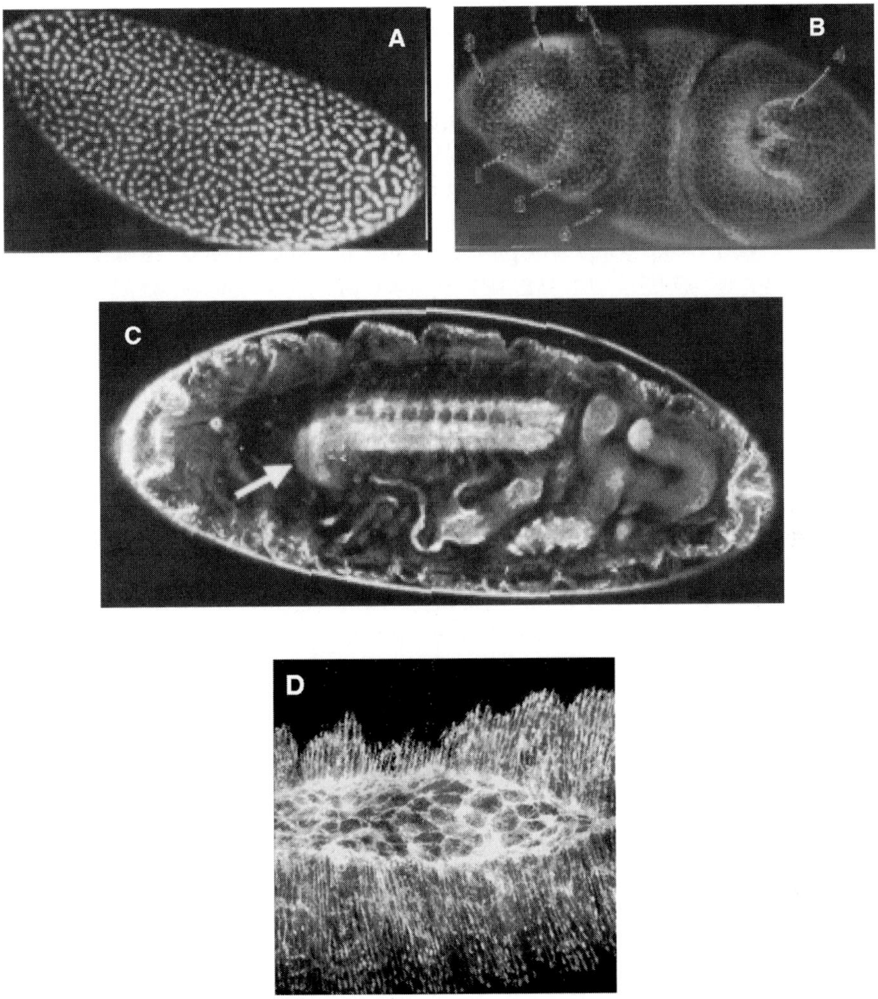

Fig. 3. Examples of *Drosophila* embryos labeled in the microtubular network (A in mitosis at an early syncytial blastoderm stage and B at a later embryonic developmental stage when complete cellularization has taken place) or in the microfilaments (C, the arrow points to the Central Nervous System. D, Microfilament involvement in dorsal closure, an embryonic operation dependent on cytoskeletal movements directed by signal transduction pathways (higher magnification).

different periods of the experiment. For hypergravity (centrifuge) and the Random Position Machine (3D clinostat) the *Drosophila* chambers developed for the ISS can be mounted and/or adapted to the Facilities. These chambers are provided with a transparent wall, facilitating the use of video observation for establishing the number of dead flies at different times during the experiment, as well as the level of motility. This parameter can be studied using the same observation capability and also using the infrared beam crossing incorporated in these containers.

The accommodation in the Magnetic Levitation Facility will be different. The flies will be incorporated in small cylinders with transparent walls. The actual size of the cylinders will have to be decided, since only the ones with tiny dimensions guarantee homogeneous levitation. The actual number of flies in the containers depends on their size, ranging from five flies in the smaller ones to 100 in the longer ones. Flies will probably not be able to feed while levitating in the Facility. For short-term experiments, of up to a few hours, this does not affect too much the survival of flies, but for long-term experiments, the experiment design will have to be modified substituting continuous levitation by periodic levitation, for instance, of a few hours a day. In any case, it is not clear whether long-term experiments will be possible in the Magnetic Levitation Facility. Effects on development will be examined by exposing embryos, larvae and/or pupae to this condition.

The experiment will be complemented by the fixation of some flies at different periods. In these fixed flies the level of defined gene expression, such as mitochondrial ones or even the overall pattern detected using microarray techniques can be followed after recovery from the facilities.

As done before, experiments will be performed with males to avoid the complication of contamination of the containers with embryos and larvae laid by the females. Nevertheless, in future Space experiments incorporating flies from successive generations in Space, they will include mixed populations of females and males. This is a consequence of the difficulty of separating males from females in Space. To test mixed female and male populations in aging and behavior experiments, we will use a different feeding medium, the so-called sugar medium made up of only sugar and agar, a medium compatible with the survival of *Drosophila* adults but not with the survival of the larvae. In this sugar-medium it is possible to maintain alive, mixed female and male populations without the complication of the progeny growth completely restricted in these conditions.

Finally, the aging experiments can be complemented with oxygen consumption measurements at defined times of the life-span curve, using an oxygen consumption measurement unit (see below for discussing the problems in this measurement).

As already mentioned, these behavior and aging experiments can be complemented with similar experiments using specific *Drosophila* strains, such as strains selected on the Ground for a different response to the gravity vector or strains showing different longevity curves. The analysis of these experiments will be critical in the decision on which strain to include in Flight Experiments with higher likelihood of giving important information on the effects of the Space environment on living beings.

Post-experimental analysis. In addition to analyze in detail some of the parameters of the experiment (video-recordings and oxygen consumption rates, food consumption rates, morphology of the preserved flies, life-spans, etc.), fixed material will be recovered both from Ground Simulation experiments and from real Microgravity ISS flights. Immunocytochemical and fluorescent observation on fixed material will be possible. Several molecular parameters can be also measured in this material such as the levels of mitochondrial RNAs and/or defined proteins such as muscle proteins. As already mentioned, it has been reported that in vertebrates, muscle proteins gel electrophoresis profiles change as a consequence of the lower weight load in Microgravity (Kraemer *et al.*, 2000). In addition, microarray analysis will establish the overall gene expression profiles in these aging populations of flies exposed to a gravity altered environment on the Ground and in Space. All this information will help and provide support in the identification of the altered mechanisms and processes.

The problems to be solved

Constraints in the ISS

At least during the next few years, the people involved in Biological Space Research should learn to live with the limitations currently imposed in the utilization of the ISS. While the crew remains restricted to three austronauts, the amount of work performed by the crew outside the normal maintenance of the ISS will be very limited. Developmental experiments will require the replacement of feeding trays and in addition, trays will have to be manipulated to prepare embryos for *in vivo* observation in flight and for fixation, once every couple of weeks. For aging experiments, the flies require less attention, making the experiment almost crew independent. Adult flies will have to be fixed at certain moments during the experiment to test the level of nuclear and mitochondrial gene expression. Obviously, these constraints do not affect the simulation facilities that will allow a better definition of the Space experiments. The type of hardware we are building (see below) is able to maintain flies during long periods of time with little or no crew involvement. Automatic embryo and adult flies preparation and fixation units are being designed and constructed as well.

A second problem will be the limitation in Storage capabilities, especially deep freeze facilities, if the experiments are to be maintained during the whole period of time between refurbishment flights to the ISS (at least 3 months). Although, a bit unclear, storage space certainly will be a scarce resource, a constraint that will emerge as soon as more experiments find their way into the ISS. Again, this problem is less important for ground simulation facilities. The optimization of the experiments in the Ground Simulation Facilities will be critical to maintain this requirement under control, since the experiments to be flown shall be fewer and better designed.

How to grow and handle Drosophila in Space

The constraints already discussed make necessary the previous study and modification of current methodological protocols to make them compatible with the Space conditions. The necessary hardware for *Drosophila* growth and handling in Space has to be also designed, built and tested.

Optimization and adaptation of the fixation methodology of Drosophila embryos and adults to the Space conditions

As far as one can tell, few attempts have been made in the past of fixing and/or preserving in an inactive state in Space the organisms to be studied afterwards on the Ground. *Drosophila* embryos, larvae and imagoes are surrounded by tight envelopes and cuticles extremely difficult for fixating fluids to penetrate. In some cases, embryos were kept frozen, not an optimal preservation method for further study after recovery on the Ground (Marco *et al.*, 1996).

Even in the case of the work in ground laboratories, the preservation of living *Drosophila* animals is a not easily solved problem. Some success has been achieved in the case of *Drosophila* embryos but the methods are complex enough not to make easy their adaptation to the Space conditions (Steponkus *et al.*, 1990; Mazur *et al.*, 1992; Ashburner, 1992).

As we have been discussing, it is almost certain that *Drosophila melanogaster*, one of the key model systems in current Biological Research, will be among the organisms accommodated in the International Space Station (ISS) or in Future Space Missions connected with Life Sciences. For example, our Madrid group is interested in the study of the *Drosophila* long-term adaptation to this unusual environment. Thus, sooner or later, the problems of fixation of the different stages of development will have to be addressed. Experiments performed in the Madrid laboratory in the past couple of years have succeeded in adapting the current techniques to the Space constraints. Designing and building an automatized fixation unit is underway (see below). With some success, current protocols have been adapted to maintain fixed animals during at least 3 months, the minimum refurbishment shuttle flight time with minimal handling and maximal automation. These protocols can be used for immunocytochemistry, morphological examinations at the level of the Optical Microscope and Molecular Biology experiments involving DNA, RNA and possibly proteins.

Simplified permeabilisation/fixation protocols for Drosophila embryos

A permeabilization-fixation protocol, simple enough to be used in the Space conditions and able to preserve embryos during at least 3 months has been developed. The first step in the protocols, dechorionization by a short treatment with a hypochlorite solution, should not give many problems when carried out in the Space Conditions. On the other hand, the following steps in the classic protocols based on formaldehyde-heptane treatments, are much more difficult to be adapted to Microgravity conditions. The much simpler Steponkus-Cadwell protocol (Steponkus *et al.*, 1996) although capable to open holes in the embryonic vitelline membrane through which a good nuclear DAPI signal can be obtained, is far from optimal for immunocytochemistry.

The more successful protocol developed in the Madrid laboratory uses a commercial non-formaline fixative, NEOFIX©, during 10 min just after two short treatments with isopropanol and hexane (permeabilization reagent). After this step the embryos can be stored in methanol at 4 °C and after sample recovery the procedure can be successfully completed with a post fixation step with Heptane/PBS and Heptane/Methanol. This protocol combines the first part of the Steponkus-Cadwell protocol and the last part of the Heptane classic protocol but without paraformaldehyde (PFA).

Using PFA as fixative, the methanol conservation of embryos can be extended more than a year without much loss of detail. The NEOFIX© Protocol can be used in the ISS, and can detect the Central Nervous System at least after 16 weeks of conservation in methanol at 4 °C (see Fig. 4). Preservation at room temperature causes significant losses of signal. Morphology is acceptable when studied at the optical microscopy level, but not surprisingly due to the use of alcohol based-fixatives, the preservation at the electron microscopical level is poor. One has to be reminded that getting good electron microscopical images from *Drosophila* embryos requires the utilization

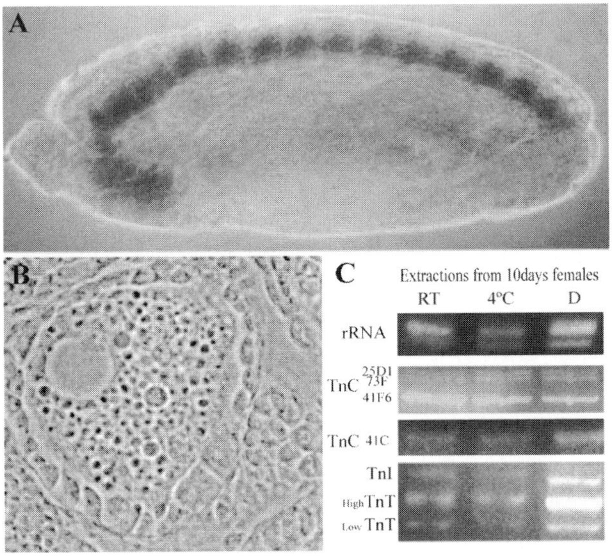

Fig. 4. NEOFIXfixation results. (A) Immunocytochemistry with ELAV antibody. (B) Phase-Contrast embryo microscopy. (C) RNA extractions &RT-PCR for Troponin cDNAs

of strong fixation procedures using high concentrations of aldehydes and osmic tetroxide (Ashburner, 1989), procedures difficult to adapt to the Space conditions and possibly incompatible with immunocytochemistry. Electron microscopical observation of *Drosophila* larval and imagoes tissues require dissection of the specimens and fixation of the isolated organs, again a procedure difficult to adapt to the Space conditions. In fact, if one examines the huge *Drosophila* literature, not many studies using electron microscopical approaches will be found.

Immunocytochemistry and molecular biology applications

In addition to the ELAV antibody, specific for the Central Nervous System, anti-Tropomyosin (TM) & anti-Troponin T (TnT) antibodies have been applied for immunodetection of the Muscular system in *Drosophila* embryos conserved in methanol for 10 weeks at 4 °C. In the case of TM the immunosignal is weak using either PFA or NEOFIX© fixation. In the case of the TnT antibody, the immunosignal is strong and similar in the two experimental conditions (data not shown). Thus, the success of long term-delayed immunocytochemistry is very much antibody dependent and the optimal conditions should be worked out in each case.

The NEOFIX© protocol tends to return a weaker and more diffuse immunosignal than PFA fixation. These results suggested a less than optimal level of morphological preservation. The same embryo fixation protocols were followed by rehydratation and LR White Resin inclusion after 3,5 months in methanol. In Figure 4B we can see a 2 μm sections of a NEOFIX© fixed embryo in phase-contrast microscopy. The embryo morphology was similarly preserved at the level of Optical Microscopy no matter the protocol used.

Apart from morphological and cell biological techniques, the studies were extended to Molecular Biology. The optimal and simpler preservation method for this purpose has been dehydration in cold anhydride acetone. Conservation at room temperature also gives acceptable results. In samples stored at −20 °C and 4 °C, conservation of rRNA is almost perfect during at least 1 month (data not shown). In Figure 4C we show that rRNA is preserved after 7 months in acetone at 4 °C or even at room temperature (amount reduced approximately two-fold). In RT-PCR experiments involving the detection of the Troponin C transcripts, all expected bands appear revealing the isoform pattern for embryos and for adults.

Preservation of living embryos, larvae or imagoes of* Drosophila*. How to recover living samples of every generation?

This is an open problem for *Drosophila melanogaster*. While some organisms, such as *C. elegans*, *Artemia franciscana*, even some other *Drosophila* and insect species have developed mechanisms of adaptation that allow the preservation of living animals or embryos during long periods of time with minimal requirements, so far this capability has not been solved in an uncomplicated way for *Drosophila melanogaster*. The methods that have been published which allow to preserve living embryos in frozen conditions are complicated and relatively inefficient and, thus, not easily applicable to the Space research. This fact has also the consequence of the maintenance of large collections of living stocks in the Stocks Centers and in the multitude of *Drosophila* laboratories around the world.

Drosophila embryos, larvae and imagoes are cold sensitive and will not survive relatively short exposures to low temperatures. The relatively cold sensitivity of *Drosophila melanogaster* (Czajka and Lee, 1990; Steponkus *et al.*, 1996) puts a limit to the possibility of using lower temperatures to increase the survival of defined generations until the ISS is refurbished, since the developmental embryonic, larval and pupation period can be slowed down only up to a month and a half (Petavy *et al.*, 2001) by keeping the animals at 12 °C, the minimum temperature compatible with relatively high survival numbers. On the other hand, preliminary experiments indicate that other Drosophila species, particularly, *Drosophila subobscura* has higher survival levels at lower temperatures. The genome sequencing of *Drosophila subobscura* is in a very advanced stage and therefore it could provide an alternative if it were necessary.

Obviously much more work has to be carried out before *Drosophila* methodologies are considered ready and adapted for long-term Space experimentation.

Hardware situation

The Biolab facility is the laboratory designed to support biological experiments on micro-organisms, cells, tissue cultures, small plants and small invertebrates on board the ISS (see the chapter by Enno Brickmann). The Biolab will be installed inside the European Columbus module which is scheduled for launch in October 2004. In addition to its general facilities as the glovebox, incubators, or freezers, the Biolab uses standard Experiment Containers mounted on two centrifuge rotors that provide either Microgravity or variable accelerations. One is typically used as 1 *g* reference, allowing to study the specific effect of weightlessness. The experiment hardware located inside the experiment containers, the Experimental Support Equipment, are provided by the investigators. Ten types of Experimental Support Equipment have been tested by ESA as bread-boards for use inside the Experimental containers for experiments with cell and tissue cultures and with small animals (see the chapter by Enno Brickmann).

Under the sponsorship of the European Space Agency, we have designed, developed and tested the Experimental Support Equipment item1 (ESE-1) for *Drosophila melanogaster*, in accordance with the Experiment Container and the

Biolab facilities requirements. For this purpose, the special constraints likely to be encountered in any ISS mission were taken into account, such as the scarcity in energy, storage room and crew-time availability. In addition, we have been taking into consideration that sample exchange and equipment refurbishment will take place not before than 3 months, when Shuttle Flights are expected to visit the Space Station. Consequently, any experiment in the ISS requires the development of specific methodologies and equipment. On the one hand, as already shown, we have modified basic handling techniques to make possible experiments with *Drosophila* in Space conditions. On the other hand, the hardware we have developed is capable of maintaining *Drosophila* multigeneration experiments in the ISS taking advantage of the Biolab/MSC facilities. In addition of this main unit we are designing devices for embryo preparation and observation in flight, fixation, preservation or storage. All the hardware presented here would give the opportunity to study the adaptation of *Drosophila* to space environment in terms of development and behavior, at the level of phenotypic, cellular and molecular analysis.

Drosophila *multigeneration and behavior studies with the ESE-1 main device*

The Experimental Support Equipment Item 1 (ESE–1) of ESA is an automated culture system for multigenerational experiments with insects. Based on a concept that maximizes automatization, ESE-1 will allow the culture in space of many generations of fruit flies to study adult behavior and embryo development with minimal participation of the crew. The main unit consists in two cultivation chambers and a set of six food trays that can be automatically transferred into the cultivation chamber at specific times to feed the flies. Photographs of an improved ESE-1 unit are shown in Figure 5. The main improvements in the current device are the following: the chambers have been lined with a white Teflon covering, the windows are now made of glass instead of the old plastic ones. These changes allow a better visualization of the flies in the Chamber. Also, the drum has been tightened and the access from the trays to the chambers has been made more hermetic to avoid fly trapping and/or escaping.

Inside the unit chamber, an infrared light and detector can be used to automatically monitor fly activity in combination with the video images obtained through the transparent wall of chambers. To complete these observations, an analysis of the adaptation of the flies to normal gravity after a long-term space flight will be performed, in term of viability, geotaxic response, mating and fertility.

In ESE-1, the tray exchange may be programmed to take place during short periods of time (about 4–8 h) to collect undeveloped embryos. The embryos and larvae in the food tray can have different fates: (a) to be discarded and killed by the specific killing actuator, that provides enough heat to the trays to kill the embryos without affecting the stability of the feeding medium (see Fig. 5B);

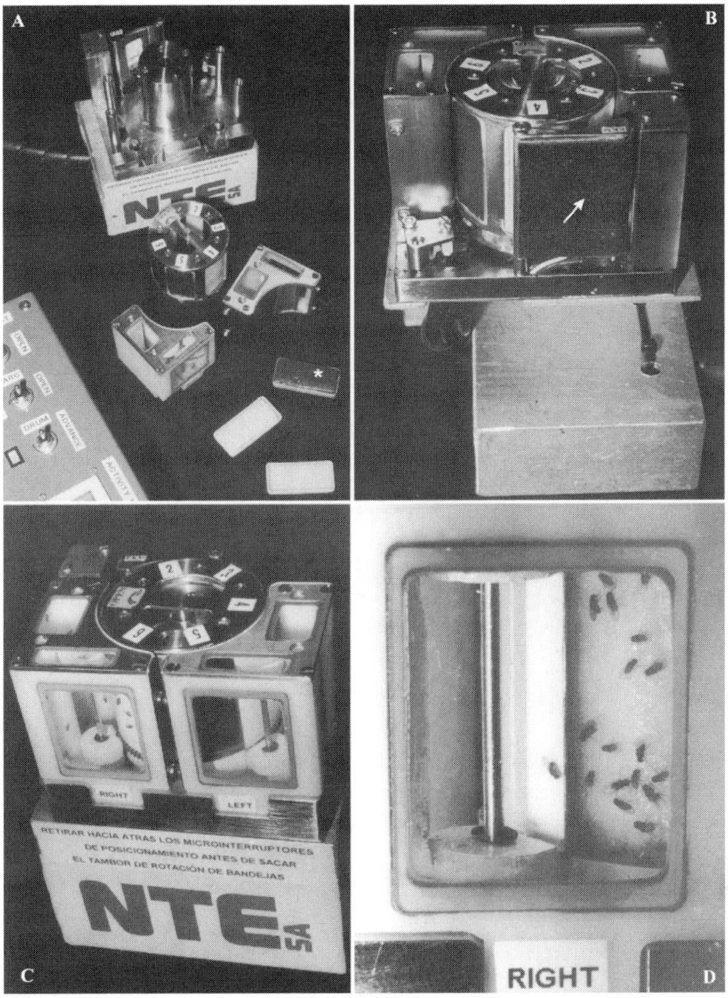

Fig. 5. Improved ESE 1 unit. (A) Dismantled, the drum and the chambers can be seen as well as the trays loaded with feeding medium (*). (B) Both chambers have been installed. View from behind with the killing actuator in the front (arrow). (C) Complete loaded with flies. Right chamber door closed, left open with the medium visible. (D) Close-up of the flies feeding on the tray.

(b) to be transferred to the second chamber to obtain a new fly generation; (c) to be removed for in-flight observation with the Biolab microscope; and (d) to be removed for fixation and storage until post flight analysis.

A prototype of a sample collection/fixation unit

As described, most fixation procedures require the use of different fluids and washing cycles between fluids. Furthermore, some of the fixation procedures

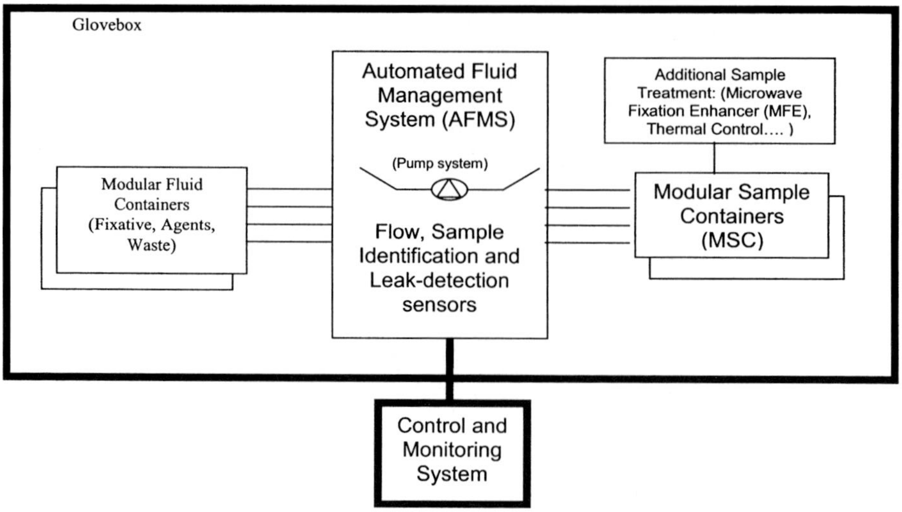

Fig. 6. Possible concept for a fixation and preservation facility. Different fluids stored into a container specifically designed to store hazardous fluids are circulated in an experiment-defined sequence through the sample. The facility has the capability of transferring samples or part of the samples from one sample container to other (carried by circulating fluids), performing filtration, concentration processes, etc. and also has thermal control and possibility of electromagnetic radiation enhancement of the fixation activity.

may also be facilitated by certain treatments such as microwave enhancement (Medina *et al.*, 1994) and mechanical operations, as filtrations or phase separations. As far as we know, no currently existing multi-user facility in the ISS allows for such a flexibility of experiment peculiar procedures, nor do they guarantee a reasonable safety level for the operations.

The approach suggested for biological sample fixation in an especially dedicated facility, consists of leaving the sample in the EC during the initial phase of sample preparation until the sample is confined within a container, which is suitable of being inserted into a common facility (this will be a requirement for the experiment developer). This facility shall be especially prepared for safe handling of hazardous fluids, and it shall be equipped with thermal control and with the possibility of microwave or UV fixation enhancement.

Figure 6 presents a basic block diagram for such a facility, which is basically a fluid handling facility. This facility could consist of two bodies. The first body would serve as storage for the different fluids, including waste fluids. Thermal control of the reservoirs does not need to be active, since this part of the facility can be stored in a cooler/incubator until use.

The second body would contain pumps and valves allowing fluids to circulate through the sample and physical stimuli to be applied to the sample. It may even be capable of taking a sample from the experiment container/culture chamber, and guiding the probe into a vial, which then undergoes the fixation or preservation treatment, or to an observation chamber capable of being mounted on an ISS microscope.

An example of a complex fixation process is tissue fixation for histological or cytological examination, or, alternatively, fixation of adhesive cells in culture. In particular, cultured cells must be detached from the surface on which they are growing before their fixation. This step requires removal of the culture medium, washing, enzymatic detachment (in most cases cells are trypsinized), and final wash of the loosened cells.

Exact amounts of chemicals and accurate time control are crucial parameters in maintaining the morphological quality of samples, as well as their molecular integrity, especially concerning the possibility of performing further investigations involving DNA.

A tentative experimental protocol could be as follows: The culture plate or the vial with the suspension of cells on micro-carrier is submitted to a continuous flow with a suitable buffer. A first flushing step replaces the old medium and washes the cells. A second sequence, in which the continuous flow is stopped, would allow exact trypsinization of the culture. Then the cells could be transported, through the second washing process, to a different chamber, in which they are stopped by a filter to be ready for the final treatment (reseeding, fixation, staining), in a similar way as described in the above example.

This device could also serve for *Drosophila* embryos permeabilisation, fixation and storage as described in a previous section. A modification of the concept will allow to store embryos and adults until post-flight for specific genetic expression analysis.

The problem of the measurement of the oxygen consumption of the adult flies in Microgravity

Another problem to be solved is the availability of a procedure for efficiently measuring oxygen consumption of flies and/or other biological samples in Space. This is a problem that remains untackled. The critical problem is guaranteeing the hermeticity of the chambers where the measurements will be made and to avoid the influence of other organisms such as those that may grow in the *Drosophila* food. An adaptation of the adult fixation unit could work. Once the flies are transferred into a hermetically sealed unit, oxygen consumption can be measured. The basis for the measurement could be either the drop in the pressure as the carbon dioxide produced in respiration is fixed with a base such as is carried out in the old Warburg respirometer, the decrease in the pH of the CO_2 fixating solution or any equivalent approach. If the unit could be mounted in the available centrifuges and in the Random Position Machines, control measurements at gravity-altered conditions would be possible.

The Future

The next step. The long-term adaptation to the altered ground simulation/ISS environment

Once the results of the current phase of research are known, the path will be open to start the preparation of a long-term goal, the establishment of a permanent colony of *Drosophila* in the ISS and the investigation of the changes that the organism may endure there.

In the case of these long-term experiments, if changes in developmental processes, in behavior and aging responses have been previously detected in Ground simulation experiments and in intermediate length experiments in Space, the study of how they will continue changing will be the focus of the experimentation. In these experiments, an important aspect will be the identification of the genes that may be changing as the experiment proceeds. In addition, the use of mutant strains may give clues on how to stop or even accelerate adaptations and changes. Isolating RNAs from the different generations after the recovery of fixed adult samples will allow to check the gene expression profiles using microarray technology: How many genes will change expression during the long-term adaptation to the Space environment? Will they be related to the possible phenotype changes that the animals are experiencing?

One problem with these experiments is the limited size of the experimental populations. Although, large in comparison with other animal systems that may be also involved in Space Research, the actual size of the populations will be limited to about 50 individuals in each container, thus, probabilistic drift effects may play a big part in what we may detect. This is an argument to support that the experiment is carried out in parallel in several containers, probably at least four or six and that populations are occasionally mixed among the containers to increase the actual effective population size. Carrying these experiments well in advance in ground simulation conditions may help us in better designing the real Space Experiments. Actually the ESE-1 container has a special tray that will allow mixing of populations in the two chambers.

What can we expect?

In short or intermediate length experiments, namely, those in which one or several generations of *Drosophila* will be grown in Space, there are two extreme distinct possibilities with regard to development: (A) Clear detectable differences may appear between the developmental mechanisms and processes in simulated and real Microgavity and those that take place under normal gravity conditions, independently of whether the overall process is finally regulated and the final outcome is a normal larva and fly. In this case, the use of hypomorphic mutants affected in the same processes probably will enhance the differences, making possible the identification of strains that are able to develop under normal gravity conditions but they may fail doing so in simulation and real Microgravity conditions. A possible complication is when the results of the ground simulation experiments also differ from the ones obtained in real Microgravity (in the ISS). If so, further experiments trying to establish the nature of the difference will be required. (B) No detectable differences appear. In this case, the conclusion is that the involvement of the gravity vector in the current developmental processes is insignificant if any and therefore, unless long-term effects appear, development will not be an obstacle to the future plans of the involvement of complex animal systems in the future Space Exploration (but see below).

With regard to aging and behavior, there are again two extreme possibilities. The motility changes in the fly behavior may continue correlating with the aging effects. Longer exposures (up till 2 months) to simulation/ real Microgravity will increase the severity of the effects. Flies selected for delayed aging may show a different response than flies selected for short life spans. Mitochondrial genes and function are affected. Oxygen consumption in altered gravity is increased. The use of flies which have lost the gravity perception mechanisms by mutation and/or selection do not show the same response as the normal wild type flies. Alternatively, the effects may become less important as the exposure to the Microgravity conditions is increased.

In genotypic terms, how many genes may change in expression? The use of microarrays will show if the group of genes expressed in the different developmental stages and during the aging process in simulated and real Microgravity are different from the normal gravity conditions. Do different strains with different gravity perception and/or life-spans show differences in the repertoire of expressed genes?

Now we will turn to the possible phenotypic responses in long-term experiments, namely, in experiments where a permanent colony of *Drosophila* will be established in Space:*Loss of graviresponse and the abnormal motility*. One possible outcome of the long-term adaptation of *Drosophila melanogaster* to the Space environment is a possible change in the abnormal motility response. It can be anticipated that the behavior of the different generations of *Drosophila* will become more normal. If that is the case, will also this occur in the Ground Simulation experiments? If the change happens only occasionally, will it be limited to certain groups of flies (will stochastic events be involved in the changes?) or will they happen more consistently? How will the patterns of gene expression change in these populations? Which will be the actual genetic, developmental and physiological mechanisms involved in these changes? Will the graviperception mechanisms, not well understood yet in *Drosophila* become altered in these animals? How will they behave when recovered and transferred back to Earth? How will the aging pattern of these animals become modified? These and other interesting questions could be addressed in these experiments.

Changes and adaptation of muscle/Motility. In mammals, changes in the isoform pattern of muscle proteins have been reported after short-term exposure to Microgravity, similar to the ones found in ground conditions where gravity effects are counteracted, just as bed laying and/or hindleg suspension in model systems. Similar information is missing in the case of invertebrates. *Drosophila* changes in motility behavior in Microgravity but no information on changes in isoform patterns in *Drosophila* muscle has been reported, although there is an increasing knowledge on the patterns of isoform types associated with the different types of musculature in the fruitfly. It is likely that some changes may be also occurring in insects. The possibility of correlating possible changes with the functional changes detected in the behavior of the fruitfly is an interesting outcome of this analysis.

Changes in flying/jumping capability. In previous Space experiments it has been noticed that fruit flies in Microgravity seldom attempt to jump, the first step in the initiation of flight. Flies cannot take off by directly flying but they jump first using a strong second leg muscle, called TDT (Tergite depressor of trochanter). In contrast, flies in the same type of containers on the Ground do attempt to jump much more often. This is an observable change in behavior that can also be studied in long-term experiments to monitor if there are changes in this kind of behavior as new generations grow in Microgravity. This means that flies probably are less prone to fly in this environment, something that has not been directly tested but it is suggested by the observations mentioned at the beginning of this paragraph.

Finally, changes in size and shapes of the fruit flies and/or some of their organs, as well as changes in the size and shapes of the embryos and on the developmental mechanisms affected by Real or Simulated Microgravity could take place after long term exposures to this abnormal environment. Both are relative simple type of observations easily monitored during these experiments. In the case something of this sort is detected, it may give ideas of how the changes may be occurring and which would be their selective advantage if any.

From a genetical point of view, several key questions may be addressed in these experiments, especially if the strains used in the Space/simulation experiments have been selected taking into account the properties of their genomes, the location of potential transposable elements, heterochromatin features, etc.

How stable is the eucaryotic genome in Space? The use of these strains will allow to detect possible long-term changes in overall chromosome architecture or more likely changes in the sites and numbers of transposable elements. It is perhaps not very likely to find many changes in these key elements of the cellular architecture of high organisms, but it may be important to confirm the stability of the genome in this new environment, as a first stage in any attempt to send other organisms, such as human beings, in long exploration or even colonization initiatives.

Are the changes due to selection, random drift or both? The identification of any change in the phenotypic/genotypic properties of *Drosophila* flies exposed during multiple generations to the Space environment, will open the way of analyzing if the changes are selected as a consequence of their adaptive nature or simply the consequence of chance (random drift). To explore this alternative, it will be necessary to run replicas of the experiments both simultaneously or successively.

Are the changes reversible? Finally, the possibility of bringing back samples of the fruit flies to more normal environments could test the capability of reversion of the changes.

The final step: Towards the colonization of another planet or celestial body. Life support systems

The problems discussed in this article have two main objectives: (1) to identify and estimate the importance of the genotypic and phenotypic changes that biological systems may endure if exposed for very long-periods of time to strange environments likely to be found during the next steps of the Space Exploration Program. In particular, if human beings participate in this enterprise, it is important to find out the possible effects that they may experience. In this respect, the use of model systems that may show the effects much faster than the relatively long-lived, slowly reproducing human species seems mandatory. (2) It is possible, or even likely, that sooner or later the scenario of a possible colonization of a near-by extraterrestrial planet or satellite may become an objective, or even the objective of Human Society. To clarify the issues raised in this article is a prerequisite before this enterprise can be initiated. There are other prerequisites as well, for example, to extend the exploration of the possible targets for a colonization effort, to obtain as much information as possible of any remains or footprints of living or pre-living organisms that may have existed on these extraterrestrial locations (Maurel and Zaccai, 2001).

Independently of the mechanism of development involved, it is clear that the living organisms thriving in any site of the Biosphere have properties that are extraordinarily well adapted to the environmental conditions that they encounter in these sites. For example, in the case of the mammals, it is obvious that the skeletal framework depends on the size and weight of the animal, something that can be clearly seen if we present the skeletons of two different organisms, such as a cat and an elephant at the same scale (Fig. 7). Similar differences are also implemented in the circulatory system, which shall be able to pump blood to and from all parts of the organism. No one should harbor any doubts that the differences in developmental characteristics driven by the differences in the genes of the two species are responsible of these differences and that environmental effects will play only marginal roles in shaping the differences in morphology and properties at least in the short and intermediate run. Longer-term experiments of development in Microgravity could verify this assertion. Cats and elephants should maintain in the short-term, similar

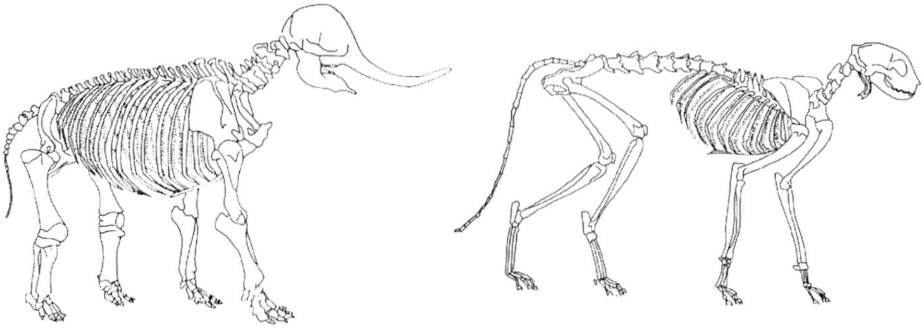

Fig. 7. Skeletons of an elephant and a cat drawn at the same scale. It can be seen how these animals have adapted their morphology and organization of the same bone repository to meet the challenges of the earth gravity.

morphologies as they show on the Ground, even if they develop in the absence of gravity. At most, they may show accentuated effects of the osteoporosis and the circulatory changes observed by Austronauts in their current experience in Microgravity. But, what will happen in the long term? Will they not adapt to the levels that a smaller gravity force will impose to them? Is not this what evolutionary forces have been producing on living organisms since they become establish in our Planet? It is true that these evolutionary changes are extremely slow and that they may be marginal in even experiments lasting many years.

Nevertheless, the knowledge of the changes that living organisms of all characteristics, but especially multicellular ones, may be experiencing will give information on the likelihood that organisms taking part in an effort to colonize another celestial body could be changing in directions undesired by us. This knowledge may prevent that the results of the operation produces undesired consequences.

Another problem that we have almost not mentioned yet but which should certainly be solved before any colonization attempts are initiated is understanding how multi-component systems are able to interact and become self-sustaining. The efforts that Space Programs are putting on the issue of Life Support Systems reflect the importance of the problem. As far as information is available, current life support systems are relatively simple and the work carried out on them concentrates on the issue of achieving self-sustainability for long period of times. Before any colonization attempts are made, these systems should be complicated and include living organisms of a higher organization. Model systems of the type discussed here, should be eventually included in these systems. Although, not always useful for the maintenance of Humans, systems incorporating additional model organisms, such as yeast, *Drosophila*, *Artemia* (a small crustacean), Amphibian, Avian and/or Mammalian model systems should be explored in a direction that eventually will define the minimal system that may be used in such a venture. The actual details of this Research Program is outside the expertise and objectives of the authors of this chapter, but it is obvious that the scientists working in the next generation of Life Support Systems in Space will need the information derived from experiments such as those that we have discussed here.

Acknowledgements

Miquel Pastor (NTE, Spain), Jutta Kramer (originally at Fokker Space, The Netherlands), and several different ESA/Shuttle Teams with whom we have collaborated in the past should be acknowledged for their involvement and support in our work during the past years.

References

Agarwal, S. and Sohal, R.S. (1994) DNA oxidative damage and life expectancy in houseflies. Proc. Natl. Acad. Sci. USA 91, 12332–12335.
Albrecht-Bühler, G. (1990) In defense of 'Nonmolecular' Cell Biology. Int. Rev. Cytol. 120, 191–241.
Amsterdam, A., Lin, S. and Hopkins, N. (1995) The *Aequorea victoria* green fluorescent protein can be used as a reporter in live zebrafish embryos. Dev. Biol. 171, 123–129.
Arbeitman, M.N., Furlong, E.E., Imam, F., Johnson, E., Null, B.H., Baker, B.S., Krasnow, M.A., Scott, M.P., Davis, R.W., White, K.P. (2002) Gene expression during the life cycle of *Drosophila melanogaster*. Science 297, 2270–2275.
Arthur, W. (2000) *The origin of Animal Body Plans*. J. Wiley & Sons, Chichester.
Ashburner, M. (1989) *Drosophila. A Laboratory Manual*. pp. 206–207. Cold Spring Harbor Laboratory Press.
Ashburner, M. (1992) Frosted flies. Science 258, 1896–1897.
Barthmaier, P. and Fyrberg, E. (1995) Monitoring development and pathology of *Drosophila* indirect flight muscles using green fluorescent protein. Dev. Biol. 169, 770–774.
Benguría, A., Grande, E., de Juan, E., Ugalde, C., Miquel, J., Garesse, R. and Marco, R. (1996) Microgravity Effects on *Drosophila melanogaster* behavior and aging. Implications of the IML-2 Experiment. J. Biotechnol. 47, 191–202.
Berry, M.V. and Geim, A.K. (1997) Of flying frogs and levitation. Eur. J. Phys 18, 307–313.
Bikle, D.D., Harris, J., Halloran, B.P. and Morey-Holton, E. (1994) Altered skeletal pattern of gene expression in response to Space Flight: comparison to the response to hind limb elevation. Am. J. Physiol 267, 822–827.
Brock, J.P. (2000) *The Evolution of Adaptive Systems*. Academic Press, London.
Burger, E.H. and Klein-Nulend, J. Mechanotransduction in bone-role of the lacuno-canalicular network, FASEB J. 1999, 13 Suppl. S101–12.
Burger, E.H. and Klein-Nulend, J. (1998) Microgravity and bone cell mechanosensitivity. Bone. 22, 5 Suppl. 127S–130S.

Calleja, M., Peña, P., Ugalde, C., Ferreiro, C., Marco, R. and Garesse, R. (1993) Mitochondrial DNA remains intact during *Drosophila* aging but the levels of mitochondrial transcripts are significantly reduced. J. Biol. Chem. 268, 18891–18897.

Chapes, S.K., Morrison, D.R., Guikema, J.A., Lewis, M.L. and Spooner, B.S. (1994) Production and action of cytokines in space. Adv. Space Res. 14(8), 5–8.

Claassen, D.E. and Spooner, B.S. (1994) Impact of Altered Gravity on Aspects of Cell Biology. Int. Rev. Cytol. 156, 301–373.

Clejan, S., O'Connor, K. and Rosensweig, N. (2001) Tri-dimensional prostate cell cultures in simulated microgravity and induced changes in lipid second messengers and signal transduction. J. Cell Mol. Med. 5, 60–73.

Cogoli, A. and Cogoli-Greuter, M. (1997) Activation and proliferation of lymphocytes and other mammalian cells in microgravity. Adv. Space Biol. Med. 6, 33–79.

Cogoli, A. (1993) The effect of hypogravity and hypergravity on cells of the immune system. J. Leukocyte Biol. 54, 250–268.

Cogoli, A., Bechler, B., Cogoli-Greuter, M., Criswell, S.B., Joller, H., Jollber, P., Hunziger, E. and Müller, O. (1993) Mitogenic signal transduction in T lymphocytes in microgravity. J. Leukocyte Biol. 53, 569–575.

Cogoli-Greuter, M., Meloni, M.A., Sciola, L., Spano, A., Pippia, P., Monaco, G. and Cogoli, A. (1996) Movements and interactions of leukocytes in microgravity. J. Biotechnol. 47(2–3), 279–287.

Cooke, J., Nowak, M.A., Boerlijst, M. and Maynard-Smith, J. (1997) Evolutionary origins and maintenance of redundant gene expression during metazoan development. Trends Genet. 13(9), 360–364.

Czajka, M.C. and Lee, R.E., Jr. (1990) A rapid cold-hardening response protecting against cold shock injury in *Drosophila melanogaster*. J. Exp. Biol. 148, 245–254.

de Groot, R.P., Rijken, P.J., Boonstra, J., Verkleij, A.J., de Laat, S.W. and Kruijer, W. (1991) Nuclear responses to protein kinase C signal transduction are sensitive to gravity changes. Exp. Cell Res. 197, 87–90.

de Groot, R.P., Rijken, P.J., Den Hertog, J., Boonstra, J., Verkleij, A.J., de Laat, S.W. and Kruijer, W. (1990) Microgravity decreases c-fos induction and serum response element activity. J. Cell. Sci. 97, 33–38.

Dobzhansky, T., Levene, H. and Spassky, B. (1972) Effects of selection and migration on geotactic and phototactic behaviour of *Drosophila*. Proc. R Soc. Lond B Biol. Sci. 180(58), 21–41.

Duprat, A.M., Husson, D. and Gualandris-Parisot, L. (1998) Does gravity influence the early stages of the development of the nervous system in an amphibian? Brain Res. Rev. 28(1–2), 19–24.

Edgerton, V.R. and Roy, R. R. (2000) Gravitational biology of the neuromotor systems: a perspective to the next era. J. Appl. Physiol. 89(3), 1224–1231.

Edwards, K.A., Demsky, M., Montague, R.A., Weymouth, N. and Kiehart, D.P. (1997) GFP-moesin illuminates actin cytoskeleton dynamics in living tissue and demonstrates cell shape changes during morphogenesis in *Drosophila*. Dev. Biol. 191, 103–117.

Endow, S.A. (1999) GFP fusions to a microtubule motor protein to visualize meiotic and mitotic spindle dynamics in Drosophila. Methods Cell Biol. 58, 153–163.

Finch, C.E. and Tanzi, R.E. (1997) Genetics of aging. Science 278, 407–411.

Foe, V. (1989) Mitotic domains reveal early commitment of cells in Drosophila embryos. Development 107, 1–22.

Gerhart, J. and Kirschner, M. (1997) *Cells, Embryos and Evolution.* Blackwell Science.

Goodwin, B.C. (1990) The evolution of Generic forms. pp. 107–117. *In* Maynard Smith, J and Vida, G. (eds.) *Organizational Constraints on the Dynamics of Evolution.* Manchester University Press, Manchester.

Grimm, D., Bauer, J., Kossmehl, P., Shakibaei, M., Schoberger, J., Pickenhahn, H., Schulze-Tanzil, G., Vetter, R., Eilles, C., Paul, M., Cogoli, A. (2002) Simulated microgravity alters differentiation and increases apoptosis in human follicular thyroid carcinoma cells. FASEB J. 16, 604–606.

Gruener, R. and Hoeger, G. (1990) Vector-free gravity disrupts synapse formation in cell culture. Am. J. Physiol. 258, c489–c494.

Gruener, R., Roberts, R. and Reitstetter, R. (1994) Reduced receptor aggregation and altered cytoskeleton in cultured myocytes after space-flight. Biol. Sci. Space 8, 79–93.

Hammond, T.G., Lewis, F.C., Goodwin, T.J., Linnehan, R.M., Wolf, D.A., Hire, K. P., Campbell, W.C., Benes, E., O'Reilly, K.C., Globus, R.K., Kaysen, J.H. (1999) Gene expression in space. Nat. Med. 5, 359.

Hemmersbach, R., Volkmann, D. and Hader, D.P. (1999) Graviorientation in protists and plants. J. Plant Physiol. 154(1), 1–15.

Hirsch, J. and Ksander, G. (1969) Studies in experimental behavior genetics. V. Negative geotaxis and further chromosome analyses in *Drosophila melanogaster.* J. Comp. Physiol. Psychol. 67(1), 118–122.

Horn, E. and Foller, W. (1998) Induction of a gravity-related response by a single receptor cell in an insect. Naturwissenschaften 85, 121–123.

Husson, D. (1998) Etude du rôle de la pesanteur dans la fécondation et l' embryogènese de l' amphibien urodele Pleurodeles waltl, Ph. D. Thesis, Toulouse.

Jacinto, A., Woolner, S. and Martin, P. (2002) Dynamic analysis of dorsal closure in *Drosophila:* from genetics to cell biology. Dev. Cell 3, 9–19.

Johnson, F.B., Sinclair, D.A. and Guarente, L. (1999) Molecular biology of aging. Cell 96, 291–302.

Kalthoff, K. (1996) *Analysis of Biological Development.* McGraw-Hill, New York.

Khaoustov, V.I, Risin, D., Pellis, N.R. and Yoffe, B. (2001) Microarray analysis of genes differentially expressed in HepG2 cells cultured in simulated microgravity: preliminary report. In Vitro Cell Dev. Biol. Anim. 37, 84–88.

Kohler, R. (1994) *Lords of the flies.* University of Chicago Press, Chicago and London.

Kraemer, W.J., Staron, R.S., Gordon, S.E., Volek, J.S., Koziris, L.P., Duncan, N.D., Nindl, B.C., Gomez, A.L., Marx, J.O., Fry, A.C., Murray, J.D. (2000)

The effects of 10 days of spaceflight on the shuttle Endeavor on predominantly fast-twitch muscles in the rat Histochem. Cell Biol. 114, 349–355.

Lewis, M.L., Cubano, L.A., Zhao, B., Dinh, H.K., Pabalan, J.G., Piepmeier, E.H. and Bowman, P.D. (2001) cDNA microarray reveals altered cytoskeletal gene expression in space-flown leukemic T lymphocytes (Jurkat). FASEB J. 15, 1783–1785.

Lewis, M.L., Reynolds, J.L., Cubano, L.A., Hatton, J.P., Lawless, B.D. and Piepmeier, E.H. (1998) Spaceflight alters microtubules and increases apoptosis in humanlymphocytes (Jurkat). FASEB J. 12, 1007–1018.

Li, X. and Noll, M. (1994) Evolution of distinct developmental functions of three *Drosophila* genes by the acquisition of different cis-regulatory regions. Nature 367, 83–87.

Lints, F.A. and Soliman, M.H. (1988) *Drosophila as a Model Organism for Ageing Studies*. Blackie and Son Ltd, London.

Malacinski, G.M. (1990) An essay on redundancy within developmental processes. pp. 123–151. *In* Malacinski, G.M. (ed.) *Cytoplasmic Organization Systems*. McGraw Hill, New York.

Marco, R., Díaz, C., Benguría, A., Mateos, J., Mas, J. and de Juan, E. (1999) The role of gravity in the Evolutionary Emergence of Multicellular Complexity. The Effects of Microgravity on arthropod. Dev. Aging Adv. Space Res 23, 2075–2082.

Marco, R., Benguría, A., Sánchez, J. and de Juan, E. (1996) Effects of the Space Environment on *Drosophila melanogaster* Development. Implications of the IML-2. Exp. J. Biotechnol. 47, 179–190.

Marco, R., González-Jurado, J., Calleja, M., Garesse, R., Maroto, M., Ramirez, E., de Juan, E. and Miquel, J. (1992) Microgravity effects on *Drosophila melanogaster* development and aging: Comparative analysis of the results of the Fly Experiment in the Biokosmos 9 URSS biosatellite flight. Adv. Space Res. 12(1), 157–166.

Marco, R., Vernós, I., González, J. and Calleja, M. (1986) Embryogenesis and aging of *Drosophila melanogaster* flown in the Space Shuttle. Naturwissenschaften 73, 431–432.

Marthy, H.J. (2002) Developmental Biology of Animal Models under varied gravity conditions: A review. Vie Milieu 52, 149–166.

Marthy, H.J., Gasset, G., Tixador, R., Eche, B., Schatt, P., Dessommes, A., Marthy, U. and Bacchieri, R. (1998) Skeletogenesis in sea urchin larvae under modified gravity conditions. Adv. Space Res. 21(8–9), 1151–1154.

Maurel, M.C. and Zaccai, G. (2001) Why biologists should support the exploration of Mars. Bioessays 23, 977–978.

Mazur, P., Cole, K.W., Hall, J.W., Schreuders, P.D. and Mahowald, A.P. (1992) Cryobiological preservation of Drosophila embryos. Science 258, 1932–1935.

Medina, F.J., Cerdido, A., Maroto, M., Manzanares, M. and Marco, R. (1994) Enhancement of the Immunochemical detection of antigens by microwave

irradiation. Benefits and limitations studied in isolated plant nuclei and *Drosophila* embryos *in toto*. Histochemistry 102, 45–50.
Mesland, D.A.M. (1992) Mechanisms of Gravity Effects on Cells: Are there Gravity-Sensitive Windows. Adv. Space Biol. Med. 2, 234–276.
Miller, M.S. and Keller, T.S. (2000) Effects of Short Duration Microgravity on *Drosophila* Melanogaster (Fruit Fly) Activity http://www.emba.uvm.edu/VSGC/rocket/rocket.htm.
Moody, S.A. and Golden, C. (2000) Developmental biology research in space: issues and directions in the era of the international space station. Dev. Biol. 228, 1–5.
Miquel, J. (1984) Effects of Microgravity and Hypergravity on invertebrate Development In *NASA Developmental Workshop* Arlington, NASA TM 86756, p 7–16.
Miquel, J. and Fleming, J. (1984) A two-step hypothesis on the mechanism of in vitro cell aging: cell differentation followed by intrinsic mitochondrial mutagenesis. Exp. Gerontol. 19, 31–36.
Miquel, J. and Fleming, J. (1986) Theoretical and experimental support for an "oxygen radical-mitochondrial injury". Hypothesis of cell aging. pp. 51–74. *In* Johnson, J.E., Jr. *Free Radicals, Aging and Degenerative Diseases*. Alan R. Liss, New YorK.
Miquel, J. (1984) Effects of Microgravity and Hypergravity on invertebrate Development. In NASA Developmental Workshop Arlington. NASA TM 86756, 7–16.
Miquel, J. and Souza, K, A. (1991) Gravity Effects on Reproduction, Development and Aging Adv. Space Biol. & Med. 1, 104–154.
Miquel, J., Economos, A.C., Fleming, J. and Johnson, J.E., Jr (1980) Mitochondrial role in cell aging. Exp. Gerontol. 15, 575–591.
Miquel, J., Lundgren, P.R. and Binnard, R. (1972) Negative geotaxis and mating behavior in control and gamma-irradiated *Drosophila* Drosophila Inf. Service 48, 60.
Miquel, J., Lundgren, P.R., Bensch, K.G. and Atlan, H. (1976) Effects of temperature on the life-span, vitality and fine structure of *Drosophila melanogaster*. Mech. Ageing Dev. 5, 347–370.
Moore, D. and Cogoli, A. (1996) Gravitational and space biology at the cellular level. pp. 1–116. *In* Moore, D., Bie, P. and Oser, H. (eds.) *Biological and Medical Research in Space*. Springer Verlag.
Newman, S.A. and Comper, W.D. (1990) Generic physical mechanisms of morphogenesis and pattern formation. Development 110, 1–18.
Orr, W.C. and Sohal, R.S. (1994) Extension of life-span by overexpression of superoxide dismutase and catalase in *Drosophila melanogaster*. Science 263, 1128–1130.
Partridge, L. and Barton, N.H. (1993) Optimality, mutation and the evolution of aging. Nature 362, 305–311.
Pearl, R. (1928) *The rate of living*. London University Press, London.

Petavy, G., David, J.R., Gibert, P. and Moreteau, B. (2001) Viability and rate of development at different temperatures in Drosophila: a comparison of constant and alternating thermal regimes. J. Therm. Biol. 26, 29–39.

Pigliucci, M. (2001) *Phenotypic plasticity. Beyond Nature and Nurture.* John Hopkins University Press, Baltimore and London.

Powell, J.R. (1997) *Progress and Prospects in Evolutionary Biology. The Drosophila Model.* Oxford Universiyt Press, New York.

Raff, R.A. (1996) *The shape of life. Genes, Development and the Evolution of Animal Form.* The University of Chicago Press, Chicago and London.

Reitstetter, R.Effects of vector-averaged gravity on single channel properties of the acetylcholine receptor in cultured myocytes NASA technical memorandum941994179.

Rose, M.R. (1991) *Evolutionary Biology of Aging.* Oxford University Press, New York.

Rose, M.R. and Lauder, G.V. (1996) *Adaptation.* Academic Press, San Diego.

Schatten, H., Lewis, M.L. and Chakrabarti, A. (2001) Spaceflight and clinorotation cause cytoskeleton and mitochondria changes and increases in apoptosis in cultured cells. Acta Astronaut. 49, 399–418.

Schatten, H., Chakrabarti, A., Taylor, M., Crosser, M. and Mitchell, K. (1998) The centriole-centrosome complex is affected by microgravity during cell division and in cilia of sea urchin embryos: results from space flight experiments. Microsc. Microanal. 2(4 Supplement), 1132–1133.

Schmitt, D.A., Hatton, J.P., Emond, C., Chaput, D., Paris, H., Levade, T., Cazenave, J.-P. and Schaffar, L. (1996) The distribution of protein kinase C in human leukocytes is altered in Microgravity. FASEB J 10, 1627–1634.

Schwarze, S.R., Weindruch, R. and Aiken, J.M. (1998a) Decreased mitochondrial RNA levels without accumulation of mitochondrial DNA deletions in aging *Drosophila melanogaster*. Mutat. Res. 382, 99–107.

Schwarze, S.R., Weindruch, R. and Aiken, J.M. (1998b) Oxidative stress and aging reduce COX I RNA and cytochrome oxidase activity in *Drosophila*. Free Radical Biol. Med. 25, 740–747.

Shigenaga, M.K., Hagen, T.M. and Ames, B.N. (1994) Oxidative damage an mitochondrial decay in aging. Proc. Natl. Acad. Sci. USA 91, 10771–10778.

Sohal, R.S. and Weindruch, R. (1996) Oxidative stress, caloric restriction, and aging. Science 273(5271), 59–63.

Souza, K.A., Black, S.D. and Wassersug, R.J. (1995) Amphibian development in the virtual absence of gravity, Proc. Natl. Acad. Sci. 92, 1975–1978.

Steponkus, P.L., Myers, S.P., Lynch, D.V., Gardner, L., Bronshteyn, V., Leibo, S.P., Rall, W.F., Pitt, R.E., Lin, T.T., MacIntyre, R.J. (1990) Cryopreservation of *Drosophila melanogaster* embryos. Nature 345, 170–172.

Steponkus, P.L., Caldwell, S., Myers, S.P. and Cicero, M. (1996) Crypreservation of *Drosophila melanogaster* embryos. Adv. Low-Temp. Biol. 3, 257–316.

Tautz, D. (2000) A genetic uncertainty problem Trends. Genetics 16(11), 475–477.

Texada, D.J., Armstrong, J.D., Carter, E., Munjaal, R.P. and Beckingham K.M. (2002) Identification of genes involved in Gravitaxis in Drosophila 43rd Annual Drosophila Research Conference San Diego Abstract 786.

Uva, B.M., Masini, M.A., Sturla, M., Prato, P., Passalacqua, M., Giuliani, M., Tagliafierro, G. and Strollo, F. (2002) Clinorotation-induced weightlessness influences the cytoskeleton of glial cells in culture. Brain Res. 934, 132–139.

Van Loon, J.J., Bervoets, D.J., Burger, E.H., Dieudonne, S.C., Hagen, J.W., Semeins, C.M., Doulabi, B.Z. and Veldhuijzen, J.P. (1995) Decreased mineralization and increased calcium release in isolated fetal mouse long bones under near weightlessness. J. Bone Miner. Res. 10(4), 550–557.

Vernós, I., González-Jurado, J., Calleja, M. and Marco, R. (1989) Microgravity effects on the oogenesis and development of embryos of *Drosophila melanogaster* laid in the Spaceshuttle during the Biorack experiment. Int. J. Develop. Biol. 33, 213–226.

Vico, L., Lafage-Proust, M.H. and Alexandre, C. (1998) Effects of gravitational changes on the bone system in vitro and *in vivo*, Bone 22 (5 Suppl): 95S–100S.

Wallace, D.C. (1999) Mitochondrial diseases in man and mouse. Science 283, 1482–1488.

Walther, I, Pippia, P., Meloni, M.A., Turrini, F., Mannu, F. and Cogoli, A. (1998) Simulated microgravity inhibits the genetic expression of interleukin-2 and its receptor in mitogen-activated T lymphocytes. FEBS Lett. 436, 115–118.

Wang, E. (1999) Age-dependent atrophy and microgravity travel: what do they have in common? FASEB J. 13(Supplement), S167–S174.

Yeh, E., Gustafson, K. and Boulianne, G.L. (1995) Green fluorescent protein as a vital marker and reporter of gene expression in Drosophila. Proc. Natl. Acad. Sci. USA 92, 7036–7040.

Developmental Biology Research in Space
H.-J. Marthy (editor)

Morphogenesis and Gravity in a Whole Amphibian Embryo and in Isolated Blastomeres of Sea Urchins

Akemi Izumi-Kurotani[1,*] and Masato Kiyomoto[2]

[1]*Space Utilization Research Center, Institute of Space and Astronautical Science, Yoshinodai 3-1-1, Sagamihara, Kanagawa 229-8510, Japan*
[2]*Tateyama Marine Laboratory, Ochanomizu University, Kou-yatsu 11, Tateyama, Chiba 294-0301, Japan*

Abstract

Fertilization and subsequent embryogenesis of newts occurred normally under microgravity in two Astronewt flight experiments. By accumulation of the results from the amphibian flight experiments including 'Astronewt', it is considered that gravity has rather small effects on the early development of amphibian eggs. However, some temporary abnormalities, which recover in the course of the further developmental process, have been observed. Some regulations may occur in whole embryos. For a thorough knowledge about the role of gravity in morphogenesis, we need to investigate the gravitational effects on a single cell in a whole embryo. We propose a new experimental system with sea urchin embryos and micromeres for further studies at a cellular level of the effects of gravity on morphogenesis.

Introduction

As one of the important objectives in space biology, the relationship between the gravity vector and morphogenesis in early amphibian development has attracted many investigators (Gualandris-Parisot *et al.*, 2002; Neff *et al.*, 1993; Souza *et al.*, 1995; Ubbels *et al.*, 1989, 1994; Ubbels, 1997; Yokota *et al.*, 1992, 1994; Young and Tremor, 1968). Here we report on one of these experiments, a series of 'Astronewt' and discuss a further approach with a new experimental system focused on studies at a cellular level.

* E-mail: kuro@pub.isas.ac.jp

Fertilization and early development of newt in space (Astronewt experiment) – Morphogenesis in a whole embryo

A series of space experiments with newts, termed 'Astronewt', was performed to study effects of the gravity force on fertilization and subsequent embryogenesis (Mogami *et al.*, 1996a,b; Yamashita *et al.*, 1997, 2001). So far two flight missions were carried out : IML-2 (The Second International Microgravity Laboratory), a Spacelab mission in 1994 (Principal Investigator: Professor M. Yamashita, ISAS) and SFU (Space Flyer Unit), an unmanned platform mission in 1995 (Principal Investigator: SFU BIO group).

Amphibian eggs as material for developmental space biology

Amphibian eggs are traditionally a useful material for general developmental biology and many basic as well as brand-new findings have been obtained. The accumulation of knowledge based on experiments with such eggs represents also a significant contribution to the research field of developmental space biology.

Within a practical context, a great advantage for space experiments is the fact that many amphibian species spend the winter by hibernating. Such hibernating animals can be kept and transported at a low temperature without the need for feeding and without frequent concerns about water quality.

Due to their relatively large size (usually 1–3 mm in diameter), amphibian has been hypothesized that they may be influenced directly by gravity. Actually there is some evidence for effects of gravity on the establishment of the embryonic axes (Fig. 1). The vegetal hemisphere of the amphibian egg cytoplasm has higher density with more yolk platelets than the animal hemisphere, whose color is darker with a lot of pigments. Once the egg is fertilized, the elevation of the fertilization membrane occurs, coming off from the exterior surface of egg by swelling of water and the perivitelline space appears. This space allows the egg to rotate freely in response to earth gravity. The pale vegetal hemisphere of the fertilized egg orientates to the gravity force vector. Thus amphibian eggs usually start their development with the

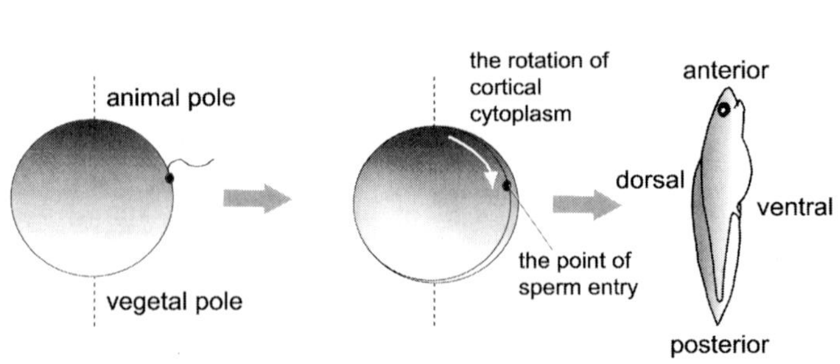

Fig. 1. The determination of embryonic axes in anuran eggs.

orientation of the animal-vegetal axis to the definite direction under 1G conditions, e.g., usually this axis roughly corresponds to the primary embryonic axis (head-tail axis). Concerning the dorso-ventral axis of embryos, another gravity orientation response is necessary: a series of asymmetric surface rotations and an internal asymmetric cytoplasm shift occurs. The cortical layer (cell membrane and the underlying cortical part of the cytoplasm) rotates against the internal cytoplasm. The direction of the rotation defines the dorso-ventral axis of the embryo. The rotation occurs before the first cleavage, while the cytoplasm is in the condition of a colloidal solution. Early stages are more sensitive to the gravity force and thus are more sensitive to gravity changes during the establishment of the embryonic axes. The direction of the rotation usually relates to the point of sperm entry in monospermic anurans. In urodeles, fertilization is physiologically a polyspermy (the nuclei fusion occurs with a single sperm nucleus). The correspondence between the direction of rotation and the sperm entry point is less clear in this order. However the mechanism for the establishment of the embryonic axis appears to be similar to that in anurans. Various experiments with anuran and urodele amphibian eggs have been performed under different gravitational conditions (microgravity in space, alteration of the direction of the gravity force vector by clinostats or by inclination, and hypergravity by centrifuges), to study whether the gravity force has actually any effect on the establishment of the embryonic axes (Neff *et al.*, 1993; Souza *et al.*, 1995; Ubbels *et al.*, 1989, 1994; Ubbels, 1997; Yokota *et al.*, 1992, 1994; Young and Tremor, 1968).

The Astronewt experiment

Newts (Cynops pyrrhogaster)

In anurans, fertilization occurs externally and is usually designed for flight experiments to be artificially *in vitro*. In contrast, in urodeles, the female receives the spermatophores from males, keeps them in her cloaca and fertilization occurs internally just before oviposition. For experiments, the oviposition of fertilized eggs is induced by a hormonal treatment. Therefore, the experiment can be performed with *in vivo* fertilized eggs and for flight experiments, females are sent into orbit, which have received spermatophores in advance from males on the ground. The body of a female is indeed the most reliable container for transport, keeping eggs and sperm in best conditions.

Newts can live in completely submerged conditions and oviposition normally occurs under such a condition. The container for keeping adult females in orbit can also serve for oviposition and embryogenesis.

In addition to the general amphibian characteristics, it appears that newts have great practical advantages in developmental space biology research (Mogami *et al.*, 1996a,b).

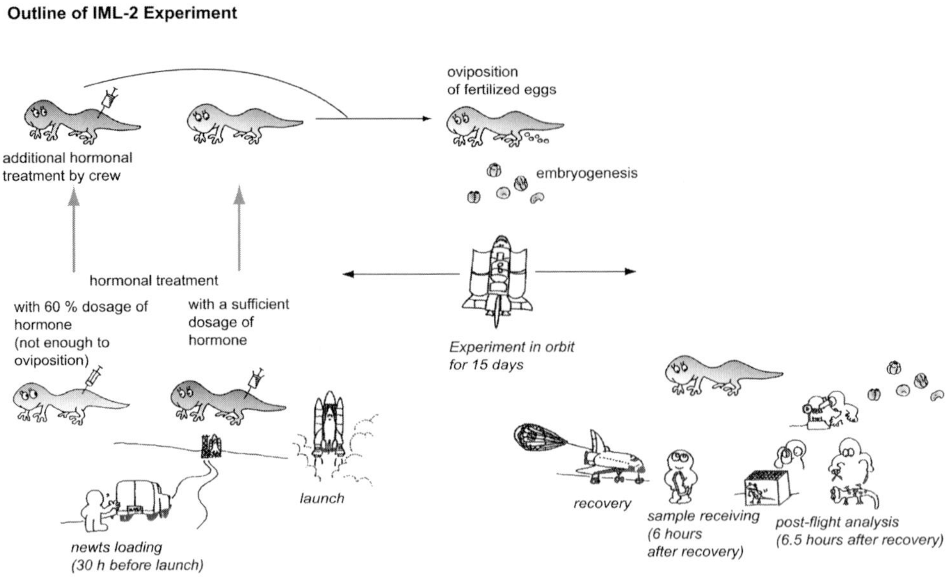

Fig. 2. Outline of IML-2 Experiment. Four female newts were sent into space in submerged conditions without hibernation. Two methods for hormone treatment were used for oviposition in orbit. 30 h before launch, two newts in a cassette (A-3) were treated with a sufficient dosage (250 IU) of hCG (human chorionic gonadotropins), which was contained in a capsule for a slow release. Another two females, which were separately housed in two cassettes (A-1, A-2), were treated with 100 IU of hCG and planned to be injected additionally by the crew. In orbit, video images of eggs and embryos were recorded by the crew. After 15 days, all females, eggs and embryos were thought to be recovered alive. However two females died in orbit, and there were neither eggs nor embryos in any cassettes after the recovery.

Outline of the experiments

Flight experiments, termed 'Astronewt', with *Cynops pyrrhogaster*, were carried out to investigate whether the *in vivo* fertilization and the subsequent development, including the establishment of the embryonic axis, can occur normally under microgravity. The outlines of the two flight experiments are shown in Figures 2 and 3.

Figure 4 shows images of embryos and of an egg obtained in these flight experiments. The macroscopic morphology of embryos developed under microgravity was as normal as on earth. The time course of development was similar to that of eggs fertilized and developed on earth at the same temperature.

Astronewt investigators have concluded therefore, that microgravity and the space environment do not affect the *in vivo* fertilization and the subsequent embryogenesis in terms of morphology and the time course of development. These results are enforced by similar findings obtained by other flight experiments with different amphibian species. Fundamental steps of development, including the establishment of embryonic axes, do not require the presence of the gravity force (Mogami *et al.*, 1996a; Yamashita *et al.*, 2001).

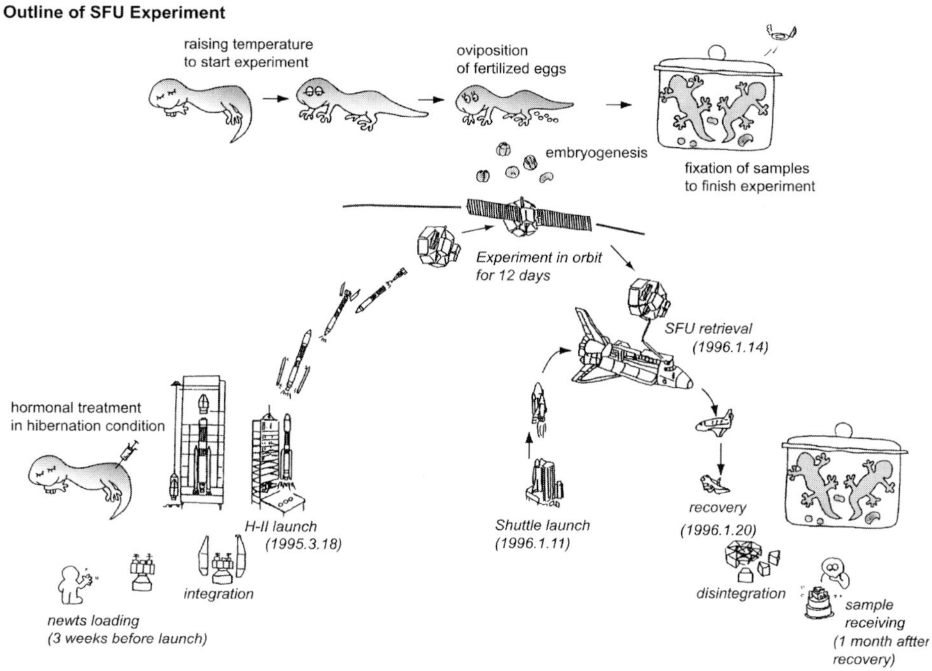

Fig. 3. Outline of SFU Experiment. Two females were sent into space for the unmanned flight in hibernating condition at a low temperature (4 °C). Three weeks before launch, two females were treated with 150 IU of hCG, contained in a capsule for a slow release and separately housed in two containers. Seven days after the arrival in orbit, the experiment was started by raising the water temperature to 20 °C. The containers were observed every 30–60 min by down-linked images. After 12 days, a fixative (gultaraldehyde) was introduced to the containers to fix the specimens for post-flight analysis. SFU was retrieved by an arm of the Shuttle and recovered after 10 months of space flight.

For the Astronewt experiment in IML-2, the AAEU (Aquatic Animal Experiment Unit) developed by NASDA (National Space Development Agency of Japan) was sent into space (Sakimura *et al.*, 1999; Uchida *et al.*, 2002). Aquatic animal experiments in space are characteristic and traditional subjects in Japanese space life sciences, and NASDA aquatic animal experiment facilities, built for the NASA Space Shuttle, had several flight opportunities carried out with various aquatic animals. The facilities are expected to promote further studies on aquatic animals as candidate hardware for the International Space Station (ISS). Experiments with newts and advanced AAEU are likely to bring new insights as to the formation of germ cells and the breeding through multiple generations (Ijiri, 1996, 1997; Wakahara, 1990).

Temporary abnormalities and further approaches

Under both continuous alteration of the gravity vector direction by clinostat rotation and real microgravity in space, abnormalities in early embryogenesis

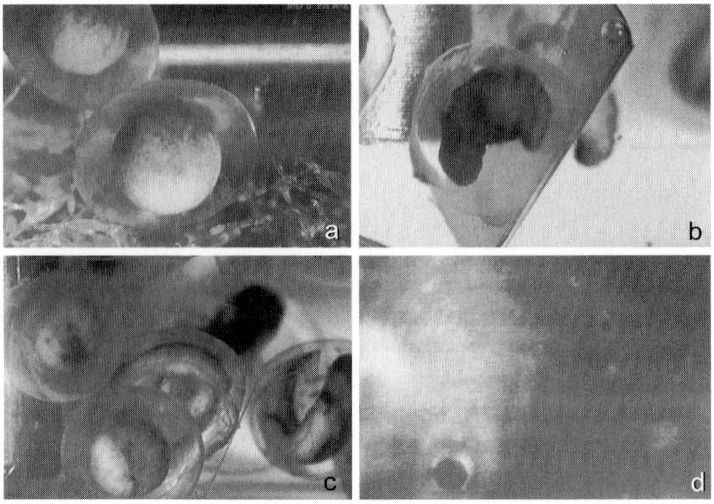

Fig. 4. Images of embryos and an egg obtained in two flight experiments of 'Astronewt'. (a) Embryos in A-2 cassette of AAEU on the mission day 5 in IML-2. (b) Embryos in A-3 cassette of AAEU on the mission day 5 in IML-2, (c) Embryos in A-3 cassette of AAEU on the mission day 12 in IML-2. (d) An egg obtained in down-linked still video images 5 days after raising of the water temperature in the SFU (Yamashita, *et al.*, 2001).

of *Xenopus* eggs have been observed (Black *et al.*, 1996; Souza *et al.*, 1995; Yokota *et al.*, 1994). The third cleavage furrow formed at a lower latitude. Blastcoeles in blastulas were located lower with a thicker and multi-layered archenteron roof. These phenomena were recovered in the course of the developmental process and no significant differences in morphology were observed at later stages.

In urodeles, evidence for effects of microgravity on embryogenesis in the salamander *Pleurodeles waltl* has been recently described (Gualandris-Parisot *et al.*, 2002): irregular segmentation, abnormal closure of the neural tube and several abnormalities corresponding to cortical cytoplasmic movements, a decrease in cell adhesion and a loss of cells. These abnormalities were temporary and subsequently reversible similar to that stated in *Xenopus* eggs.

The accumulation of such observations in whole embryos of amphibians seem to allow just one conclusion: gravity does not seriously influence the early development of amphibian embryos, and the abnormalities produced by the different gravitational conditions are 'temporary' and are regulated in the course of the subsequent developmental process.

The next flight opportunity for the Astronewt project will clarify whether embryos of *Cynops pyrrhogaster* developed under microgravity will also show such temporary abnormalities.

Generally, the recovery might occur by the potency of regulation or by the homeostatic activity of the embryo. Mogami *et al.* (1996b) discussed possible explanations for 'temporary' abnormalities.

We may argue that the homeostatic activity is particularly efficient in a whole embryo. Moore and Cogoli (1996) reviewed gravitational biology including embryology from a cellular point of view. Is there any effect of gravity on individual cells of the embryo, but masked in a whole embryo? Search for the evidence of the homeostatic activity in the embryos and further investigations of a gravitational effect on this activity will be one of the most interesting subjects in future developmental biology. It is necessary to investigate the effect of a space environment on both, isolated single cells (blastomeres) of an embryo and the whole embryo. By research at the cellular level, we may reveal the hidden effects of gravity on blastomeres and may learn to understand the mechanism of the homeostatic activity in whole embryos.

Effect of gravity on skeletogenesis in isolated micromere cultures of sea urchin embryos-Morphogenesis in isolated blastomeres

As said above, research for morphogenesis in space, at the cellular level, is expected to be a promising approach to investigate the role of gravity, which is possibly masked in a whole embryo.

Our new subject, 'skeletogenesis in a sea urchin embryo' is thought to be an efficient approach to further explore that role of gravity. Skeletogenesis in both the whole embryo and cultured isolated blastomeres, can be observed easily.

In this section, this experimental approach, especially the culture system with isolated blastomeres will be described. Preliminary results were obtained under various gravity conditions such as hypergravity on a centrifuge, continuous alteration of the gravity vector direction by a vertical clinostat, and a short-term microgravity (20 s) with a gravity force change by a series of parabolic flights. The possibility of a space experiment will also be discussed.

Sea urchins

Gametes of sea urchins, that are the eggs and sperm, are traditional materials for research on early animal development and numerous papers deal with developmental biology aspects of sea urchins, ranging from classical famous papers on experimental morphology to recent approaches for molecular mechanism analysis of differentiation and morphogenesis.

Artificial control of spawning in all seasons of sea urchins

There is only one disadvantage in sea urchins as biological material: there is a defined breeding season in Japanese sea urchins as in most other species. We cannot obtain gametes out of this season and it is impossible to use the same species throughout the year. This problem is serious, especially for a flight experiment, since space missions can be shifted for unexpected reasons, and launch windows do not open only in a breeding season of sea urchins.

Therefore, we tried to control gonadal maturation artificially. *Hemicentrotus pulcherimus*, whose breeding season is usually in winter in the Kanto region of Japan, was selected for the following reasons: (1) this species is common at the seaside on the periphery of the Honshu and Kyushu islands in Japan and we can collect easily many animals; (2) an adult is small with short spines, and we can keep the animals in a water tank as a relatively large population; (3) it is easy to observe the morphology of the egg due to its high transparency; (4) a high elevation of the fertilization membrane allows removing it easily for collecting the micromeres; (5) the spicule formation rate in cultured micromeres is high; (6) cultured micromeres form long simple-shaped which are suitable for observation and image analysis.

Since breeding of most of Japanese sea urchin species depends mainly on the annual temperature change of the seawater, the gonadal maturation can be induced simply by a temperature control (Sakairi *et al.*, 1989; Yamamoto *et al.*, 1988). The best temperature profile for the artificial induction of the gonadal maturation was defined, on the base of the criteria of the fertilization rate, the hatching rate and the spicule formation rate. The best temperature profile for summer and autumn use is obtained as follows: animals were kept at the summer temperature from the autumn of the preceding year. Although, we could not completely suppress the natural maturation during winter, the breeding periods were shorter than in natural conditions. By cooling to winter temperature, before use, the maturation could be induced. In animals kept under a temperature control, the rates for fertilization, hatching and spicule formation were equally high as in animals matured naturally during winter (Kiyomoto *et al.*, 2001).

Preservation of eggs and sperms

Eggs and sperm were obtained by injection of 0.5 ml of 10 mM acetylcholine chloride into the adult sea urchins. After rinsing with artificial seawater, the eggs in the presence of antibiotics (100 U/ml of penicillin and 100 μg/ml of streptomycin) were kept cool in a refrigerator (4 °C). Such eggs could be fertilized during 2 days and developed normally. For the preservation of sperm, the removed testis was put in a small sample tube without seawater and stored in a refrigerator (4 °C or on ice). Such sperm fertilized the eggs at least for four days. Anthony *et al.* (1996) reported on the preservation of gametes, fertilized eggs and embryos of sea urchins for space experiments. Further studies on cryo-preservation are in progress (Marthy, pers. comm.).

The effects of microgravity on fertilization and early development in sea urchins

Marthy (1990), Marthy *et al.* (1994a,b, 1996) and Marthy (1997) have carried out several series of experiments to investigate the effects of microgravity on fertilization and early development in whole eggs of the sea urchins

Paracentrotus lividus and *Sphaerechinus granularis*. The results indicate that fundamental processes necessary for fertilization, the subsequent embryogenesis and the bio-mineralization of spicules occurred normally in the absence of the gravity force. This suggests that the influence of microgravity on a whole egg is rather small.

Marthy *et al.* (1996) have also shown the effects of environmental factors, potentially including microgravity, on the process of the positioning of the skeletogenic cells (determining primarily shape and size of the skeletal rods).

Skeletogenesis in micromere cultures

Spicules in sea urchins

The structural plan of a sea urchin larva is symmetric while that of an adult has a pentamerous radial symmetry (Fig. 5). The structure of the larva is supported by two symmetrical spicule arrangements, and the spicules are formed by the primary mesenchyme cells (PMCs), which are originating from the micromeres at the 16-cells stage.

A spicule consists of $CaCO_3$ crystals in the form of calcite. Spicules are easy to observe *in vivo* in the larva, even as the smallest granules at the beginning of their formation. It is particularly easy to observe by a polarization microscope due to its single crystal property (Okazaki and Inoué, 1976).

The research subject of the skeletogenesis process includes very interesting topics such as the metabolism of the calcium ion (the transportation and accumulation), as well as the mechanism of the spicule formation and the regulation of the shape and length of the spicules.

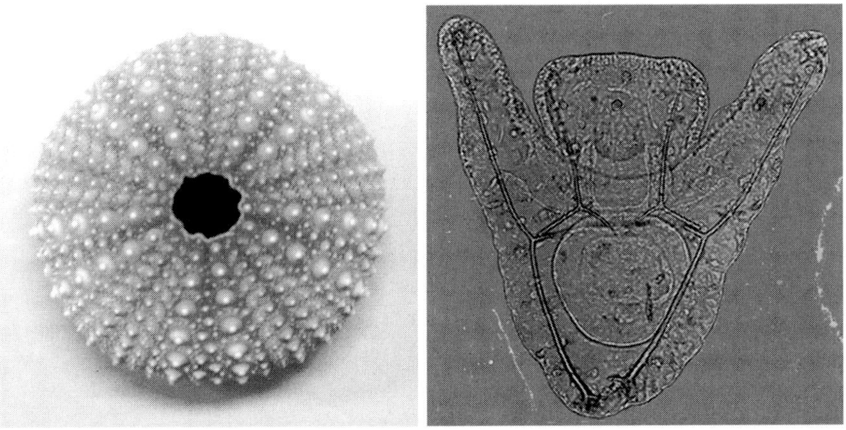

Fig. 5. The structural plans of a sea urchin adult and a larva. Left: the adult test of *Strongylocentrotus nudus*, note the pentamerous radial symmetry. Right: the early stage of the pluteus larva of *Strongylocentrotus nudus*.

Protocols for isolation and culture of micromeres

When isolated micromeres are cultured in seawater with 4% horse serum or 10 times diluted blastocoelic fluid of sea urchin blastulae (Kiyomoto *et al.*, 1991; Okazaki, 1975), they alter into migratory cells after proliferation and cellular aggregation. Spicule formation begins about 25 h after the start of the culture (20 °C) (Fig. 6). For further details on skeletogenesis *in vivo* and *in vitro* one is referred to several reports (e.g., Harkey and Whiteley, 1980, 1985; Decker and Lennarz, 1988a,b; Ettensohn and Malinda, 1993; Marthy and Bacchieri, 1999).

Our procedure for isolation and culture of micromeres is as follows:

The fertilization membrane is removed in Ca, Mg-free seawater just after the complete elevation of fertilization membrane. The eggs are allowed to develop in Ca-free seawater up to 16-cells stage. Under this condition, blastomeres lose adhesion to each other and dissociate easily. Micromeres are isolated from macromeres and mesomeres by a sucrose density gradient.

Hundred micromeres (for morphological observation and measurement) or 500 micromeres (for extraction of mRNA) are introduced into each well of a 96-wells plate. The culture medium was seawater containing 4% horse serum or 10 times diluted blastocoelic fluid (Kiyomoto *et al.*, 1991; Okazaki, 1975). The wells were coated in advance with fibronectin for a more stable adhesion of the cells. The index of spicule formation activity is represented as 'Relative Length' in Figures 8 and 9. Digital images of spicules were transformed to two-valued black and white patterns by NIH image and the outline length of each spicule was measured. The 'Total length of spicules in a well of the multi-well plate'

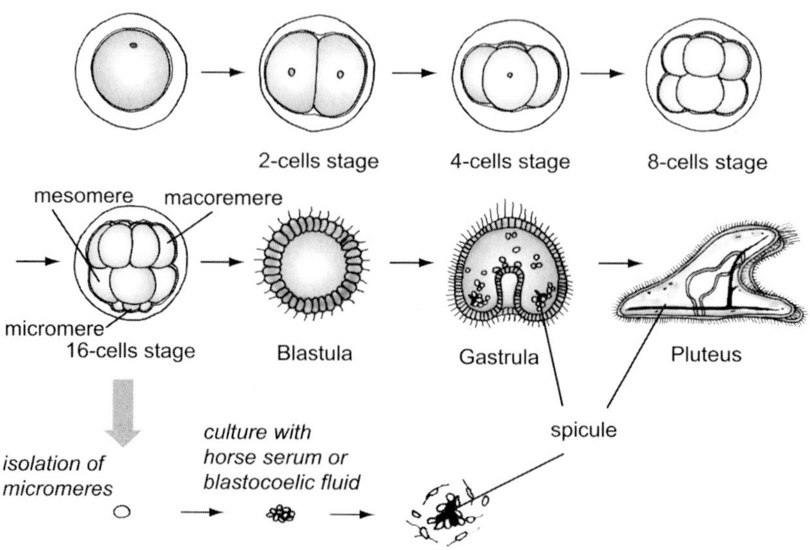

Fig. 6. Spicule formation in an embryo and in a culture of isolated micromeres.

was calculated as a half value of the total outline length of all spicules in a well. The 'Relative Length' is a relative value of the 'total length in a well' compared with the value of control. We also checked the expression of spicule matrix proteins (SM30, SM50) by RT-PCR.

Skeletogenesis in cultured micromeres under various conditions of gravity

Preliminary results from experiments on the role of the gravity force in skeletogenesis by using this new sea urchin experimental system can be summarized as follows.

Hypergravity on a centrifuge

When the micromeres were exposed to hypergravity (55G) from the migrating cell period (16 h after fertilization) up to the spicule elongation period (50 h after fertilization), skeletogenesis was almost suppressed. The result of RT-PCR also showed a decrease in mRNA expression for a spicule matrix protein, SM30. Exposure to hypergravity after the beginning of spicule formation (29 h after fertilization) or during the migrating cell period (between 16 and 29 h after fertilization) did not completely suppress the spicule formation (Fig. 7). It indicates that the complete suppression of skeleton formation occurs only under a continuous exposure to hypergravity (from the migrating cell period up to the spicule elongation period).

Clinostat experiment

Micromeres cultured from the migrating cell period (16 h after fertilization) were installed in a vertical clinostat and kept in this condition up to the spicule elongation period (48 h after fertilization). The lengths of the spicules after 50 h of clino-rotation were smaller than those in control samples (Fig. 8). The result suggests that simulated microgravity suppresses the spicule formation.

Parabolic flight experiment

Although, short term microgravity (20 s) and alterations of the gravity force by a parabolic flight are less efficient in a slow process as is the spicule formation, the sensitivity of the cells may be detected by some effects during post-flight cultures on the ground.

Micromeres cultured just before the migration (12 h after fertilization), before the beginning of spicule formation (24 h after fertilization) and during spicule elongation (36 and 48 h after fertilization) were exposed to the alteration of the gravity force during a parabolic flight and then further cultured for 48 h on the ground. After 24 h post-flight, the length of the spicules was shorter than that in the ground control (Fig. 9). The expression of SM30 mRNA

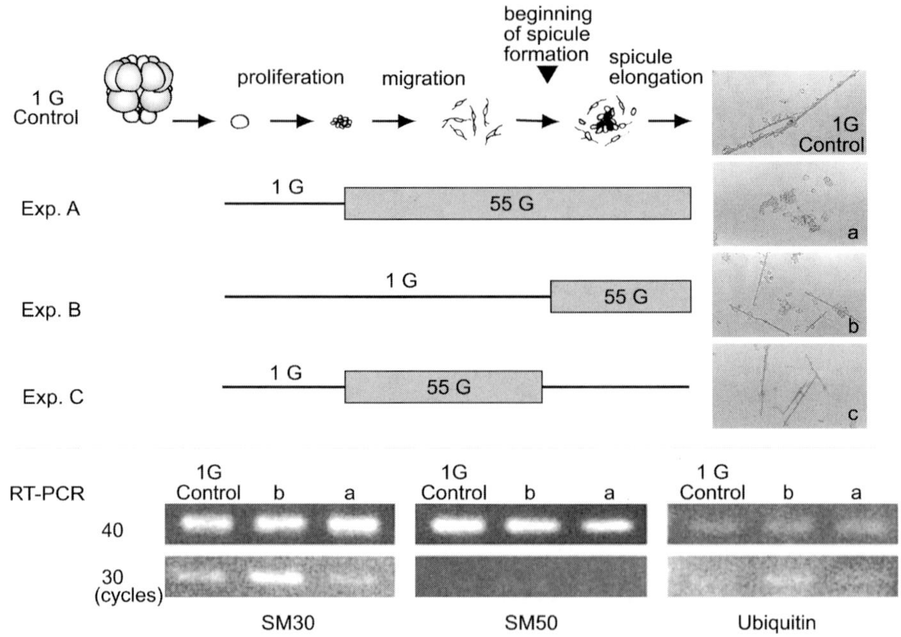

Fig. 7. The effects of hypergravity, by a centrifuge, on the spicule formation in micromere cultures. Upper part: the time course of developmental stages under 1G condition and the time course of the alteration of gravity in three experiments. The images of spicule formation in each experiment are shown on the right side. Lower part: Gene expressions for the spicule matrix proteins SM30 and SM50 and ubiquitin by RT-PCR. The letters a, b and 1G-control correspond to those in the upper images.

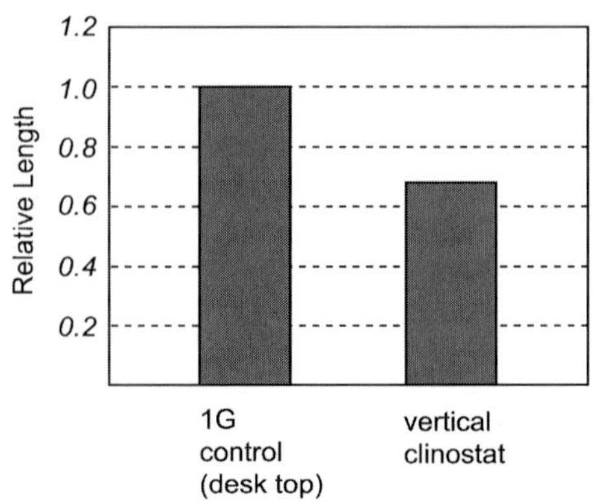

Fig. 8. The effect of a vertical clinostat on spicule formation activity of micromere cultures.

also decreased. Both results show a suppressive effect of the parabolic flight on the spicule elongation. It appears that the cells recover from possible flight effects since no significant differences in the spicule length are observed after 48 h in culture.

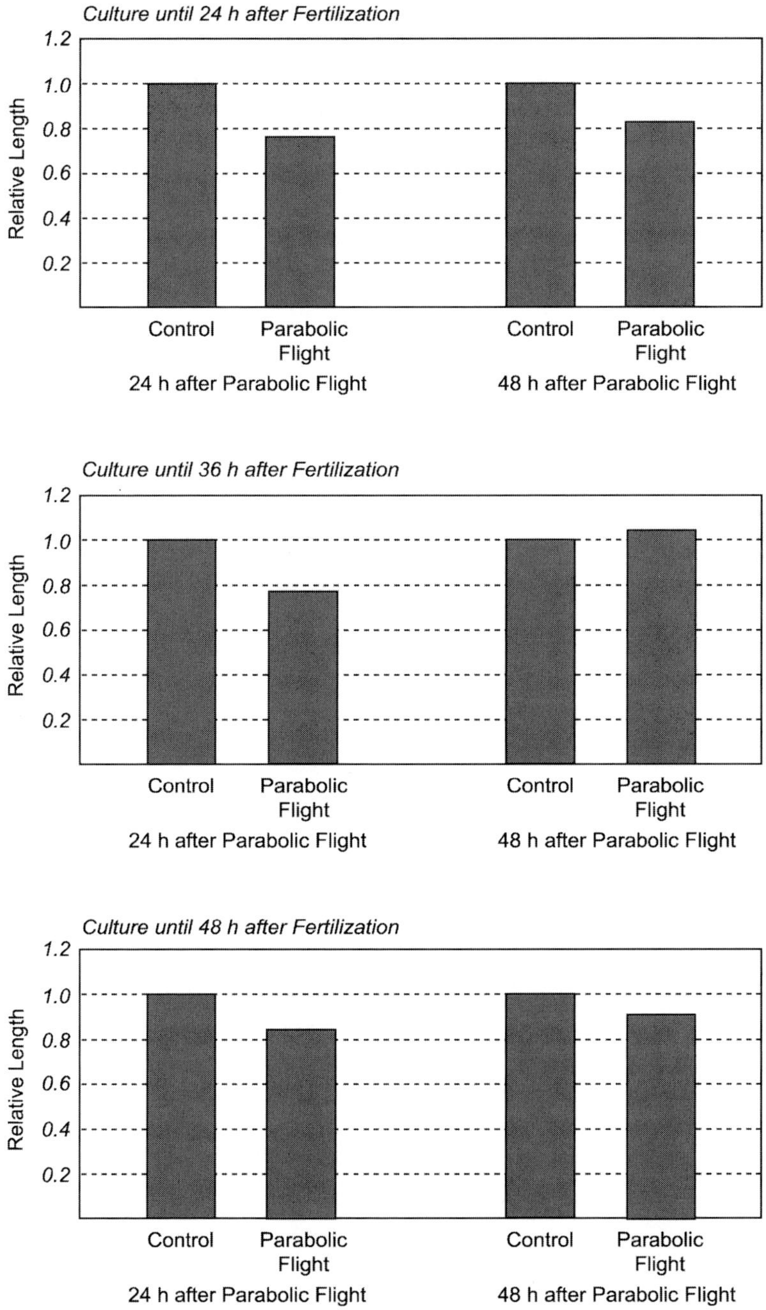

Fig. 9. The effect of a series of parabolic flights on spicule formation activity of micromere cultures. Micromere cultures 24, 36 and 48 h after fertilization were submitted to the parabolic flight experiment and spicule formation activities in three samples were measured 24 and 48 h, respectively after the flight.

Possibilities for space experiments

It appears that the various gravity conditions influence the spicule formation of cultured micromeres and their elongation. Especially, the negative effect on the bio-mineralization process seems to be consistent with the response of the mammalian skeleton to microgravity (LeBlanc *et al.*, 1996). A period-specific sensitivity for gravity changes appears partially. All these findings are important when planning an actual space experiment.

In summary, there are many interesting topics on possible effects of gravity in skeletogenesis of sea urchin larvae. One of the most attractive subjects is certainly the identification of the role of gravity in the mechanisms of skeletogenesis, which includes cell fusion processes, morphology and motility of the migrating cells, the regulation of spicule size and shape, and the time course of the gene expression for spicule matrix proteins. Furthermore, the relationship between gravity and the transport and/or accumulation of calcium ions during skeletogenesis is important. Microgravity-induced decalcification known in vertebrates is a serious problem. It has also not yet been studied in invertebrates. One has to understand in fact both mechanisms since 'bio-mineralization' includes in a common process, the calcium ion transport and the accumulation. Our new experimental system with micromeres of sea urchins appears promising to investigate such problems at a cellular level. A comparison with results from experiments with whole embryos will further enhance a better understanding of the role of gravity in animal morphogenesis.

Acknowledgements

This study was supported partially by 'Ground Research for Space Utilization' promoted by NASDA and Japan Space Forum. A. Izumi-Kurotani the author thanks S. Uchida (Mitsubishi Heavy Industries, Ltd.) for information of AAEU. We greatly appreciate the technical supports of Mr. M. Yamaguchi and Mr. H. Eguchi. We express a special thank to Dr. S. Nemoto. This manuscript would not be completed without the cooperation of Dr. H.-J. Marthy and Dr. N. West.

References

Anthony, P., Ausseil, J., Bechler, B., Benguria, A., Blackhall, N., Briarty, L.G., Cogoli, A., Davey, M.R., Garesse, R., Hager, R., Loddedkemper, R., Marchant, R., Marco, R., Marthy, H-J, Perry, M., Power, J.B., Schiller, P., Ugalde, C., Volkmann, D. Wardrop, J. (1996) Preservation of viable biological samples for experiments in space laboratories, J. Biotech. 47, 377–393.

Black, S.D. and Gerhart, J.C. (1985) Experimental control on the site of embryonic axis formation in *Xenopus laevis* eggs centrifuged before cleavage. Dev. Biol. 108, 310–324.

Black, S., Larkin, K., Jacqmotte, N., Wassersug, R., Pronych, S. and Souza, K. (1996) Regulative development of *Xenopus laevis* in microgravity. Adv. Space Res. 17, 209–217.

Decker, G.L. and Lennarz, W.J. (1988a) Skeletogenesis in the sea urchin embryo. Development 103(2), 231–247.

Decker, G.L. and Lennarz, W.J. (1988b) Growth of linear spicules in cultured primary mesenchymal cells of sea urchin embryos is bi-directional. Dev. Biol. 126(2), 433–436.

Ettensohn, C.A. and Malinda, K.M. (1993) Size regulation and morphogenesis: a cellular analysis of skeletogenesis in the sea urchin embryo. Development 119(1), 155–167.

Gualandris-Parisot L., Husson, D., Bautz, A., Durand D., Kan, Paulette, Aimar, C., Membre H., Duprat, A.-M. and Dournon, C. (2002) Effects of space environment on embryonic growth up to hatching of salamander eggs fertilized and developed during orbital flights, Biol. Sci. in Space, 16(1), 3–11.

Harkey, M.A. and Whiteley, A.H. (1980) Isolation, culture and differentiation of echinoid primary mesenchym cells. Roux' Arch. Dev. Biol. 189, 111–122.

Harkey, M.A. and Whiteley, A.H. (1985) Mass isolation and culture of sea urchin micromeres. In Vitro Cell Dev. Biol. 21(2), 108–113.

Ijiri, K. (1996) Fish mating behavior and development of their eggs in space. . In: Plant in Space Biology (H. Suge ed.) pp. 1–11, *Institute of Genetic Ecology*, Tohoku University, Sendai.

Ijiri, K. (1997) Development of space-fertilized eggs and formation of primordial germ cells in the embryos of medaka fish. Adv. Space Res. 21, 1155–1158.

Kitajima, T. and Matsuda, R. (1982) Specific protein synthesis of sea urchin micromeres during differentiation. Zool. Mag. 91, 200–205.

Kiyomoto, M. and Tsukahara, J. (1991) Spicule formation-inducing substance in sea urchin embryo. Dev. Growth Differ. 33, 443–450.

Kiyomoto, M., Yamaguchi, M., Nemoto, S., Eguchi, H. and Izumi-Kurotani, A. (2001) Artificial control of sea urchin spawning in all seasons. Zool. Sci. 18(Supplement), 86.

LeBlanc, A., Schneider, V., Shackelfold, L., West, S., Ogavov, V., Bakulin, A. and Veronin, L. (1996) Bone mineral and lean tissue loss after long duration spaceflight. J. Miner. Res. 11, S323.

Marthy, H.-J. (1990) Microgravity and development of aquatic animals, pp. 255–257, ESA SP-307, ESA, Paris.

Marthy, H.-J., Schatt, Ph. and Marthy, U. (1994a) Sounding rocket flights, suitable opportunities for developmental biology studies in space, pp. 193-196, ESA SP-355, ESA, Paris.

Marthy, H.-J., Schatt, P. and Santella, L. (1994b) Fertilization of sea urchin eggs in space and subsequent development under normal conditions, Adv. Space Res. 14(8), (8)197–(8)208.

Marthy, H.-J., Gasset, G., Tixador, R., Schatt, P., Eche, B., Dessommes, A., Giacomini, T., Tap, G. and Gorand, D. (1996) The sea urchin larva, a suitable model for biomineralisation studies in space (IML-2 ESA Biolack experiment '24-F urchin'). J. Biotech. 47, 167–177.

Marthy, H.-J. (1997) Sea urchin eggs under microgravity conditions. In: Life Science Experiments performed on Sounding Rockets —1985–1994 (A. Cogoli, U. Friedrich, D. Mesland and R. Demets, eds.), pp. 81–91, ESA SP-1206, ESA, Paris.

Marthy, H.-J. and Bacchieri, R. (1999) In vitro spiculogenesis of sea urchins: an in vitro model system for studying biomineralization processes. ELGRA News 21, 135–136.

Mogami, Y., Imamizo, M., Yamashita, M., Izumi-Kurotani, A., Wiederhold, M.L., Koike, H. and Asashima, M. (1996a) AstroNewt: Early development of newt in space. Adv. Space Res. 17(6/7), 257–263.

Mogami, Y., Koike, H., Yamashita, M., Izumi-Kurotani, A. and Asashima, M. (1996b) Early embryogenesis of amphibians in space: Astronewt for the space embryology in IML-2 and SFU. Adv. Astron Sci. 91, 1089–1097.

Moore, D. and Cogoli, A. (1996) Gravitational and space biology at the cellular level. *In:* D. Moore, P. Bie and H. Oser, (eds.) *Biological and Medical Research in Space*. pp. 1–106, Springer-Verlag, Berlin.

Neff, A.W., Yokota, H., Chung, H.-M., Wakahara, M. and Malacinski, G.M. (1993) Early amphibian anuran morphogenesis is sensitive to novel gravitational fields. Dev. Biol. 155, 270–274.

Okazaki, K. (1975) Spicule formation by isolated micromeres of the sea urchin embryo. Amer. Zool. 15, 567–581.

Okazaki, K. and Inoué, S. (1976) Crystal property of the larval sea urchin spicule. Dev. Growth. Differ. 18(4), 413–434.

Sakairi, K., Yamamoto, M., Ohtsu, K. and Yoshida, M. (1989) Environmental control of gonadal maturation in Laboratory-reared sea urchins. *Anthocidaris crassispina and Hemicentrotus pulcherimus*. Zool. Sci. 6, 721–730.

Sakimura, T., Suzuki, T., Matsubara, S., Uchida, S., Kato, M., Tanemura, R. and Honda, S. (1999) NASDA aquatic animal experiment facilities for space shuttle. Biol. Sci. Space 13(4), 314–320.

Souza, K.A., Black, S.D. and Wassersug, R.J. (1995) Amphibian development in the virtual absence of gravity. Proc. Natl. Acad. Sci. USA 92, 1975–1978.

Uchida, S., Masukawa, M. and Kamigaichi, S. (2002) NASDA aquatic animal experiment facilities for space shuttle and ISS. Adv. Space Res. 30(4), 797–802.

Ubbels, G.A., Berendsen, W. and Narraway, J. (1989) Fertilization of frog eggs on a sounding rocket in space. Adv. Space Res. 9, 187–197.

Ubbels, G.A., Reijnen, M., Meijerink, J. and Narraway, J. (1994) *Xenopus laevis* embryos can establish their spatial bilateral symmetrical body pattern without gravity. Adv. Space Res. 14, 257–269.

Ubbels, G.A. (1997) Establishment of polarities in the oocyte o *Xenopus laevis*: the provisional axial symmetry of the full-grown oocyte of *Xenopus laevis*. Cell Mol. Life Sci. 53, 382–409.

Wakahara, M., Neff, A.W. and Malacinski, G.M. (1985) Development of delayed gastrulae and permanent blastulae from inverted *Xenopus laevis* eggs. Acta Embryol. Morph. Exp. 6(3), 193–209.

Wakahara, M. (1990) Germ plasm: its role in species continuity on the earth and in space. *In:* M. Asashima and G.M. Malacinski (eds.) *Fundamentals in Space Biology*. pp. 139–151, Japanese Science Press, Tokyo.

Yamamoto, M., Ishine, M. and Yoshida, M. (1988) Gonadal maturation independent of photic conditions in laboratory-reared sea urchins, *Pseudocentrotus depressus* and *Hemicentrotus pulcherimus*. Zool. Sci. 5, 979–988.

Yamashita, M., Izumi-Kurotani, A., Imamizo, M., Mogami, Y., Koike, H., Komazaki, S. and Asashima, M. (1997) BIO experiment on space flyer unit: Effects of gravity on early development of Japanese red-bellied newt. J. Space Tech. Sci. 13(2), 12–17.

Yamashita, M., Izumi-Kurotani, A., Imamizo, M., Koike, H., Pfeiffer, C.J., Komazaki, S., Sasaki, F., Ohira, Y., Kashima, I., Kikuyama, S., Ohnishi, T., Mogami, Y., Asashima, M. (2001) Japanese red-bellied newts in space—Astronewt experiment on space shuttle IML-2 and space flyer unit. Biol. Sci. Space 15(Supplement), 96–103.

Yokota, H., Neff, A.W. and Malacinski, G.M. (1992) Altering the position of the first horizontal cleavage furrow of the amphibian *Xenopus* egg reduced embryonic survival. Int. J. Dev. Biol. 36, 527–535.

Yokota, H., Neff, A.W. and Malacinski, G.M. (1994) Early development of *Xenopus* embryos is affected by simulated gravity. Adv. Space Res. 14, 249–255.

Young, R.S. and Tremor, J.W. (1968) The effect of weightlessness on the dividing egg of *Rana pipiens*. Bioscience 18, 609–615.

Developmental Biology Research in Space
H.-J. Marthy (editor)

Developmental Biology of Urodele Amphibians in Microgravity Conditions

Christian Dournon*

Laboratory of Experimental Biology-Immunology, EA 3442: Genetic, Signaling, Differentiation, Henri Poincaré University-Nancy 1, B.P. 239, 54506 Vandoeuvre-lès-Nancy cedex, France

Abstract

Among the urodele amphibians, only *Cynops pyrrhogaster* and *Pleurodeles waltl*, two species of the Salamandridae family, were used in space experiments. The advantages for using urodeles reside (i) in reproduction: a few months after natural breeding, females can lay eggs in absence of males after a hormonal treatment, because spermatozoa were preserved in the cloacal pelvic glands of matted females, (ii) in the rate of development which is slower in *Cynops* and *Pleurodeles* than in the anuran *Xenopus*, (iii) in their physiological properties: they can live in a closed water container or in a moisturized environment, and they can fast during several days. Moreover, urodeles have an important phylogenetic interest. Many biological phenomena differ from those of anurans, such as fertilization events, the germ cell origin and the migration toward the differentiating gonads, and their regeneration capabilities.

The main goals of the space experiments were to answer the following questions. On the one hand, does fertilization occur normally in microgravity? Is subsequent embryonic development normal in microgravity? Is further development and reproduction normal after return to Earth? On the other hand, does microgravity affect the organs in adult animals? Does microgravity affect the regeneration of organs?

Fertilization in space is clearly demonstrated. However, subsequent embryonic development appears to be altered in microgravity. In *Pleurodeles*, abnormalities such as cortical cytoplasmic movements, decrease of cell adhesion, and loss of cells were observed. Although, early development was not strictly normal as a consequence of embryological regulation phenomena, young hatching larvae had normal morphological phenotypes and swimming behavior. After landing, no differences were observed between born-in-space

* E-mail: Christian.Dournon@scbiol.uhp-nancy.fr

animals to standard ones during the embryonic development to adulthood. The analyses of their offspring demonstrated that the percentages of fertilization and development were in accordance with the control animals. No genetic abnormalities were detected during the analysis of the offspring. The development of their progenies were also without characteristic differences compared to control *Pleurodeles*.

Microgravity seems to have effects on the morphological and histological structures of organs of flight adults. However, as was the case in several experiments the number of analyzed adults was low, and it is too early to conclude on specific effects of microgravity. Moreover, in certain flights the temperature was not regulated, and an increase in temperature occurred. Conditions of these space flights had certainly influenced the samples, and consequently the interpretations of results.

Space flights have clear effects on organs in regeneration. But more specifically, they have long term effects that last several weeks after the return of the animals to Earth. A similar result was also obtained for otoconia several months after landing. So far, however, no clear hypothesis could be proposed to interpret these observations.

Introduction

Amphibians are subdivided into anurans, urodeles and gymnophiona. Most of these vertebrates are characterized by metamorphosis and living outside of water with the exception of egg laying that occurs in water. In anurans, tadpoles have inner gills located in lateral cavities where the forelimbs growing, and have a long tail to swim. After metamorphosis, the frog or toad has no tail and jump or swim with their long hind limbs. Anurans are classified into 4,381 species (Lecointre and Le Guyader, 2001) and only a few of them were used for space experiments. Some examples include *Rana pipiens* (Young and Tremor, 1968), *Rana temporaria* (Vinnikov *et al.*, 1973), *Rana catesbiana* (Bracchi *et al.*, 1975), *Xenopus laevis* (Vinnikov *et al.*, 1976, 1980; Ubbels *et al.*, 1989, 1994; Souza *et al.*, 1995; Black *et al.*, 1996; Ubbels, 1997), and *Hyla japonica* (Feuilloley *et al.*, 1993; Yamashita *et al.*, 1997; see chapter on anurans in this book). In urodeles, larvae have external gills and fore- and hindlimbs grown externally and not located in a cavity during their development. After metamorphosis, the animals retain their tail, which is used to walk or swim. They are classified in 429 species (Lecointre and Le Guyader, 2001). Among the urodeles, only *Cynops pyrrhogaster* and *Pleurodeles waltl*, two species of the Salamandridae family, were used in space experiments. The gymnophiona have a worm aspect and are characterized by the absence of legs. These animals are classified into 165 species (Lecointre and Le Guyader, 2001). No one has used them for experimentation in space flight.

Cynops pyrrhogaster originates from Japan and is called "Imori". It is a fire-bellied newt living temporally in summer at 15–23 °C. The genus *Cynops* is closely related to the European *Triturus* and is aquatic most of its life. This species breeds from April to early July after hibernation. It inhabits ponds, rice fields, ditches, swamps and streams. Animals can be collected from a water creek near rice paddies in late autumn or winter after natural mating. Embryos, larvae, and adults are routinely used in laboratories. Embryos and larvae are organized according to a table of development in which the stages are established following various specific morphological landmarks, such as the development or disappearance of gills or the development of the limbs (Okada and Ichikawa, 1947; Okada, 1989).

Pleurodeles waltl originates from Spain, Portugal and Morocco (Rudolf *et al.*, 1996). Adults have a laterally flatted tail and bulging ribs. This species was introduced in laboratory rearing around 1950 and crossed under laboratory conditions at 16–25 °C, where the breeding period occurs between September and April. Embryos and larvae are staged according to the table of development of Gallien and Durocher (1957). Genotypic females are heterogametic ZW and genotypic males are homogametic ZZ (Gallien, 1954; Collenot, 1973; Dournon and Houillon, 1984).

The first advantage of urodeles for space experiments concerns reproduction (Gualandris-Parisot *et al.*, 2002b). A few months after natural breeding, *Cynops* and *Pleurodeles* females can lay eggs in absence of males after a preparatory hormonal treatment or by injection of hormone. Spermatozoa were preserved in the cloacal pelvic glands of matted females. During an induced egg laying, spermatozoa passed the jelly coat and penetrated the oocytes in a natural manner as in normal *in vivo* fertilization. On board space vehicles, only inseminated urodele females were used to obtain embryos. In contrast, only anuran females were used in space, but as fertilization is external, spermatozoa were formerly prepared on ground and then used in space for *in vitro* fertilization. The advantage for using urodeles resides in the rate of development, which is slower in *Cynops* and *Pleurodeles* than in *Xenopus*. For these species reared at 18 °C, hatching occurs after about 20, 12 and 2 days, respectively. Consequently, it is easier to study embryogenesis in detail on urodele species than on *Xenopus*. On the other hand, it is more convenient to use *Xenopus* to study advanced larval development, since *Xenopus* grow more rapidly than *Cynops* or *Pleurodeles*. Another advantage is linked to the physiological properties of urodeles. Under experimental conditions, *Cynops* females can live and lay eggs in a closed water container, and *Pleurodeles* females can survive and reproduce in a moisturized environment (Yamashita *et al.*, 2001; Husson *et al.*, 2001). Urodeles have slower movements than anurans, and are easier to manipulate in microgravity conditions. They can fast during several days (up to 4 weeks for *Pleurodeles*). During the fasting period, they produce little excrement, and consequently this limits material and time for their maintenance. Among the disadvantages, the molecular biology and genetics are less developed in these urodeles than in *Xenopus*. However, urodeles have an important phylogenetic interest. Many biological phenomena differ from those of anurans, such as fertilization events, the germ cell origin and the migration toward the differentiating gonads, and their regeneration capabilities.

Feeding and rearing methods in microgravity conditions

Methods of feeding were defined to studying larvae and adults during long space missions. Under laboratory conditions, *Pleurodeles* larvae were reared in basins with clean tap water at room temperature and submitted to a daylight variation. They were periodically fed with nauplii of *Artemia salina* developed from cystae in salted water. During the 1999 space mission "Perseus" aboard Mir, larvae were continuously reared at 18 °C in hermetic cylindrical boxes and in darkness except during the phases of feeding and during medium changes. On board Mir, live food i.e., *Artemia* nauplii was produced in an adequate quantity for the *Pleurodeles* larvae depending on their stages of development. *Artemia salina* cystae and sea salt were dry stocked in syringes until 3 months. Cystae were activated with the modified physiological medium of Steinberg (Dournon *et al.*, 2001) 3 days before use. Management of syringes, food distribution and water cleaning were performed over 21 days by a French cosmonaut aboard MIR (Durand *et al.*, 2000).

For *Pleurodeles* adults, an initial protocol of force-feeding was successfully performed for 5 months aboard Mir during the French mission "Perseus". A force-feed method had to be defined since adults could not take food themselves under microgravity conditions. A cosmonaut introduced into the mouth a stomach catheter connected to a food containing syringe, and dosed the food quantity using a dispenser. The selected food was a complete aliment for dogs and cats called "Prescription diet Canine/Feline a/d" made in USA and produced by the Hill's Pet Nutrition SNC (Sofia-Antipolis, France). It was diluted, sterilized and conserved aboard Mir at ambient temperature (20–30 °C). After the space mission, two females survived. Both laid fertilized eggs twice, once in flight, and subsequently 1 week after landing (Veynachter *et al.*, 2002).

Experiments performed in space using urodele amphibians

Due to French-Russian collaborations, *Pleurodeles waltl* were previously introduced into USSR and Russian laboratories. Animals were first launched to space on board the satellite Bion 7 during the Russian Cosmos 1667 flight in 1985, from July 10–17 (Mitashov *et al.,* 1987). Until 2001, several experiments using eggs, embryos, larvae, juveniles and adults of *Pleurodeles waltl* were performed by Russian or French groups on board Russian space vehicles. Japanese groups in collaboration with other groups used *Cynops pyrrhogaster* on board the American shuttle Columbia in 1994 and on board a Japanese space platform put in orbit in 1995. The known space missions in which newts were embarked are gathered Table 1.

Table 1

Known space missions in which urodele amphibians were embarked

Dates	Duration	Flights	Experiments	Instruments	Species	Principal investigators
1985, July 10–17	7 days	Cosmos 1667 Bion 7	Triton	Triton	*Pleurodeles waltl*	Mitashov *et al.*
1987, September 29–October 12	12.5 days	Cosmos 1887 Bion 8	Triton	Triton	*Pleurodeles waltl*	Mitashov *et al.*
1988, April 14–23	13.5 days	Foton 4	Triton	Triton	*Pleurodeles waltl*	Mitashov *et al.*
1989, April 26–May 10	14 days	Foton 5	Triton	Triton	*Pleurodeles waltl*	Mitashov *et al.*
1989, September 15–29	14 days	Cosmos 2044 Bion 9	Triton	Triton	*Pleurodeles waltl*	Mitashov *et al.*
1990, April 11–26	15 days	Foton 6	Triton	Triton	*Pleurodeles waltl*	Mitashov *et al.*
1991, October 4–19	5.5 days	Foton 7	Triton	Triton	*Pleurodeles waltl*	Mitashov *et al.*
1992-93, December 29–January 10	12 days	Cosmos 2229 Bion 10	Triton	Triton	*Pleurodeles waltl*	Mitashov *et al.* Dournon *et al.* Duprat *et al.*
1994, June 14–July 2	17.5 days	Foton 9	Triton	Triton		Mitashov *et al.*
1994, July 8–23	15 days	Columbia Shuttle STS 65-IML-2	Astronewt	AAEU	*Cynops pyrrhogaster*	Yamashita *et al.* Wiederhold *et al.*
1995–96, March 19–January 20	Activated during 12 days	Japanese H-II rocket – Endeavour Shuttle STS 72	Astronewt	NEU	*Cynops pyrrhogaster*	Yamashita *et al.*
1995, February 16–March 3	14 days	Foton 10	Triton Torcol Célimène	Triton IBIS	*Pleurodeles waltl*	Mitashov *et al.* Dournon *et al.* Duprat *et al.*
1996, August 17 –September 2	14 days	Soyuz TM 24 and-23-Mir Cassiopée Mission	FERTILE	Triton CTA and Fertile	*Pleurodeles waltl*	Mitashov et al. Dournon *et al.* Duprat *et al.*
1996–97, December 24–January 7	13.5 days	Bion 11	Triton	Triton	*Pleurodeles waltl*	Mitashov *et al.*
1997, October 9–23	13.5 days	Foton 11	Torcol Célimène	IBIS	*Pleurodeles waltl*	Dournon *et al.* Duprat *et al.*
1998, January 29–February 19	20 days	Soyuz TM 27 – 26-Mir Pégase Mission	FERTILE	CTA and Fertile	*Pleurodeles waltl*	Dournon *et al.* Duprat *et al.*
1999, February 20–August 28	5 months	Soyuz TM 29 – 28-Mir Perseus Mission	Genesis	CTA and Fertile	*Pleurodeles waltl*	Dournon et al. Duprat *et al.*
2001, October 21–31	9 days	Soyuz TM 33 – 32-ISS Andromède Mission	Aquarius	CTA	*Pleurodeles waltl*	Dournon *et al.*

Hardware used for experiments

A box called Triton. Russian groups worked on juveniles or adults of *Pleurodeles* and used metal boxes closed with transparent plastic covers. Boxes were 300 mm in length, 200 mm in width and 90 or 80 mm in height. Water-absorbent paper covered the internal metal surfaces, and 500 ml of water were put in the containers. Holes were made through the cover for passive aeration throughout a gas changing filter. This hardware was used for short space missions. Each box contained up to 10 adult females. Boxes were not thermally regulated; consequently, the rearing temperature was ambient to the space vehicles.

The aquatic animal experiment unit (AAEU). This hardware was provided by three Japanese organizations: ISAS (Institute of Space and Astronautical Science), MITI (Ministry of International Trade and Industry) and NASDA (National Space Development Agency of Japan) (Chee *et al.*, 1993; Sakimura *et al.*, 1999). The essential part of the AAEU was a controlled life support system, used to keep aquatic animals alive for at least 19 days in the Spacelab. The AAEU consisted of a main unit Aquarium Package that had an autonomous life support system. Four adults and approximated 144 young embryos of *Cynops* were kept in three cassette-type water tanks in the Aquarium Package. They were completely confined in the cassettes throughout the experiment. A special life-support system supplied oxygen, removed carbon dioxide, ammonia and organic substances, and regulated temperature at 24 °C. The animals could be viewed through a window and a video system was attached to the viewing port for recording observations of animal behavior. No 1G centrifuge and no fixative injection devices had been integrated in the AAEU. It was previously used in 1994 on board the American shuttle Columbia on the Second International Microgravity Laboratory (IML-2) during the STS 65 mission (Mogami *et al.*, 1996a,b; Wiederhold *et al.*, 1997a; Yamashita *et al.*, 2001).

The newt experiment unit (NEU). This hardware developed by ISAS, MITI and NASDA was used on board the first Space Flyer Unit (SFU), a Japanese space platform put in orbit on March 18, 1995 by a H-II rocket. After 10 months of space flight, the SFU was retrieved by the American space shuttle Endeavor during the STS-72 mission in 1996. Two *Cynops* females were kept in two rectangular boxes (30 × 27 × 140 mm). Close up images of the newt box was taken by a CCD camera and downlinked to the ground. Circulated water was filtered, and a sensor unit monitored dissolved oxygen and temperature. Mixed pellets of potassium peroxide and calcium peroxide were placed in the air feeding line to generate oxygen and absorb carbon dioxide. A fixative solution could be introduced in the boxes by electric valves opened at the end of the flight. At launch site, temperature was kept low for hibernation (4 °C). In orbit, the water temperature was raised to 20 °C. NEU was mounted on the main panel of the Special Payload Unit of the SFU. Experimental computer and power unit were installed on the side panel. Electrical contacts to the newt system on ground were made through an umbilical connector at the fairing of the H-II rocket (Mogami *et al.*, 1996a,b; Yamashita *et al.*, 2001).

The IBIS. The Instrument for BIology in Space (IBIS) is a thermally controlled instrument conceived by CNES and a French manufacturer COMAT SA. It was used aboard the automatic Russian biosatellites Bion or Foton. This instrument is subdivided into three parts. One part is a refrigerator used to block the cell metabolism or the embryo development of amphibians, plants or other model organisms. The second part is an incubator held at 18–37 °C for cell cultures or small organisms, and the third part is a thermo-regulated 1G-centrifuge. Twin cassettes-samples are automatically transferred from the refrigerator to the incubator and onto the centrifuge, activated by temperature. Fixations can be automatically performed at any time. At the end of the experiment, the twin cassettes-samples are transferred to the refrigerator.

The CTA and Fertile instrument. The instruments called CTA and Fertile were conceived by CNES and COMAT in collaboration with French biologists (Gualandris-Parisot *et al.,* 1998; Husson *et al.*, 2001). The CTA (in French: Container de Transport Aller) was a thermally controlled container used for the transportation of animals to space stations. It was connected to the electrical power supply of the vessel Soyuz. The CTA mass was 9.6 kg. The trip toward the Mir station or ISS took 2 days. During this period, the crew had access to animals. The CTA can receive three boxes of transport containing *Pleurodeles* adults. In this configuration, it was set at 18 ± 2 °C and was an air pump ventilated (0.2 l/min) in the boxes. Each was 192 mm in length, 152 mm in width and 47 mm in height, and the inner volume was 1 l. In each box, two adults were maintained without free water, but instead with a damp towel that covered the two largest internal surfaces to moisturize their skin. In another configuration, the CTA can receive twenty miniaquaria in 11.25 l available volume (30 cm in length, 25 cm in width and 15 cm in height). The CTA was set at 20 ± 1 °C and was ventilated inside by an air pump (0.2 l/min).

The Fertile instrument was put aboard Mir in May 5, 1996 by the Progress M-31 and used during three space missions in 1996, 1998 and 1999. The unit contained an incubator set at 18 ± 1 °C. Dimensions were $400 \times 412 \times 677$ mm with a mass of 47 kg. This instrument had several different functions. Three "egg-laying boxes" (226 mm in length, 164 mm in width and 76 mm in height) allowed spawning by females in good condition. The internal volume was one liter and each box was ventilated by an air circulation system (0.2 l/min). Two polycarbonate portholes allowed visual control of the interior of the box. The internal surface was covered with a water-absorbent tapestry covering a sponge and allowing permanent hydration of the skin of the adults or the jelly coat of laid eggs. The tapestry and sponge were saturated with a filtered (0.22 μm) modified physiological medium of Steinberg (Dournon *et al.*, 2001). A system of water distribution and absorption was designed for liquid control before and after each opening of the box, to prevent water spilling into the space station and to avoid drowning of the animals when the boxes were closed. This component consisted of the egg-laying boxes, with two trays storing 28 egg boxes in microgravity. Three types of egg boxes were used: (1) boxes allowing the development of embryos until the end of the space mission for live recovery; (2) boxes containing a cartridge of fixative liquids such as formaldehyde or glutaraldehyde for in-flight fixation at selected stages of the embryonic development; and (3) one large box for storing eggs which were not immediately used. The component also contained a centrifuge providing 1G aboard Mir and stored 10 egg boxes used as controls, a video camera recorded the development of 20 μG-embryos located in a special egg box equipped with a lighting system, and a reservoir containing 2.8 l of a filtered (0.22 μm) modified physiological medium of Steinberg to hydrate adults and eggs. Four other 2.8 l reservoirs were available to stock water for more than 8 months without bacteriological and physico-chemical alterations and used during longer space missions.

A third unit was a passive container. Inside, live animals and fixed biological samples were placed for return to Earth. It consisted of expansion moss covered with resin and "Nomex" material to limit variations of temperature in the absence of electrical supply on board Soyuz. The mass was 4 kg and its design was adapted to the small volume available in Soyuz. The return trip took was approximately 6 h, but biological materials remained inside for only 1 day.

Miniaquaria. Two types of miniaquaria were used for *Pleurodeles* embryos placed inside a CTA before the launch. Miniaquaria, developed by EADS-Astrium, the German Space Establishment (DLR) and German biologists were used for recovering living larvae after landing. They were previously used for *Xenopus* tadpole experiments (Sebastian and Horn, 1998; Horn and Sebastian, 1999). Each miniaquarium was 80 mm in length, 40 mm in width and 20 mm in height and had two transparent faces (bioFOLIE 25/Heraeus a 25 μm thick plastic foil, which is O_2- and CO_2-permeable). For protection and handling, the miniaquaria were introduced in Type 1 containers from ESA Biorack facility. The Type 1 container has two gas permeable faces (Horn and Sebastian, 1999).

Particular miniaquaria developed by the CNES were used to fix *Pleurodeles* embryos or larvae during space flights with the help of the crew. These boxes were rectangular and 80 mm in length, 40 mm in width and 20 mm in height. Inside each box, miniaquarium (10 ml) could receive at least five embryos in tap water, which were separated by a porous membrane (5 μm) from an air chamber (20 ml). The miniaquarium also included a cartridge of a fixative liquid. For protection, each box was introduced in a hermetic Type 1 container. These miniaquaria put into the CTA were used on board the Soyuz TM 33 and then on the I.S.S during the "Andromède Taxi Mission" (2001).

Results

The results reported in this paper are those communicated by international journals, reviews or at congresses. However, not all the published results are accessible. For example, previous papers written in Russian language appear newly and regularly on international servers. The results obtained on urodeles during space mission concern:

- microgravity effects on fertilization, embryogenesis and early organogenesis. Post-flight development and reproduction on ground were also analyzed;
- microgravity effects on intact or regenerating organs of adult animals during space conditions and after return to Earth.

Development under microgravity conditions and postflight development

The main goals of the space experiments, which used *Cynops pyrrhogaster* and *Pleurodeles waltl*, were to answer the following questions. Does fertilization occur normally in microgravity? Is subsequent embryonic development normal in microgravity? Is further development and reproduction normal after return to Earth?

Does fertilization occur normally in microgravity?

Obtaining fertilized eggs under space conditions
In the AAEU Astronewt experiment and the duplicate FERTILE experiments, eggs were laid by preinseminated females injected with human chorionic gonadotrophins (HCG) for *Cynops* (Yamashita *et al.*, 2001) and human luteinizing hormone-releasing hormone (LH-RH) for *Pleurodeles* (Aimar *et al.*, 2000). In the NEU Astronewt experiment, the inseminated females were launched in hibernation at 4 °C and were treated with HCG enclosed in a drug delivery system inserted in the females (Imamizo *et al.*, 1994). An increase in temperature during space flight, coupled with the hormonal treatment induced egg laying (Mogami *et al.*, 1996b; Yamashita *et al.*, 2001). For the first FERTILE experiment, inseminated females were kept at 8 °C during 16–12 weeks after natural mating and were placed at 18 °C at launch. For the second FERTILE experiment, the natural mating occurred 8–2 weeks before the space experiment and the inseminated females were maintained at 13–16 °C and placed at 18 °C at launch (Dournon *et al.* 2001). In flight, *Cynops* were reared at 24 °C (Mogami *et al.*, 1996b; Yamashita *et al.*, 2001) and *Pleurodeles* at 18 °C (Gualandris-Parisot *et al.*, 2002a). In all the experiments, synchronous controls were performed at the ground laboratory but some days shifted to reproduce unexpected events occurring during the flight experiment.

For the two Astronewt experiments, 6 μG- and six ground control-inseminated females were used. Five μG-females provided 71 embryos. At the end of the experiments, two females were fixed in flight, two died, and two returned alive. On ground, all the control females survived except two, which were fixed according to flight events.

For both FERTILE experiments, 12 μG- and the 12 ground control-inseminated females were used. Ten μG-females provided 1620 eggs placed in μG-containers and 341 eggs placed on the 1G-centrifuge aboard Mir. Eleven 1G-ground females provided 1576 eggs. At landing, all the *P. waltl* females were alive.

Fertilization in space is demonstrated

In *Pleurodeles waltl*, 1–14 spermatozoa enter the oocyte, but only one male pronucleus fuses with the female pronucleus (Labrousse, 1959; Picheral, 1977). The occurrence of fertilization in microgravity was demonstrated by the presence of spermatozoa in the perivitellin space of eggs laid aboard Mir, spermatic impacts on the surface of μG-eggs, and one female pronucleus and several male pronuclei in the examined μG-eggs and control eggs. Moreover, the demonstration of an actual fertilization was completed by the analysis of the distribution of peptidase-1 genes, from parents to progenies obtained in μG conditions. Peptidase-1 is a dimeric and polymorphic enzyme. Each monomer is coded by a gene linked with a sex chromosome, the gene *Pep 1A* with the sex chromosome Z and the gene *Pep 1B* or *1β* with the sex chromosome W. Genetic analyses ruled out parthenogenesis or gynogenesis and proved that fertilization did occur in microgravity (Aimar *et al.*, 2000). Previous work aboard Bion 10 had shown that microgravity does not modify the peptidase-1 expression (Bautz *et al.*, 1996b).

The determination of the symmetrical plan position differs between urodeles and anurans

In amphibian eggs, the gravity center is located on the animal-vegetative axis in the vegetative hemisphere. During fertilization, eggs are freed from the surrounding envelopes and turn inside to place the gravity center in concordance with the ground gravity vector. Then, the rotation of symmetrisation occurs, materialized by the grey crescent that indicates the dorsal region of the future embryos. The superficial cytoplasm, including the pigments of the animal hemisphere, rotates inside the egg. Early work using anuran amphibians suggested that gravity could be necessary to determine the position of the symmetrical plan in amphibians (Pflüger, 1883; Roux, 1884; Kathariner, 1901; Morgan, 1904; Pasteels, 1941). Ancel and Vintemberger (1948) had shown that if eggs were slowly rotated by 360°, the imposed plan of rotation coincides with the future plan of symmetry. These data raised the question of the role of Earth gravity on the determination of the symmetrical plan and on the subsequent development in amphibians. Later, it was demonstrated on *Xenopus laevis* eggs that were centrifuged between fertilization and the first cleavage, that hypergravity induced twin-duplicated embryos (Black 1989; Black and Gerhart 1986; Neff *et al*., 1990, 1993). But during the same period, space experiments using *Xenopus* stated that embryos, produced from *in vitro* fertilization performed in microgravity, were normal with a symmetrical plan (Ubbels *et al*., 1989, 1994; Ubbels 1997; Souza *et al*., 1995).

For all anuran species studied in ground or space experiments, fertilization is external and monospermatic. At present in anurans, it is well known that the spermatic impact point with the animal-vegetative pole axis determine the position of the symmetrical plan, with the exception of particular experimental conditions (Ancel and Vintemberger, 1948).

Most urodeles, as *Pleurodeles*, are polyspermatic under natural conditions. Several spermatozoa penetrate the eggs and only one male pronucleus merges with the female pronucleus. Moreover, in the absence of spermatozoa, the surrounding membranes naturally come unstuck from the egg cells laid in water, and a rotation of equilibration occurs. Consequently, for space experiments, the potential role of the egg rotation or the superficial cytoplasm displacement were not pertinent questions in urodeles as in anurans. As expected, all embryos of *Cynops pyrrhogaster* and *Pleurodeles waltl*, laid and developed in microgravity, displayed normal bilateral symmetry (Yamashita *et al*., 2001; Gualandris-Parisot *et al*., 2002a).

The cortical reaction of fertilization is altered in microgravity

Pleurodeles eggs stored in μG for a minimum of 2 h showed the cortical reaction i.e., the migration of the cortical pigment toward the animal pole after fertilization and demonstrated the success of the egg activation in μG. However, the cortical reaction is altered. On board, eggs exhibited large unpigmented areas in the animal hemisphere, and the ascent of the pigment toward the animal pole was amplified in comparison to control 1G-eggs. This perturbation was unrelated to the orientation of the cleavage planes and was still visible in μG-eggs up to the morula stages. 34.4% of μG-embryos were affected. Data revealed a significant difference between μG and 1G-ground embryos, and no difference between 1G-centrifuge on board and 1G-ground embryos (Gualandris-Parisot *et al*., 2002a).

Egg nucleus location

In natural conditions, the zygotic nucleus resulting from the fusion of the two pronuclei is located on the animal-vegetal axis, thus determining the meridian position of the furrow. Consequently, a furrow extended besides the animal-vegetal axis corresponds to a lateral position of the nucleus. With one exception, all examined *Pleurodeles* eggs (No = 25) exhibited normal location of the first furrow. This indicated that microgravity did not impair the migration of the pronuclei and did not provoke an off-axis location of the zygotic nucleus

(Aimar *et al.*, 2000; Gualandris-Parisot *et al.*, 2002a). This observation does not preclude a displacement of the zygotic nucleus from the animal-vegetal axis toward the vegetal pole, as observed in experiments using *Xenopus laevis* in microgravity (Souza *et al.*, 1995).

Evidence of effects of μG, gravity and hypergravity on microvilli at the beginning of the first cleavage

In ground development, microvilli of *Pleurodeles waltl* eggs were easily observed through SEM. Microvilli covered the egg surface and displayed characteristic changes after fertilization. Microvillus transformations, observed in the vicinity of the animal pole of eggs, were classified in five types corresponding to five stages of egg transformation during the first cycle (Aimar, 1991). At fertilization, microvilli are hemispherical or conical-shaped and uniformly distributed (type 1). About 1 h after fertilization, the microvilli progressively transform, organizing between 3–4 h post-fertilization a discontinuous network of convoluted crests in association to deep depressions (type 2). This sponge-like structure regresses after 4 h post-fertilization (type 3). The egg surface is then smooth, exhibiting some flattened or spherical microvilli (type 4). At the beginning of the first division (6 h post-fertilization), finger-like microvilli elongate on the edges of the forming furrow (type 5) while, on the other areas of the egg surface, microvilli are still of type 4. On the furrow base between the furrow edges, rounded microvilli resembling type 4 microvilli develop.

The main characteristics of microvilli of fertilized eggs (No = 150) submitted to gravitational forces were compared during the first division cycle. In the experiment, four batches of *Pleurodeles* eggs were used. Eggs provided from fertilization at μG were subdivided in two batches. In one batch, they developed continuously in μG, and in the other they were immediately put in the 1G-centrifuge on board Mir. Other eggs from spawning in the ground laboratory were reared in 1G ground gravity or were immediately put on the centrifuge to be submitted to 1.4G, 2G or 3G. As a rule, microvilli alterations increased with the duration of exposure to a constant gravitational force, and changes in the microvilli pattern appeared to depend on the magnitude of the gravitational forces. A minimum of 2 h was necessary to induce an effect on egg microvilli. Thus, alterations were mostly observable at the time of the first cleavage, which occurred 6 h after fertilization.

For example, at fertilization, the conical microvilli had a uniform distribution and density of 20 microvilli per μm^2. Afterwards, density decreased but did not differ significantly between eggs submitted to different gravitational forces. However, at the time of the first cleavage and by comparison to μG-eggs, the density of microvilli of 1.4G-eggs and 3G-eggs was reduced by 4 and 8-fold, respectively. Changes of the microvilli in shape and diameter seemed to be also correlated to the gravity level applied to the eggs. At 2–4 h post-fertilization, the respective percentages of round or elongated microvilli were different when μG-eggs up to 3G-eggs were compared. At the beginning of the first cleavage, the difference in patterns of microvilli of the furrow sides became more evident. The percentage of rounded microvilli decreased by 30% from μG-eggs to 1.4G-eggs whereas the percentage of elongated microvilli increased. A 12% decrease was observed for μG-eggs and reached 38% for 1.4G-eggs. Finally, 96% of the microvilli of the 3G-eggs produced a typical voluminous lobed shape.

Microvilli elongation was the most evident microvilli feature, which appeared to depend on the gravitational forces and on the time that these forces were applied to the eggs. Although, the chronological schedule of microvilli pattern transformations occurred identically to that described for 1G-ground control eggs, microgravity and hypergravity provoked alterations on the microvilli characteristics. As a rule, microvilli elongation, compared to 1G-ground control eggs, was reduced on μG-eggs similar to 1G-eggs on board Mir as well as on 1.4G-eggs on ground, and amplified on eggs submitted to 2G and 3G. At about 4 h of development, μG-eggs displayed crests, 0.25–1 μm long. In contrast in 1G, ridges were approximately 1 μm. They reached 1.5 μm or more in size in 2G and 3G (Fig. 1). No lethal consequences of these alterations on early development of μG-eggs were observed (Aimar *et al.*, 2000).

After fertilization, is subsequent embryonic development normal in μG?

Embryogenesis

Up to now, *Cynops pyrrhogaster* and *Pleurodeles waltl* were models for studying embryonic development in space conditions. Concerning *Cynops,* five females provided 71 embryos video recorded at late morula, early blastula, gastrula, neurula, tail bud stage and before the first gill ramification. The morphology and the time course of the development were analyzed by close up video images taken by the crew during the flight. All embryos were judged as normal. In respect to this method, the authors concluded that microgravity and a space environment do not affect early developmental processes in *Cynops* (Yamashita *et al.*, 2001). Concerning *Pleurodeles*, twelve females provided a total of 977 developed embryos obtained during two space experiments. The first experiment had been achieved beyond and the second within the reproductive period. The percentage of development (Number of developed embryos at the 2-cell stage or at later stages/total number of eggs) was 21 and 82% for the eggs laid by the μG-on board females, and 23 and 78% for the eggs laid by the 1G-ground control females for the first and second space mission respectively. The percentage of developed embryos was 13 and 87% for the 1G centrifuged eggs on board MIR, and 10 and 73% for the 1.4G centrifuged eggs in the ground laboratory for the first and second space missions, respectively (Table 2). The embryonic development is detailed hereafter.

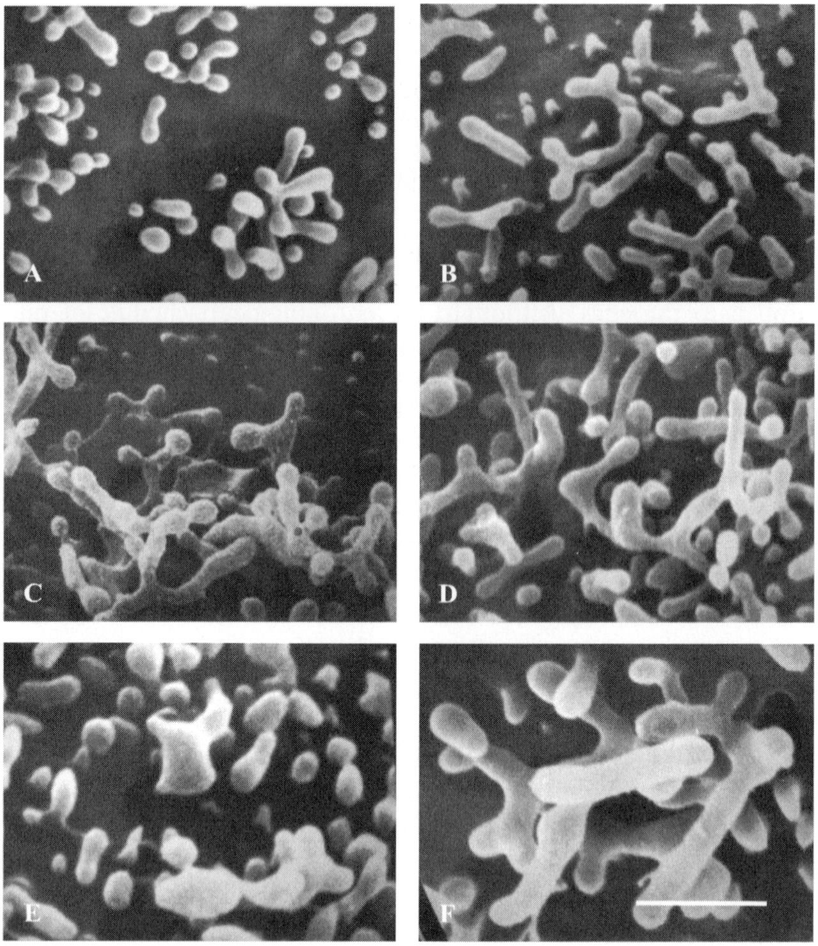

Fig. 1. Microgravity and hypergravity effects on microvillies of *Pleurodeles* eggs. SEM aspects of finger-like MV of the first furrow edges of *Pleurodeles* eggs submitted to various gravitational strengths. The size and extension of the MV increase progressively from μG to 3G. A: μG. B: 1G (centrifuge in space). C: 1G (ground gravity). D, E and F: centrifugation in ground laboratory at 1.4, 2 and 3G, respectively. Bar = 1 μm. Reproduced by permission of "Society for the Study of Reproduction"; from: Aimar, Bautz, Durand, Membre, Chardard, Gualandris-Parisot, Husson & Dournon – Biol. Reprod. 63, 551–558 (2000).

Cleavage period

Following the first furrow, μG-embryos of *Pleurodeles* displayed an irregular cleavage pattern. For example, three, five or seven cells were observed in the animal hemisphere (Fig. 2). Irregular cleavages were seen in 31.6, 7.7 and 2.3% of μG-, 1G-on board and 1G-ground embryos, respectively. Using χ^2 test with the Yates correction χ_c^2, the difference between μG- and 1G-ground embryos

Table 2

Embryos of Pleurodeles layed aboard the Mir space station or in the ground laboratory

	Embryos fixed during the flight			Embryos kept alive during the flight			Larvae reared after the flight		
	No of eggs used to study the embryonic development	No of developed and fixed embryos	% of developed and fixed embryos	No of eggs used to obtained in live embryos	No of embryos in live after landing	% of in live embryos after landing	No of larvae kept alive after landing	No of in live larvae tested with peptidase-1 assays	No of reproductive adults[1]
Cassiopée Mission									
In μG on board MIR	504	108	21.4	194	43	22.2	22	13	9
In 1G MIR centrifuge	147	19	12.9	21	14	66.7	3	3	3
In 1G on ground	527	120	22.8	139	1	0.7	0	0	0
Pégase Mission									
In μG on board MIR	719	592	82.4	122	51	41.8	29	13	5
In 1G MIR centrifuge	173	150	86.7	0	0	0	0	0	0
In 1G on ground	680	528	77.6	230	93	40.4	66	12	12
Total missions									
In μG on board MIR	1223	700	57.2	397	94	23.7	51	26	14
In 1G MIR centrifuge	320	169	52.8	21	14	66.7	3	3	3
In 1G on ground	1207	648	53.7	369	94	25.5	66	12	12

[1]At left, number (No) and percentage (%) of developed embryos fixed in flight at different stages of development. In the middle, number (No) and percentage (%) of embryos obtained alive after delivery of samples on ground. Some embryos were used for histological analyses and *enzymatic* assays. At right, number of animals kept alive and reared after landing, and crossed when mature.

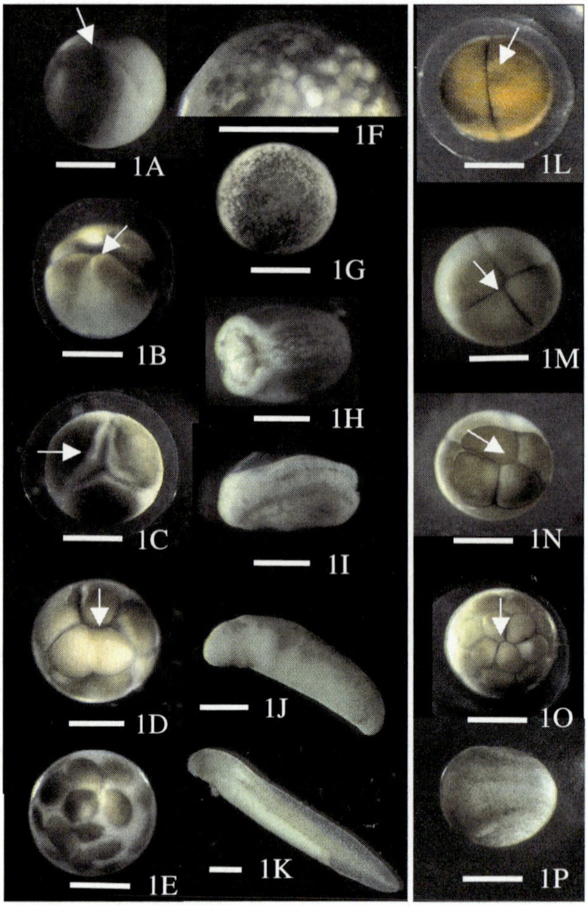

Fig. 2. *Pleurodeles* development in space conditions. **A** to **K**: μG-embryos fertilized, developed and fixed on board MIR. **L** to **P**: 1G- ground control embryos. Arrows indicate the animal pole. **A**: beginning of the first cleavage. Abnormal pigmented area. **B**: beginning of the second cleavage. Abnormal pigmented area. Enlarged intercellular space. **C**: abnormal three-cell stage. **D**: Five blastomeres instead of 4 at the animal hemisphere. Enlarged intercellular space. **E**: morula stage. Yolk platelets between blastomeres. **F**: detail of a blastula. Showing detached cells. **G**: blastula at stage 7. **H**: neurula at stage 15. Loss of neurectodermal cells. **I**: neurula at stage 20. Neural tube not closed in the dorsal region. **J**: embryo at tail bud stage 25. Normal aspect. **K**: embryo at stage 32. Normal aspect. **L**: two-cell embryo. **M**: four-cell embryo. **N**: eight-cell embryo. Four pigmented and four unpigmented cells of the animal and vegetative hemispheres, respectively. **O**: morula stage. **P**: neurula at stage 15. Bar = 1 mm. Reproduced by permission of "Japanese Society for Biological Sciences in Space"; from: Gualandris-Parisot, Husson, Bautz, Durand, Kan, Aimar, Membre, Duprat & Dournon – Biol. Sci. Space. 19, 3–11 (2002).

was statistically significant, but there was no difference between 1G-centrifuge and 1G-ground embryos (Gualandris-Parisot *et al.*, 2002a).

In *Xenopus laevis*, Souza *et al.* (1995) indicated that under μG conditions, the cleavage furrow was positioned normally at the two-cell stage, but the mitotic asters showed a slight displacement toward the vegetal pole. Using a clinostat on the ground to obtain virtual microgravity, Neff *et al.* (1993) and Yokota *et al.* (1994) observed that the third cleavage shifted toward the embryo equator.

During cleavage in *Pleurodeles*, the intercellular space was significantly enlarged at the animal surface in 40.4% of μG-embryos. This was observed in 40.4, 10.7 and 1.5% of μG-, 1G-on board and 1G-ground embryos, respectively. The differences were statistically significant (using χ^2 test with the Yates correction χ_c^2) between μG- and 1G-centrifuge or 1G-ground embryos. There was no difference between 1G-centrifuge and 1G-ground embryos. This morphological aspect, which could be interpreted as a decrease in cell adhesion, was also observed at the beginning of gastrulation and during neurulation.

Histological observations of embryos at advanced blastula stage indicated that the blastocoel was located inside the animal hemisphere, as it was in control embryos. The blastocoel roof made up of two layers of cells had an identical thickness for the μG- and 1G-blastulae (Gualandris-Parisot *et al.*, 2002a). On both *Xenopus laevis* and *Pleurodeles waltl* embryos, μG has effects at the beginning of cleavage.

Gastrulation period

Only young gastrulae of *Pleurodeles* were obtained in flight. Histological observations showed that the blastocoel roof was thicker in μG- than in 1G-ground gastrulae. However, as in the case of old blastulae, the roof was always composed of two layers of weakly cohesive cells in μG-embryos (Gualandris-Parisot *et al.*, 2002a). The thickness of the μG-embryo roof was related to a reduction of cohesion between ectoderm cells and not to an increase in the number of cell layers, as observed by Ubbels *et al.* (1994) for μG-embryos of *Xenopus laevis* or with cell proliferation as proposed by Black *et al.* (1996). However, the data of Neff *et al.* (1993) which used a clinostat, suggested that the thickening of the blastocoel roof of *Xenopus* embryos was due to a change in the latitude of the third cleavage plane rather than a modification of cell proliferation. Moreover, De Mazières *et al.* (1996) observed in μG *Xenopus* that the blastocoel was shifted toward the vegetal hemisphere. These differences between frogs and salamanders could be due to modifications in the osmoregulation and the volume of the blastocoel liquid (Ubbels, 1997).

Neurulation

In *Pleurodeles* at the end of the neurulation period, 81.3% of the μG-embryos showed an incomplete closure of the neural tube at the cephalic and trunk levels (Fig. 2). During neural tube closure, cells were shed from the neurectoderm and were observed outside the tube. In 1G-embryos, only 8.3% of the 1G-centrifuge and 0% of the 1G-ground embryos exhibited an abnormal neurulation. The differences were statistically significant (using χ^2 test with the Yates correction χ_c^2) between μG- and 1G-centrifuge or 1G-ground embryos. There was no difference between 1G-centrifuge and 1G-ground embryos.

In contrast with the neurectodermal cells, the epidermal cells that differentiated from ectoderm showed normal adhesion. In the course of development, the epidermal cells acquired cilia, the movement of which progressively induced a rotation of the embryos within the egg membranes. This movement was very slow and could only be observed by time-lapse video-recording. Under μG conditions, video-movies showed that the rotation began as it did in control embryos (Gualandris-Parisot *et al.*, 2002a).

Organogenesis and swimming behavior

In *Pleurodeles* under μG conditions after the tail bud stages, the closure of the neural tube was completed but abnormal at the cephalic level. Moreover, 25.4, 18.5 and 9.4% respectively of the μG- and 1G-centrifuged and 1G-ground embryos produced a small head. But the differences were not statistically significant between μG- or 1G-centrifuge or 1G-ground embryos. From fertilized eggs put in a video-recording box, 87.5% developed or regulated and reached the hatching stages just before the end of the space flight. The video-recording showed that regulation phenomena occurred after neurulation. Moreover, it indicated that the time of development from egg laying to the hatching stages was identical for the animals reared at 18 °C aboard Mir or in the ground laboratory (Dournon *et al.*, 2001 Gualandris-Parisot *et al.*, 2002a).

Histological and immunohistochemical analyses performed 5 h after landing showed no differences between μG- and 1G ground young larvae in the central nervous system, eyes, somites, pronephros and gut (Gualandris-Parisot *et al.*, 2002a,b). These results are in accordance with previous studies obtained during a 12-day space flight. Cultured *in vitro,* isolated neurectodermal cells or isolated mesodermal cells differentiated in a normal manner into neurons and glial cells or muscle cells (Duprat *et al.*, 1998; Husson *et al.*, 1997). The video-recording displayed that the first spontaneous contractions of the trunk musculature occurred at the same stages in μG- and 1G-ground embryos. A few hours after landing, the young-μG larvae displayed a swimming behavior identical to that of controls, with no looping or somersaulting as observed in *Xenopus laevis*.

Inner ear and otoconia

In *Cynops pyrrhogaster* reared on ground, a small number of otoconia was first observed at the gill bud stage (St. 33) (Koike *et al.*, 1995; Wiederhold *et al.*, 1995a,b). Otoconia split into separate utricular and saccular otoliths at cylindrical forelimb bud stage (St. 40). In *Cynops*, 148 embryos were launched before any of the otoconia was formed in the inner ear. After 15 days in microgravity during the IML-2 mission, 62 surviving larvae were recovered with a nearly adult inner ear (Wiederhold *et al.*, 1997a,b). In the endolymphatic sac, the otoconia were much larger and developed earlier in μG-larvae. At later developmental stages, aragonitic otoconia enter and fill the saccule. In the saccule and utricle, the otoliths were of the same size in μG- and 1G-ground larvae. Subsequently, one μG-animal was reared for 9 months. The size of the saccular otolith, as well the volume of the otoconia within the endolymphatic sac, was considerably larger than in control animals. This suggests that rearing in μG induces a process that continues for several months after introduction to 1G. Such an observation was made on regenerating forelimbs or lenses of *Pleurodeles* adults after space flight (Mitashov *et al.*, 1996).

In *Pleurodeles*, ground experiments were performed to determine the stages of the first appearance of otoconia in different regions of the inner ear. The crystallographic structures of otoconia and otoliths were defined during development and at adulthood. The saccular otoconia are first in calcite, and then in aragonite, even though in the utricle and endolymphatic sacs they are permanently in calcite and aragonite (Oukda *et al.*, 1999a,b).

During two space flights, one two weeks and the other 5 months long, *Pleurodeles* larvae were reared aboard Mir and ISS to compare the crystallographic structure of the otoconia with those of ground animals. In adults, after a 5-month space flight, biological crystals were altered like those of elderly people described by Ross *et al.* (1976) (Oukda et al., 2001). The analysis of the samples is in progress.

Is the further development and the reproduction normal upon return to Earth?

Development from hatching stage to sexual maturity

At landing, *Pleurodeles* larvae (No = 26) that developed up to the hatching stages in μG, and ground control larvae that developed in 1G (No = 12) were kept alive and underwent assays to determine their sexual genotypes. Fifteen flight larvae were genetic ZW females and 11 were genetic ZZ males. Four control larvae were genetic ZW females and 8 genetic ZZ males. After recovery of these *Pleurodeles* larvae in the laboratory, the development of both flight and control animals was similar up to metamorphosis. The development was comparable with the development of progenies derived from the natural crosses obtained before the space mission and with couples of females and males. Moreover, development was also comparable with the development of progenies reared in the laboratory during the corresponding annual period. After metamorphosis, *Pleurodeles* grew without any observable abnormalities. Twelve months after landing, the males had callosities and turgescent cloacal lips. Five months later, the other animals acquired the female phenotype. The sexual phenotype of the ground controls became evident during the same periods. Laparotomy confirmed a normal aspect of gonads that was in accordance with the sexual genotype previously detected (Dournon *et al.*, 2001).

Analysis of offspring derived from animals born and developed up to hatching in space

In the first mating, μG-males were crossed with standard ground females. In the second mating, some μG-males were crossed with μG-females. In the third mating, one μG-male was crossed with a female born in μG, and then was developed in the 1G-centrifuge aboard MIR. These females laid fertilized eggs. For all progenies, the percentage of fertilization (number of embryos at 2-cell stage/total number of eggs) and the percentage of development (number of embryos at 2-cell stage or later stages/total number of eggs) were in accordance with those obtained for standard progenies in the laboratory. Moreover, the morphology of these larvae observed at periodic intervals during all embryonic and larval development was without any particular distinguishable abnormality (Dournon *et al.*, 2001).

Microgravity effects on intact or regenerating organs of adults under space conditions and after landing

The goals of these experiments were to answer the following questions. Does microgravity affect the organs in adult animals? Does microgravity affect the regeneration of organs?

Does microgravity affect organs of adults?

The status of various organs was analyzed after different space missions in metamorphosed or adult *Cynops pyrrhogaster* and *Pleurodeles waltl* animals whose external aspect appeared normal. For *Cynops* and *Pleurodeles* most the flights were approximately 2 weeks long, except for the "Perseus mission" in which two *Pleurodeles* females were reared for 5 months aboard Mir.

In livers of *Cynops* compared to ground controls, osmiophilic granular cells were vacuolated. These cells might function to scavenge free radicals similar to Kuppfer cells in mammals. These cells might be present due to a reaction against toxins. In hepatic cells, the fine structure of the rER was deformed and morphological changes were observed in the nucleolus and in mitochondria, suggesting an arrest of protein synthesis in the liver during the flight. Similar findings were reported in the liver of treefrogs flown in 1990 aboard Mir (Yamashita *et al.*, 1997). HSp was not detected (Yamashita *et al.*, 2001).

The stomach was ulcerated in one of the two *Cynops* females and had an edema in the lamina propria. Ulcers are commonly interpreted as a stress reaction observed in animals (Yamashita *et al.*, 2001). In *Pleurodeles*, the stomach of two adult females appeared also ulcerated with edema after 5 months in space (Membre and Dournon, in preparation). The intestine was normal in *Cynops* (Yamashita *et al.*, 2001). In *Pleurodeles*, the analyzes are in progress.

In *Cynops* lungs, pulmonary cytoplasmic changes were observed by TEM. Epithelial cells were degenerating and the epithelium was disrupted (Pfeiffer *et al.*, 1995a,b; Yamashita *et al.*, 2001).

One strong effect of microgravity on living organisms is the loss of calcium from the skeleton. In amphibians, the ultimobranchial glands secrete calcitonin, which controls the metabolism of calcium. In *Pleurodeles* after the Bion 8 space flight, an increase of free calcium in the blood plasma and an increase of the functional activity of the calcitonin producing cells of the ultimobranchial glands were detected. These glands were hyperdeveloped; most of their cells died and 50-90% of the follicles were destroyed. After landing, the glands became gradually restored (Besova et Savelyev, 1993; Besova *et al.*, 1993; Mitashov *et al.*, 1996). After the Bion 10 flight, a histological study of the bones did not reveal any specific differences in the morphology of the femur, tibia and fibula. The most remarkable feature in intact animals was the increased number of polynuclear cells on the endosteal surface of both flight samples and synchronous ground controls (Berezovska *et al.*, 1997, 1998). *Cynops* muscles and bones, analyzed in their fine structure, appeared normal. HSp was not detected in the limb and the vertebra (Yamashita *et al.*, 2001).

In *Pleurodeles* blood, after the Bion 10 space flight, Michurina *et al.* (1996) reported that space flight led to an increase in the proportion of neutrophils. For flight females, the ratio between young cells of a neutrophil lineage and mature neutrophils was very high. The number of lymphocytes decreased and the ratio of lymphocytes to neutrophyls became lower than in animals of a control group. Number of eosinophyls also decreased, whereas the proportion of basophyls and monocytes remained unchanged. After a 5-month space flight, the level of immunoglobulines in two *Pleurodeles* females, two ground control females (all force-fed) and two standard females fed with *Chironoma* larvae and meat were compared. The IgM levels were normal; IgY levels were three times higher than in the control and in standard females. Moreover, space flight also affected some families of VH genes expressed with the isotype IgM (Frippiat and Dournon, 2002). This phenomenon is still being studied on larvae from a 2-week flight in 2001 (mission "Andromède").

Post-flight progeny of adult females

In 1992–93, one of the objectives of the experiment "Triton" was to analyze the progenies of *Pleurodeles* females after a 12-day space flight (Bion 10). Ten females, which had forelegs, lens and part of the tail removed, were launched. At the end of the flight, a rough and temporary increase of temperature occurred and led to the death of some animals. After landing, five females survived and were sacrificed for various goals. In the aim to analyze their progeny, ovaries freshly tacked off were immediately grafted into castrated juvenile males. The grafting was successful and males became phenotypic females. After mating, these males carrying ovaries and laid few unfertilized eggs (Bautz *et al.*, 1996a; Dournon *et al.*, 1997). Later, the consequence of space flight on the progeny was clearly analyzed in the FERTILE experiments. In this double experiment, four flight females of *Pleurodeles* were crossed with ground standard males after 2-week space flights. Post-flight progenies were obtained and the percentages of fertilization and development were similar to those obtained for standard progenies. No particular abnormalities were observed (Dournon *et al.*, 2001). Two *Cynops* females were also crossed after the landing and no abnormalities were observed (Yamashita *et al.*, 2001).

Does microgravity affect the regeneration of organs during space flights and after landing?

The regeneration phenomenon was studied in eyes, forelimbs and tails of *Pleurodeles waltl* during space flights, the duration of which was between 7 and 16 days. Lenses were tacked off between 13 and 5 days and forelimbs between 24 and 5 days before the launch. The forelimbs were removed at the ulna/radius level. Regeneration of flight animals was studied at landing and during the following weeks and compared with synchronous ground controls.

Lens and retina regeneration

An increase in regeneration rate and a synchronization of the stages of regeneration (the dispersion of the obtained stages appeared reduced) were observed in juvenile *Pleurodeles* during and after space flights. After a 7- or 14-day space flight, the stages of lens regeneration were more advanced in temperature controls and transport controls than in flight animals. However, 14 days after landing, no differences in regeneration stages were observed anywhere between flight newts and controls. The lens regeneration appeared faster in space animals than in controls (Mitashov *et al.*, 1989). After several space experiments, the rate of lens regeneration during and after the flight does not differ significantly from controls (Brushlinskaia *et al.*, 1997). However, the number of proliferating cells in lens regenerates is greater than in controls. It holds also true for the ciliary zone of the iris and the growth zone of the retina (Brushlinskaia, 1995). According to the authors, the data seems statistically significant (Mitashov *et al.*, 1996). An enlargement of the new formed lens (in diameter), described by the authors after 14 days and 4.5 months after landing, could be explained by such proliferations (Mitashov *et al.*, 1989, 1996).

Retina of *Pleurodeles* were operated 2 weeks prior to the Bion 11 launch. After a 2-week space flight, the recovery of the retina proceeded more intensively than in ground controls; the differentiation of regenerate layers in flight newts was also more rapid in process (Grigorian *et al.*, 1999). Similar findings were obtained when using a radial clinostat (Grigorian and Anton, 1994; Anton *et al.*, 1996; Grigorian *et al.*, 1998).

Limb regeneration
In *Pleurodeles* after a 2-week flight, as in the case of the lens, a synchronization of the stages of regeneration was observed, and the dispersion of the regeneration stages seemed reduced. Moreover, the rate of the forelimb regeneration was significantly higher if the experiment started when the animals already had a blastema in formation. If the experiment started earlier (when newts had a blastema at wound healing stage) no difference to controls was detected immediately after a one or two-week flight. However, nine days after landing, a higher rate of limb regeneration was detected in flight-animals. In the regenerate of these animals, differentiated structures appear earlier than in controls. However, no disturbances of the regenerate morphogenesis were observed after 4.5 months from landing (Mitashov *et al.*, 1989, 1996).

Tail regeneration
Tail regeneration was studied during the Bion 10 space flight. Fifteen days before the launch the tail was sectioned at the third posterior part level in order to allow a blastema to form. Fixed after landing, the regenerates had some differences with the regenerates from ground control females. If the neural tube, muscles, and cartilage formed similarly in space and on ground, the pigmentation of the blastema of flight-females was much stronger than in controls. These were black in color, contrasting to blastema of controls, which were very light-colored, translucent. A histological analysis indicated that melanocytes and melanosomes had extended under the epidermis layer of flight females than in controls, where only a few melanocytes and melanosomes were found (Grinfeld *et al.*, 1994, 1996). Moreover, GABA-positive cells were detected in greater number in connective tissues of experiment blastemas than in control blastemas. As the temperature increased in a rough manner at the end of the space flight, it was difficult to conclude firmly on specific microgravity effects.

Conclusion

After fertilization, subsequent embryonic development appears to be altered in μG. Although, early development was not strictly normal as a consequence of embryological regulation phenomena, young hatching larvae had normal morphological phenotypes and swimming behavior after landing. Video-movies clearly confirm this point. The ontogenetic studies in μG, obtained with *Cynops pyrrhogaster* and *Pleurodeles waltl,* complete the results obtained with the anuran *Xenopus laevis* (Ubbels *et al.*, 1989, 1994; Souza *et al.*, 1995; Black *et al.*, 1996) and the fish *Medaka* (Ijiri, 1995, 1997). In *Pleurodeles*, observed abnormalities such as cortical cytoplasmic movements, decrease of cell adhesion, and loss of cells concern the cell membrane. Modifications of the membrane were suggested by the results of Schatz *et al.* (1990); Claassen *et al.*, (1996) and Aimar *et al.* (2000). Modifications or alterations of physico-chemical properties of the membrane may involve the cytoskeleton (Spooner *et al.*, 1994; Tabony 1994; Lewis *et al.*, 1998). It will be interesting using embryos or cell cultures to determine whether abnormalities involve modifications of the cell membrane or of cell junctions.

After landing, the behavior and the morphology of urodeles born in space seem normal. In *Pleurodeles* no differences were observed between born-in-space animals to standard ones during the embryonic development to adulthood. The analyses of their offspring demonstrated that the percentages of fertilization and development were in accordance with the control animals. No genetic abnormalities were detected during the analysis of the offspring.

On-ground development of animals born in space and the development of their progenies were without characteristic differences compared to control *Pleurodeles*. Depending on methods and limits of analyses, results clearly indicate that this urodele born in space is able to live and reproduce after return to Earth. The results fit with those obtained in fish *Oryzias latipes* (Ijiri, 1997) and *Xenopus* (Souza *et al.*, 1995). However, development occurred in space up to the hatching stage, and consequently it is not yet possible to know if the animals born in microgravity conditions could live normally on Earth without perturbation of equilibrium when they arrive to Earth. It is now necessary to perform long flight experiment with animals in development.

Microgravity seems to have effects of the morphological and histological structures of organs of flight urodeles. However, as in several experiments the number of analyzed adults was low, and it is too early to conclude on specific effects of microgravity. Moreover, in certain flights the temperature was not regulated, and increases in temperature occurred. Conditions of these space flights had influenced the samples, and consequently some interpretations of the results. On the immunological aspect, the cell defense of the immune system can be stimulated in microgravity conditions.

Space flights have clear effects on organs in regeneration. But more specifically, they have long term effects that last several weeks after the return of the animals to Earth. Such a result was also obtained for otoconia several months after landing. However, no clear hypothesis could be proposed to interpret the observations.

At the beginning of this new century, space experiments must address particular attention for studies on genetic or molecular levels.

Acknowledgements

I thank Dr. Michele Nishiguchi from the New Mexico State University, Las Cruces, NM, for her suggestions and corrections.

References

Aimar, C. (1991) Cyclic changes in amphibian egg microvilli occur during the division cycle: implication of MPF. Development 112, 91–98.

Aimar, C., Bautz, A., Durand, D., Membre, H., Chardard, D., Gualandris-Parisot, L., Husson, D. and Dournon, C. (2000) Microgravity and hypergravity effects on fertilization processes in the salamander *Pleurodeles waltl*. Biol. Reprod. 63, 551–558.

Ancel, P. and Vintemberger, P. (1948) Recherches sur la détermination de la symétrie bilatérale dans l'oeuf des Amphibiens. Bull. Biol. Fr. Belg. Suppl 31, 1–182.

Anton, H.J., Grigoryan, E.N. and Mitashov, V.I. (1996) Influence of longitudinal whole animal clinorotation on lens, tail, and limb regeneration in urodeles. Adv. Space Res. 17, 55–65.

Bautz, A., Houillon, C., Aimar, C., Mitashov, V. and Dournon, C. (1996a) Fate of the grafted ovaries from female salamander Pleurodeles waltl embarked on the Cosmos 2229 flight. Adv. Space Res. 17, 269–270.

Bautz, A., Rudolf, E., Mitashov, V. and Dournon, C. (1996b) Peptidase-1 expression in some organs of the salamander Pleurodeles waltl submitted to 12-day space flight. Adv. Space Res. 17, 271–274.

Berezovska, O.P., Rodionova, N.V., Grigorian, E.N. and Mitashov, V.I. (1997) The role of osteoclasts in maintaining calcium homeostasis in newts. Izv Akad Nauk Ser Biol. 4, 495–499.

Berezovska, O.P., Rodionova, N.V., Grigorian, E.N. and Mitashov, V.I. (1998) Changes in the numbers of osteoclasts in newts under conditions of microgravity. Adv. Space Res. 21, 1059–1063.

Besova, N.V. and Savelev, S.V. (1993) The effect of weightlessness on amphibians. The ultimobranchial body. Biull Eksp Biol. Med. 116(7), 87–90.

Besova, N.V., Savelev, S.V. and Chernikov, V.P. (1993) The effect of weightlessness on amphibians. The skeleton and mineral metabolism. Biull Eksp Biol. Med. 116(7), 90–92.

Black, S.D. (1989) A step in embryonic axis specification in Xenopus laevis is stimulated by cytoplasmic displacements elicited by gravity and centrifugal forces. Adv. Space Res. 9, 159–168.

Black, S.D. and Gerhart, J.C. (1986) High frequency twinning of *Xenopus laevis* embryos from eggs centrifuged before first cleavage. Dev. Biol. 116, 228–240.

Black, S., Larkin, K., Jacqmotte, N., Wassersug, R., Pronych, S. and Souza, K. (1996) Regulative development of *Xenopus laevis* in microgravity. Adv. Space Res. 17, 209–217.

Bracchi, F., Gualierotti, T., Morabito, A. and Rocca, E. (1975) Multiday recordings from the primary neurons of the statoreceptors of the labyrinth of the bull frog. The effect of an extended period of "weightlessness" on the rate of firing at rest and in response to stimulation by brief periods of centrifugation (OFO-A orbiting experiment). Acta Otolaryngol Suppl. 334, 1–27.

Brushlinskaia, N.V. (1995) The effect of space flight factors on the cellular proliferative activity of different eye tissues during lens regeneration in Spanish newt *Pleurodeles waltlii*. Izv Akad Nauk Ser Biol. 2, 149–156.

Brushlinskaia, N.V., Grigorian, E.N., Tuchkova, S.I., Anton, H.J. and Mitashov, V.I. (1997) Organ and tissue regeneration in lower vertebrate animals during space flight and after its completion. Ontogenez 28(3), 198–210.

Chee, O.K., Maese, A.C., Nagaoka, S., Yamashita, M., Schaefer, R.L. and Wiederhold, M.L. (1993) AAEU on IML-2: a new space hardware for experiments using aquatic animals. ASGSB Bull. 7(1), 86.

Claassen, D.E. and Spooner, B.S. (1996) Liposome formation in microgravity. Adv. Space Res. 17, 151–160.

Collenot, A. (1973) Obtention, par la méthode des greffes de gonades embryonnaires, d'une femelle à descendance unisexuée femelle, chez le Triton *Pleurodeles waltl* Michah. Experientia 29, 885–887.

De Mazières, A., Gonzales-Jurado, J., Reijnen, M., Narraway, J. and Ubbels, G.A. (1996) Transcient effects of microgravity on early embryos of *Xenopus laevis*. Adv. Space Res. 17, 19–223.

Dournon, C. and Houillon, C. (1984) Genetic demonstration of functional sex inversion in Pleurodeles waltlii Michah. (urodele amphibian) under the effect of temperature. Reprod. Nutr. Develop. 24, 361–378.

Dournon, C., Bautz, A., Aimar, C., Ropars, A. and Houillon, C. (1997) Grafts of ovaries in males and females of *Pleurodeles waltl* (urodele amphibia): evidence of a long term tolerance of allografts; application following a space biology experiment. Reprod. Nutr. Develop. 37, 637–649.

Dournon, C., Durand, D., Tankosic, C., Membre, H., Gualandris-Parisot, L. and Bautz, A. (2001) Effects of microgravity on the larval development metamorphosis and reproduction of the urodele amphibian *Pleurodeles waltl*. Develop. Growth Differ. 43, 315–326.

Duprat, A.M., Husson, D. and Gualandris-Parisot, L. (1998) Does gravity influence the early stages of the development of the nervous system in an amphibian?. Brain Res. Rev. 28, 19–24.

Durand, D., Tankosic, C., Chaput, D., Gualandris-Parisot, L., Membre, H., Bautz, A. and Dournon, C. (2000) Feeding and rearing techniques used for larvae of *Pleurodeles waltl* (urodele amphibian) onboard the MIR space station. Scand. J. Lab. Anim. Sci. 27, 193–201.

Feuilloley, M., Yon, L., Kawamura, K., Kikuyama, S., Gutkowska, J. and Vaudry, H. (1993) Immunocytochemical localization of atrial natriuretic factor (ANF)-like peptides in the brain and heart of the treefrog *Hyla japonica*: effect of weightlessness on the distribution of immunoreactive neurons and cardiocytes. J. Comp. Neurol. 330(1), 32–47.

Frippiat, J.P., Dournon, C. 2002 Effects of five-month stay on board the Mir space station on immunoglobulin production in the urodele amphibian *Pleurodeles waltl, ASGSB Bull.* 18th meeting, abstract, in press.

Gallien, L. (1954) Inversion expérimentale du sexe sous l'action des hormones sexuelles chez le Triton *Pleurodeles waltlii* Michah. Analyse des conséquences génétiques. Bull. Biol. Fr. Belg. 88, 1–51.

Gallien, L. and Durocher, M. (1957) Table chronologique du développement chez *Pleurodeles waltlii* Michah. Bull. Biol. Fr. Belg. 91, 97–114.

Grigorian, E.N. and Anton, H.J. (1994) The characteristics of eye regeneration in newts after complete retinal detachment induced by a change in the effect of gravitation in an experiment using a clinostat. Izv Akad Nauk Ser Biol. 3, 342–352.

Grigorian, E.N. Anton, H.J. and Mitashov, V.I. (1998) Microgravity effects on neural retina regeneration in the newt. Adv. Space Res. 22, 293–301.

Grigorian, E.N., Tuchkova, S.I., Poplinskaia, V.A. and Mitashov, V.I. (1999) Retinal regeneration after dissection of the optic nerve in newtw exposed on board the Bion-11 biosatellite. Aviakosm Ekolog Med. 33, 15–20.

Grinfeld, S., Foulquier, F., Mitashov, V., Bruchlinskaia, N. and Duprat, A.M. (1994) Tissue regeneration in space (spinal cord, muscle and bone) in the amphibian *Pleurodeles waltl*. ESA SP 366, 181–184.

Grinfeld, S., Foulquier, F., Mitashov, V., Bruchlinskaia, N. and Duprat, A.M. (1996) Amphibian tail regeneration in space: effect on the pigmentation of the blastema. Adv Space Res. 17, 237–240.

Gualandris-Parisot, L., Bautz, A., Chaput, D., Husson, D., Durand, D. and Dournon, C. (1998) Mises au point technologiques en vue d'étudier le développement du Pleurodèle (Amphibien Urodèle) à bord de la station spatiale MIR. Récents Progrès en Génie des Procédés 62, 37–48.

Gualandris-Parisot, L., Husson, D., Bautz, A., Durand, D., Aimar, C., Membre, H., Duprat, A.M. and Dournon, C. (2002a) Effects of space environment on the embryonic development up to hatching of salamander eggs fertilized and developed during orbital flights. Biol. Sci. Space. 16, 3–11.

Gualandris-Parisot, L., Husson, D., Foulquier, F., Kan, P., Davet, J., Aimar, C., Dournon, C. and Duprat, A.M. (2002b) *Pleurodeles waltl*, amphibian urodele, is a suitable biological model for embryological and physiological space experiments on vertebrate. Adv. Space Res. 28, 569–578.

Horn, E. and Sebastian, C. (1999) A comparison of normal vestibulo-ocular reflex development under gravity and in the absence of gravity. ESA SP 1222, 127–138.

Husson, D., Gualandris-Parisot, L., Foulquier, F., Grinfeld, S., Kan, P. and Duprat, A.M. (1997) Differentiation in microgravity of neural and muscle cells of a vertebrate (amphibian). Adv. Space Res. 22, 303–308.

Husson, D., Chaput, D., Bautz, A., Davet, D., Durand, D., Dournon, C., Duprat, A.M. and Gualandris-Parisot, L. (2001) Design of specific hardware to obtain embryos and maintain adult urodele amphibians aboard a space station. Adv. Space Res. 27, 433–445.

Ijiri, K. (1995) The first vertebrate mating in space: A fish story, *Ricut (ed.), Tokyo, Japan,* 57p.

Ijiri, K. (1997) Explanation for a video version of the first vertebrate mating in space – a fish story. Biol. Sci. Space. 11, 153–167.

Imamizo, M., Yoshida, M., Fujioka, K., Takada, Y., Hisada, A., Izumi-Kurotani, A. and Yamashita, M. (1994) Sustained release of hCG minipellet for newt experiment in space. Biol. Sci. Space. 8, 226–230.

Kathariner, L. (1901) ber die bedingte Unabhängigkeit der Entwicklung der polar differenzierten Eiere von der Schwerekraft. Arch. Entwick. Mech. Org. 12, 597–609.

Koike, H., Nakamura, K., Nishimura, K., Kashima, I., Wiederhold, M.L. and Asashima, M. (1995) Non-invasive assessment of otolith formation during development of the Japanese red-bellied newt. Cynops pyrrhogaster. Hear. Res. 88, 206–214.

Labrousse, M. (1959) Phénomènes cytologiques de la fécondation chez *Pleurodeles waltl* Michah. Bull. Soc. Zool. Fr. 84, 493–498.

Lecointre, G., Le Guyader, H. (2001) Classification phylogénétique du vivant, Publisher : Belin, Paris, France.

Lewis, M.L., Reynolds, J.L., Cubano, L.A., Hatton, J.P., Lawless, B.D. and Piepmeier, E.H. (1998) Spaceflight alters microtubules and increases apoptosis in human lymphocytes (Jurkat). FASEB J. 12, 1007–1018.

Michurina, T.V., Domaratskaya, E.I., Nikonova, T.M. and Khrushchov, N.G. (1996) Blood and clonogenic hemopoietic cells of newts after the space flight. Adv. Space Res. 17, 295–298.

Mitashov, V.I., Brushlinskaya, N.V., Grigoryan, E.N., Tuchkova, S.Ya. and Anton, H.J. (1996) Regeneration of organs and tissues in lower vertebrates during and after space flight. Adv. Space Res. 17, 241–255.

Mitashov, V.I., Grigoryan, E.N., Tuchkova, S. and Cherdanzeva, E. (1987) Organs and tissues regeneration in amphibia under the space flight conditions. ESA SP 271, 299–303.

Mitashov, V.I., Grigoryan, E.N., Tuchkova, S.Ya., Oigenblik, E.A. and Malchevskaya, I.E. (1989) Lens and limb regeneration in newt during and after 13-day long spaceflight. ESA SP 1123, 85–92.

Mogami, Y., Imamizo, M., Yamashita, M., Izumi-Kurotani, A., Wiederhold, M.L., Koike, H. and Asashima, M. (1996a) Astro-newt": early development of newt in space. Adv. Space Res. 17, 257–263.

Mogami, Y., Koike, H., Yamashita, M., Izumi-Kurotani, A. and Asashima, M. (1996b) Early embryogenesis of amphibians in space: "Astro-newt" for the space embryology in IML-2 and SFU, *In* Bainum, P.M., May, G.L. and Ohkami, Y.) Strengthening Cooperation in the 21st Century. pp. 1089–1097. American Astronautical Society, San Diego. (Advances in the astronautical sciences, vol. 91).

Morgan, T.H. (1904) The dispensibility of the constant action of gravity and of a centrifugal force in the development of the toad's egg. Anatomischer Anzeiger 25, 94–96.

Neff, A.W., Wakahara, M. and Malacinski, G.M. (1990) Bifircation of the embryo's axis: analysis of variation in response to egg centrifugation. Int. J. Dev. Biol. 34, 391–398.

Neff, A.W., Yokata, H., Chung, H.M., Wakahara, M. and Malacinski, G.M. (1993) Early amphibian morphogenesis is sensitive to novel gravitational fields. Dev. Biol. 155, 270–274.

Okada, T. (1989) Development of vertebrates, *Baifukan, Tokyo*, pp. 333–335.

Okada, Y.K. and Ichikawa, M. (1947) Normal table of Triturus pyrrhogaster. Jap. J. Exp. Morphol. 3, 1–65.

Oukda, M., Bautz, A., Membre, H., Ghanbaja, J., François, M. and Dournon, C. (1999) Appearance and evolution of calcitic and aragonitic otoconia during Pleurodeles waltl development. Hear. Res. 137, 114–126.

Oukda, M., François, M., Membre, H., Bautz, A. and Dournon, C. (1999) Crystallographic and chemical composition of otoconia in the salamander Pleurodeles waltl. Hear. Res. 132, 85–93.

Oukda, M., Membre, H., Bautz, A., Durand, D., Dournon, C. (2001) Otoconia during the inner ear development and adulthood in *Pleurodeles waltl* (urodele amphibian): ground study and preliminary results in microgravity. *International Symposium on "International Scientific Cooperation onboard MIR", CNES, Lyon, France. Abstract, 4p.*

Pasteels, P.J. (1941) Recherches sur les facteurs initiaux de la morphogenèse chez les Amphibiens anoures. V-Les effets de la pesanteur sur l'œuf de Rana fusca maintenu en position anormale avant la formation du croissant gris. Arch. Biol. 52, 321–339.

Pfeiffer, C.J., Yamashita, M., Asashima, M., Izumi-Kurotani, A. and Koike, H. (1995a) Cytopathology of lungs of newts on-board Columbia space shuttle STS-65. ASGSB Bull. 9(1), 88.

Pfeiffer, C.J., Yamashita, M., Izumi-Kurotani, A., Koike, H. and Asashima, M. (1995b) Cytopathologic observations of the lung of adult newts (Cynops pyrrhogaster) on-board the space shuttle, Columbia, during the Second International Microgravity Laboratory experiments. J. Submicrosc. Cytol. Pathol. 27(4), 501–509.

Pflüger, E. (1883) Über den Einfluss der Schwerkraft auf die Teilung der Zellen. Zweite Abhandlung. Arch. für die Gesammte Physiologie des Menschen und der Tiere Arch 32, 1–79.

Picheral, B. (1977) La fécondation chez le Triton Pleurodèle: II-La pénétration des spermatozoïdes et la réaction locale de l'oeuf. J. Ultrastruct. Res. 60, 181–202.

Ross, M.D., Johnsson, L.G., Peacor, D. and Allard, L.F. (1976) Observations on normal and degenerating human otoconia. Ann. Otol. 85, 310–326.

Roux, W. (1884) Beitrag 2 zur Entwicklungsmechanik der Organismen. Über die Entwicklung des Froscheies bei Aufhebung der richtenden Wirkung der Schwere. Breslauer ärztlicher Zeitschrift 22, 57–62.

Rudolf, E., Dorazi, R., Chesnel, A., Durand, D., Bautz, A. and Dournon, C. (1996) Pleurodeles waltl and Pleurodeles poireti: correlations between their iberic or north-african origin and genes B or (of peptidase-1, a sex-linked enzyme. Bull. Soc. Zool. Fr. 121, 21–29.

Sakimura, T., Suzuki, T., Matsubara, S., Uchida, S., Kato, M., Tanemura, R. and Honda, S. (1999) NASDA aquatic animal experiment facilities for Space Shuttle. Biol. Sci. Space. 13(4), 314–320.

Schatz, A., Reitstetter, R., Linke-Hommes, A. and Briegleb, W. (1990) Gravity perception by non specialized structures. ESA SP 307, 313–314.

Sebastian, C. and Horn, E. (1998) The minimum duration of microgravity experience during space flight which affects the development of the roll induced vestibulo-ocular reflex in an amphibian (Xenopus laevis). Neurosci. Lett. 253, 171–174.

Souza, K.A., Black, S.D. and Wassersug, R.J. (1995) Amphibian development in the virtual absence of gravity. Proc. Natl. Acad. Sci. USA. 92, 1975–1978.

Spooner, B.S., Hardman, P. and Paulsen, A. (1994) Gravity in mammalian organ development: differentiation of cultured lung and pancreas rudiments during spaceflight. J. Exp. Zool. 269, 212–222.

Tabony, J. and Job, D. (1992) Gravitational symmetry breaking in microtubular dissipative structures. Proc. Natl. Acad. Sci. USA 89, 6948–6952.

Ubbels, G.A. (1997) Establishment of polarities in the oocyte of Xenopus laevis: the provisional axial symmetry of the full-grown oocyte of Xenopus laevis. Cell Mol. Life Sci. 53, 382–409.

Ubbels, G.A., Berendsen, W. and Narraway, J. (1989) Fertilization of frog eggs on a sounding rocket in space. Adv. Space Res. 9, 187–197.

Ubbels, G.A., Reijnen, M., Meijerink, J. and Narraway, J. (1994) Xenopus laevis embryos can establish their spatial bilateral symmetrical body pattern without gravity. Adv. Space Res. 14, 257–269.

Veynachter, L., Tankosic, C., Viso, M., Chaput, D., Gualandris-Parisot, L. and Dournon, C. (2002) A new method to force-feed and rear newt adults on board a space station. Scand. J. Lab. Anim. Sci. 29, 100–108.

Vinnikov, Ia.A., Gazenko, O.G., Titova, L.K., Govardovskii, V.I. and Gribakin, F.G. (1973) Development of the vestibular apparatus (labyrinth) of the frog Rana temporaria in conditions of weightlessness. Zh Evol. Biokhim. Fiziol. 8(3), 343–350.

Vinnikov, I.aA., Gazenko, O.G., Titovo, L.K. and Govardovskii, V.I. (1976) The development of the vestibular apparatus under conditions of weightlessness. Arkh. Anat Gistol Embriol. 70(1), 11–17.

Vinnikov, I.aA., Lychakov, D.V., Pal'mbakh, L.R., Govardovskii, V.I. and Adanina, V.O. (1980) Vestibular apparatus study of the toad, *Xenopus laevis*, and rats under prolonged weightlessness. Zh Evol. Biokhim. Fiziol. 16(6), 574–579.

Wiederhold, M.L., Gilpatric, K.D. and Hejl, R.J. (1995a) Otolith systems in newt larvae reared in microgravity on IML-2. Gravit. Space Biol. Bull. 9(1), 88.

Wiederhold, M.L., Yamashita, M., Larsen, K.A., Batten, J.S., Koike, H. and Asashima, M. (1995b) Development of the otolith organs and semicircular canals in the Japanese red-bellied newt Cynops pyrrhogaster. Hear. Res. 84, 41–51.

Wiederhold, M.L., Gao, W.Y., Harrison, J.L. and Hejl, R. (1997a) Development of gravity-sensing organs in altered gravity. Gravit. Space Biol. Bull. 10(2), 91–96.

Wiederhold, M.L., Pedrozo, H.A., Harrison, J.L., Hejl, R. and Gao, W.Y. (1997b) Development of gravity-sensing organs in altered gravity conditions: opposite conclusions from an amphibian and a molluscan preparation. J. Gravit. Physiol. 4, 51–54.

Yamashita, M., Izumi-Kurotani, A., Mogami, Y., Okuno, M., Naitoh, T. and Wassersug, R.J. (1997) The Frog in Space (FRIS) experiment onboard Space Station Mir: final report and follow-on studies. Biol. Sci. Space. 11(4), 313–320.

Yamashita, M., Izumi-Kurotani, A., Imamizo, M., Koike, H., Okuno, M., Pfeiffer, C.J., Komazaki, S., Sasaki, F., Ohira, Y., Kashima, I., Kikuyama, S., Ohnishi, T., Mogami, Y., Asashima, M. (2001) Japanese Red-Bellied Newts in Space-AstroNewt Experiment on Space Shuttle IML-2 and Space Flyer Unit. Biol. Sci. Space 15 (Suppl), S96–S103.

Yokota, H., Neff, A.W. and Malacinski, G.M. (1994) Early development of Xenopus embryos in affected by simulated gravity. Adv. Space Res. 14, 249–255.

Young, R.S. and Tremor, J.W. (1968) The effect of weightlessness on the dividing egg of Rana pipiens. Bioscience 18, 609–615.

Developmental Biology Research in Space
H.-J. Marthy (editor)

The Development of Gravity Sensory Systems During Periods of Altered Gravity Dependent Sensory Input

Eberhard R. Horn*

Gravitational Physiology, University of Ulm, 89081 Ulm, Germany

Abstract

Gravity related behavior and the underlying neuronal networks are the most suitable model systems to study basic effects of altered gravitational input on the development of neuronal systems. A feature of sensory and motor systems is their susceptibility to modifications of their adequate physical and/or chemical stimuli during development. This discovery led to the formulation about critical periods, which defines the period of susceptibility during post-embryonal development. Critical periods can be determined by long-lasting modifications of the stimulus input for the gravity sensory system (*GSS*). Techniques include: (1) destruction of the gravity sense organ so that the gravity cannot be detected any longer and the central neuronal network of the GSS is deprived of gravity related information, (2) loading or deloading of parts of the body by weights or counterweights, respectively, which compensates for the gravitational pull, and (3) absence or augmentation of the gravitational environment *per se* by the exposure of organisms to microgravity during spaceflights or to hypergravity by centrifugation.

Most data came from studies on compensatory eye or head movements in the clawed toad *Xenopus laevis*, the cichlid fish *Oreochromis mossambicus*, and crickets (*Acheta domesticus*, *Gryllus bimaculatus*). The responses are induced by a roll or pitch stimulation of the gravity sense organs, but are also affected by sensory inputs from proprioreceptors and eyes. The development of these compensatory eye and head responses reveals species-specific time courses. Based on experiments using spaceflights, centrifugation, lesion and loading or deloading, all species revealed a significant susceptibility to modifications of the gravity sensory input during development. Behavioral responses were depressed (*Xenopus*) or augmented (*Xenopus*, *Oreochromis*) by microgravity, and depressed by hypergravity except in crickets. In *Acheta*, however, the sensitivity of its position sensitive neuron PSI was reduced by microgravity. After termination of the period of modified gravity sensory input, all behavioral and

* E-mail: eberhard.horn@biologie.uni-ulm.de

physiological modifications disappeared, in some preparations such as the PSI of *Acheta* or the eye response in *Xenopus*, however, delayed after exposure to hypergravity. Irreversible modifications were rare; one example were malformations of the body of *Xenopus* tadpoles caused by lesion induced deprivation.

Several periods of life such as the period of hatching or first appearance of gravity related reflexes revealed a specific sensitivity to altered gravity. Although all studies gave clear evidences for a basic sensitivity of developing GSSs to long-lasting modifications of the gravity sensory input, clear arguments for the existence of a critical period in the development of the sense of gravity are still missing. It has to take into consideration that during long-term exposures, adaptation processes take place which are guided by central physiological and genetically determined set points. The International Space Station (ISS) is the necessary platform of excellence if biological research is focussed on the analysis of long-term space effects on organisms.

Introduction

Gravity related behavior and the underlying neuronal networks are the most suitable model systems to study basic effects of altered gravity (*AG*) on the development of neuronal systems. Gravity sense organs are well described in structure and function. Within the central nervous network, identified gravity sensitive neurons are known, and function can be tested by well-defined behavior such as equilibrium control during swimming, geotaxis, or compensatory eye or head movements. *Equilibrium control during swimming* causes that the back of a swimming animal is directed upwards; the response is supported by the dorsal light reaction (von Holst, 1950). *Geotactic behavior* is the orientation with respect to gravity. Most animals crawl up- or downwards on an inclined surface. Statistically, the preferred direction of orientation is parallel to the gravitational vector. Modifications of the sensitivity of the animal's gravity sensory system (*GSS*) correlate with the accuracy of maintaining the course parallel to the *g*-vector. A high GSS sensitivity is demonstrated by straight-lined trajectories which deviate little from the direction of the *g*-vector direction while large deviations and curved trajectories are a sign of a low GSS sensitivity.

An alternative method to study the GSS sensitivity during walking and swimming is the additional stimulation by a laterally placed light. The method was applied to analyze the function of GSSs in insects (cf. Horn, 1985b) and fish (von Holst, 1950). The light forces the animal to react by a compromise orientation between the directions of the gravity vector and the light vector, i.e., between geotaxis and phototaxis. A decreasing GSS sensitivity directs the animal more towards the light while a high GSS sensitivity moves the orientation towards the direction of the *g*-vector (cf. Horn, 1985c).

Thus, the shift of the compromise orientation can be used to study sensitivity modifications within the GSS caused by AG conditions (Fig. 1, upper part).

Many animal species perform *compensatory head or eye movements* if they are rolled around their longitudinal or transverse axes. Eye responses are common in animals with movable eyes such as vertebrates (roll-induced static vestibuloocular reflex (*rVOR*); cf. Cohen, 1974), crustaceans (Schöne, 1954) or cephalopodes (Budelmann, 1970). Insects have no movable eyes; but species such as honey bees, bumble bees, flies or crickets move their head in response to a body roll (roll-induced static compensatory head response (*rCHR*); Horn and Lang, 1978; Horn and Bischof, 1983). The gain G of compensatory eye or head movements, i.e., the ratio between the response angle α and the roll angle γ ($G = \alpha/\gamma$) is in the range of $G = 0.1$–0.5. This value is smaller than $G = 1$ (i.e., $\alpha = \gamma$) which would be necessary to project the visual surroundings onto the same area of the retina as during normal body posture. The reason for the low gain G is not yet known; but the corresponding small compensatory head or eye movements are large enough to give men and animals the imagination of a stable visual environment, the most likely meaning of these responses. The relation between response and stimulus can be approximated by a sine-like function (Fig. 1, lower part). Compensatory head and eye movements are excellent responses to study sensitivity modifications of a GSS during and after exposure to altered gravity.

A feature of sensory and motor systems is their susceptibility to modifications of their adequate physical and/or chemical stimuli during development (Dews and Wiesel, 1970; Hubel and Wiesel, 1970; Knudsen *et al.*, 1984; Oakley, 1993; van der Loos and Woolsey, 1973; Walton, 1998). Typical is a *critical (sensitive) period* of life which is characterized: (1) by a sensitivity of the developing system to modifications of the adequate environment, (2) by a clearly defined time window of this sensitivity (cf. Wiesel, 1982), and (3) by the irreversibility of anatomical, behavioral or physiological modifications induced by the altered environmental conditions. The most effective approach for the analysis of critical periods is sensory or motor deprivation. This implies that for the sense of gravity, spaceflights are necessary.

In vertebrates, the vestibular otolith (macular) organs are activated by shearing forces which cause a deflection of the sensory cilia by the dislocation of the heavy mass, the otolith. Shearing forces occur either during a static roll which is geometrically related to the direction of the gravitational vector, or during accelerated linear translatory movements which are not related to the gravity vector (Lichtenberger *et al.*, 1982). Thus, only the roll-induced input to the otolith organ is eliminated during spaceflights, while its stimulation by linear translatory movements during accelerated floating, walking or swimming still exists (Clément and Reschke, 1996). These physical conditions have to be considered in order to determine whether the development of otolith vestibular systems is susceptible to modifications of g-levels.

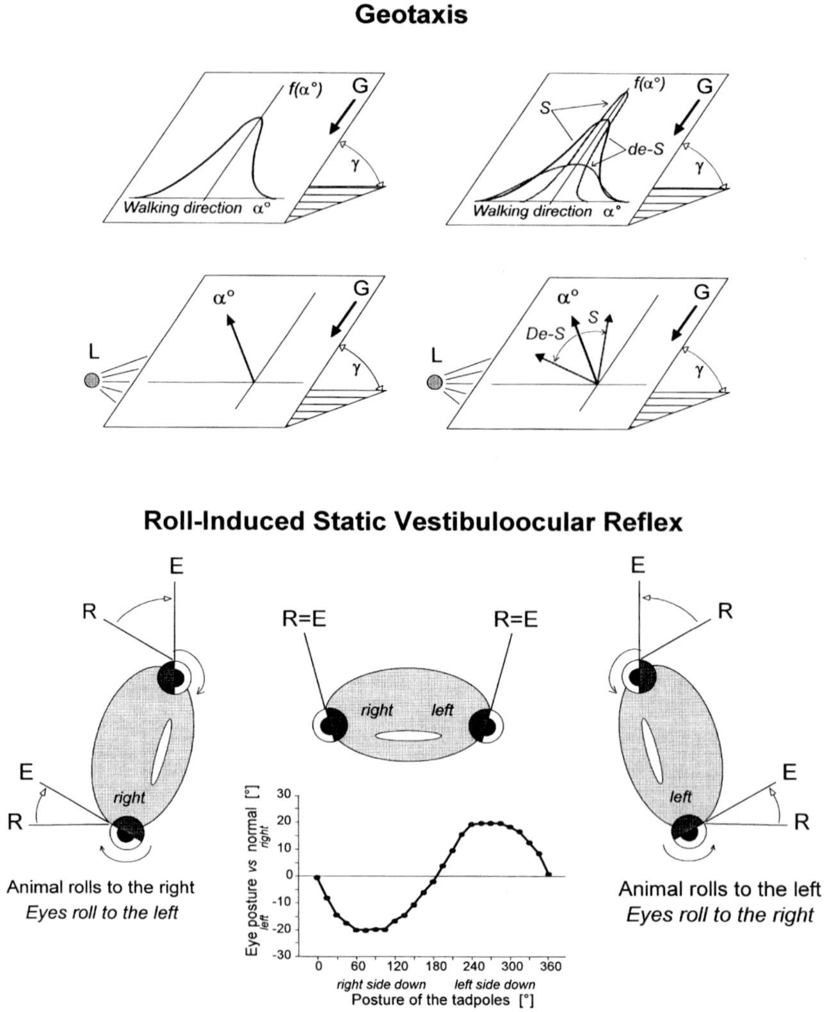

Fig. 1. Gravity related behavior in animals. Upper part: Geotaxis. Geotactic orientation on an inclined surface in an otherwise homogenous environment (upper row) or during simultaneous stimulation by a light source (lower row). Under 1g-condition, the distribution of the walking directions can be approximated by a Gaussian function. After adaptation to altered gravity, a depression of this function indicates desensitization (de-S), a concentration indicates sensitization (S) of the GSS. During additional stimulation with light, animals prefer a compromise direction. After adaptation to altered gravity, a shift of the walking trajectory towards the light direction indicates desensitization (De-S), a shift towards the direction of the gravity vector (G) sensitization (S) of the GSS. Lower part: Roll-induced vestibuloocular reflex. The eyes are rolled opposite to the body. R, orientation of the reference line (the margin of the eye cup) before the lateral roll; E, orientation of this reference line after the lateral roll. The angle between both lines is the compensatory eye response (rVOR). The inset shows the typical sine-like response characteristic for a complete 360° roll.

Standard development of gravity related behavior

Determination of developmental characteristics is useful to understand mechanisms of information processing within the adult central nervous system. These studies are mandatory for the analysis of critical periods and have to be performed at a high resolution of larval stages to detect and define the precise time window of sensitivity to AG conditions. In addition, developmental characteristics of structural components and physiological mechanisms including behavior are not always in synchrony (for *X. laevis*, cf. Horn, 1985a, 1987). The developmental characteristics of both rVOR and rCHR were determined with a high resolution only for two lower vertebrate species (*X. laevis*, *O. mossambicus*) and in one insect species, respectively (Fig. 2). Staging of developing fish and amphibia is based on external markers such as the size of the yolk sac, pigmentation, fin structures, shape of the gut, or appearance and shape of extremities (for *Xenopus*: cf. Nieuwkoop and Faber, 1967; for *Oreochromis*: cf. Anken *et al.*, 1993). Post-embryonal developmental stages in crickets are defined by the number of the respective molt.

Compensatory head and eye movements occur early during the development. In the first cricket instar (*G. bimaculatus*), a head response is induced by stimulation of a single gravity sensory cell. Thus, not only in protozoa but also in a phylogenetically advanced animal with a central nervous system and a specialized gravity sense organ, stimulation of a single gravity sensory cell

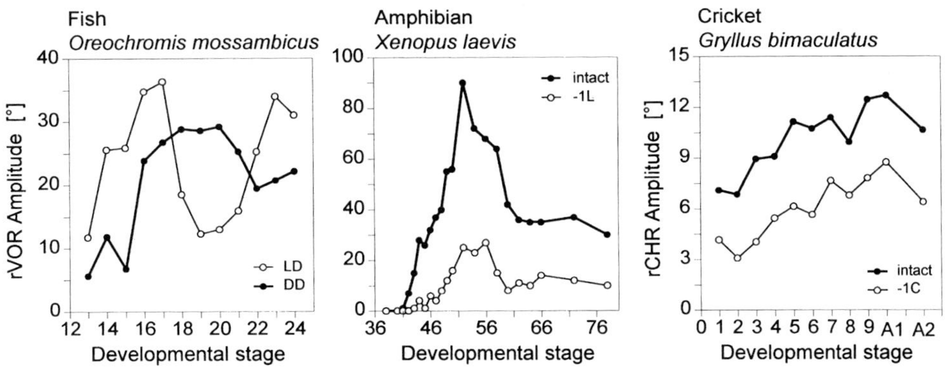

Fig. 2. Standard development of compensatory eye and head responses in lower vertebrates (fish: *O. mossambicus*; amphibian: *X. laevis*) and an insect (cricket: *G. bimaculatus*). For *Oreochromis*, characteristics are shown for animals reared under light–dark conditions (LD) and for animals reared either during early periods of life or during a 9-day period under permanent darkness (DD). For *Xenopus*, characteristics are shown for intact animals (intact) and for animals with only one labyrinth (−1L). For *Gryllus*, characteristics are shown for intact animals (intact) and for animals with only one cercal GSS (−1C). Definition of developmental stage: *Oreochromis*, cf. Anken *et al.* (1993); *Xenopus*, cf. Nieuwkoop and Faber (1967); cricket stages are defined by the number of molts; A1, A2, 4–10 days and 3–4 weeks old adults, respectively. Modified from Horn *et al.* (1986a,b), Sebastian and Horn (1999) and Horn and Föller (2001).

induces a significant response. In *X. laevis*, about 40 hair cells in the macula epithelium exist at the time when the response appears for the first time (Horn and Föller, 1998). During further development, the sine-like shape of the response characteristic is maintained. Modifications include the roll angle at which the compensatory eye or head roll obtains its maximum value; in the clawed toad (*X. laevis*), the response peak shifted from a body roll of 60° to approximately 120° during the first 5 days of development (Horn *et al.*, 1986a; Sebastian and Horn, 1999; Horn and Föller, 2001).

Fish: In *O. mossambicus*, the rVOR can be observed for the first time at developmental stage 13; thereafter, it increases until stage 17. During the stages 18 and 21, the rVOR is depressed but increases again until the animals become premature at stage 25. The rVOR development is sensitive to modifications of light conditions; dark-rearing during early periods of life retards the rVOR development (Fig. 2, left) (Sebastian and Horn, 1999). Obviously, this light sensitivity of rVOR development is an expression of the intensive visual–vestibular integration within the vestibular nuclei (Dichgans and Brandt, 1978) which supports the control of body posture (von Holst, 1950; Meyer *et al.*, 1976; Berthoz *et al.*, 1975). The rVOR depression might be caused by a transient destabilization of the rVOR system by growing projections from the visual to the vestibular system and a reorganization of rVOR related nuclei.

Amphibian: In the clawed toad *X. laevis*, the rVOR appears in 3 days after fertilization (stage 42). Both rVOR amplitude and gain increase up to stage 52, and then decrease to an almost constant value at stage 60 and older tadpoles. This basic shape of the developmental characteristic was not affected by a unilateral labyrinthectomy; only the response level decreased by approximately 60% of the values obtained from intact tadpoles (Fig. 2, middle). After surgical elimination of the utricular macula of one labyrinth, the response decrease was at the same level as after a complete destruction of only one labyrinth; destruction of the saccular macula was ineffective (Horn *et al.*, 1986a,b). Light deprivation during early development retarded the rVOR development (Sebastian *et al.*, 1996).

Cricket: Crickets possess gravity sense organs on their abdominal cerci. They are composed of pin-shaped cuticular sensilla, each linked to one sensory cell. Stimulation of these sense organs by a lateral roll induces a compensatory head roll (rCHR) (Fig. 3, left). First instars of crickets possess only two club-shaped sensilla, one on each cercus (Fig. 3, lower). In *G. bimaculatus* (Horn and Föller, 2001) as well as in the house cricket *A. domesticus* (Förster, 2001), the number of clavate sensilla increased from one for the first instar to approximately 150 on each cercus in adults during post-embryonal development. Similarly, rCHR amplitude and gain increase in both species after each molt and reach their maximum in adults. Amputation of one cercus has no effect on the sinusoidal shape of the rCHR characteristics but depresses both rCHR gain and

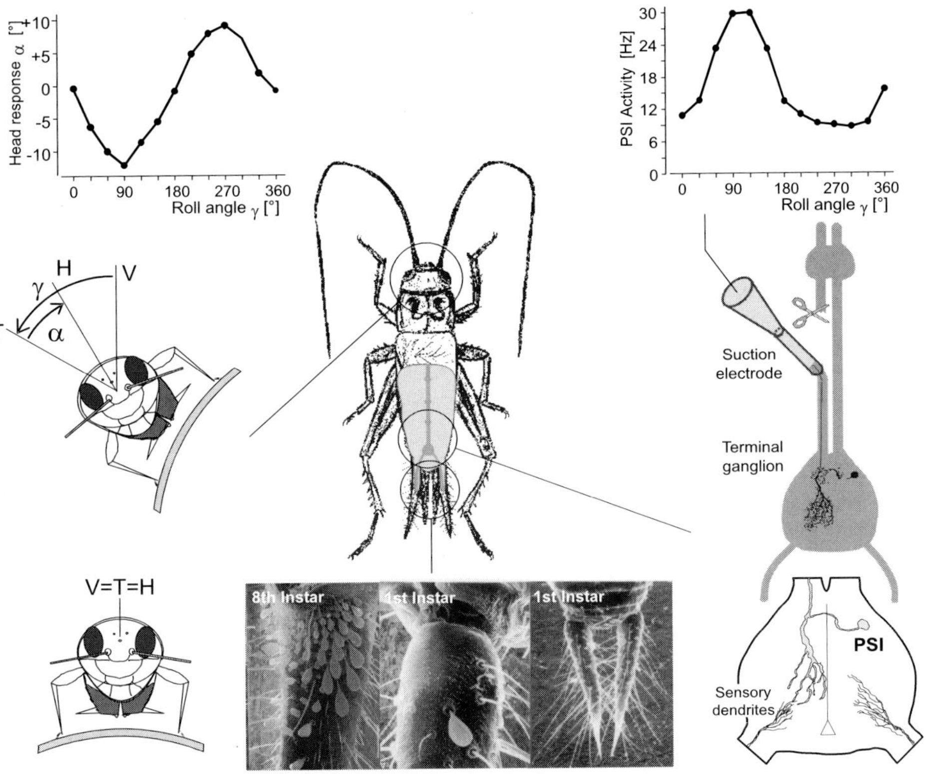

Fig. 3. The cercal GSS of crickets. It is composed of a field of club-shaped sensilla located at the base of the abdominal cerci; 1st instars possess only one sensillum on each cercus (lower); after each subsequent molt, the number increases by approximately 6–8 sensilla. Stimulation of this sense organ by a lateral roll induces a compensatory head response (rCHR). The angle α of this response depends from the roll angle γ in a sine-like manner (left). H, dorsoventral axis of the head; T, dorsoventral axis of the thorax; V, Space vertical. Each sensillum is innervated by a sensory cell. Its axon projects to the dendritic area of the large interneuron PSI. The cell body of the PSI is located in the cercal (terminal) ganglion of the ventral nerve cord; its axon projects towards the subesophageal ganglion (right). The activity of the PSI can be recorded from the connectives between two adjacent ganglia. The frequency of the PSI activity depends on the roll angle γ.

amplitude by 30–40% (Fig. 2, right). After bilateral destruction of the sense organ, the rCHR disappeared.

Modifications of the behavior by sensory deprivation or stimulus augmentation

Long-lasting stimulus modifications for the peripheral and central GSS include absence or augmentation of the stimulus *per se*, i.e., the gravitational environment, or any change of physiological input to the central nervous GSS structures.

Thus, three approaches can be used to deprive the developing GGS from gravity related information:

(1) destruction of the sense organ with the consequence that the gravitational field cannot be detected any longer and the central neuronal network of the GSS is deprived of gravity related information,
(2) deloading of parts of the body by counterweights which compensates for the gravitational pull, and
(3) exposure of the organism to microgravity (*syn.* weightlessness) during spaceflights which leaves the sense organ intact but deprives it from being stimulated by gravity related body displacements.

Many animals rely on more than one GSS in the control of their behavior. Insects use mechanoreceptors of their antennae or legs or specialized sense organs located on their cerci (Horn, 1985b); vertebrates use their vestibular macula sense organs but, in addition, gravity also affects various mechanoreceptors of the joints and muscles by the gravity-induced weight of the body (Massion *et al.*, 1995). Thus, the lesion technique is a selective deprivation approach while exposure to weightlessness acts on all GSS subsystems.

In contrast to deprivation, gravity augmentation provides the developing and adult animal with a gravitational reference direction throughout their life because the development of the sensory epithelium is not blocked by hypergravity (Vinnikov *et al.*, 1983). If hypergravity starts before the first appearance of the anatomical structure of GSS and its physiological excitability, the new structure experiences immediately the reference sensory field. Thus, if hypergravity exposure affects the development of gravity induced responses only during specific periods of life, the existence of a sensitive period to this specific environmental conditions is obvious. Basically, two techniques can be used for stimulus augmentation:

(1) loading the animal with a weight, and
(2) stimulation by centrifugation (hypergravity). Similar to the lesion and deloading techniques, loading affects specific sense organs of the complex GSS such as proprioceptors within joints of the legs or antennae, while exposure to hypergravity acts on all these components. Loading is less useful than exposure to hypergravity because animals increase their weight during development so the load has to be exchanged regularly.

Sensory deprivation by lesioning the gravity sense organ

The lesion technique is a powerful tool to determine consequences of sensory deprivation on the development of sensory systems. In rodents, it was used for the analysis of critical periods in the somatosensory system (van der Loos and Woolsey, 1973). In many animals, lesions activate regeneration mechanisms, in particular during development, by which structure and/or function of a sensory system recovers to normal. Both complete and partial replacement of the nervous system has been documented in invertebrate phyla including coelenterates, flatworms, annelids, gastropods and tunicates. Some vertebrates exhibit cyclical replacement or facultative repair in sensory systems, other vertebrates are able to replace discrete portions of the central nervous system (fish, amphibia), retina (amphibia), hair cells of the auditory (birds) or vestibular organ (mammals) or olfactory neurons (mammals) (cf. Moffett, 1996).

Regeneration capacities within the GSSs of lower vertebrates are low; however, a physiological normalization is activated by labyrinthine lesions (*vestibular compensation*, cf. Schaefer and Meyer, 1974). In contrast, larval crickets regenerate a lesioned cercal gravity sense organ; the extent of this regeneration depends on the stage at which the lesion was performed and on its repetition rate. The effects of sensory deprivation by lesion technique on the development of gravity related behavior was intensively studied in the swimming behavior and the rVOR of *X. laevis* and in the rCHR of the cricket *G. bimaculatus*.

Amphibian: Partial deprivation by the destruction of one labyrinth induces a behavioral asymmetry in *X. laevis* tadpoles characterized by a roll of the body to the lesioned side during swimming and by an asymmetrical posture of the eyes if the tadpole is lying horizontally. The direction of view of the eye on the lesioned side is directed downward while the eye on the intact side looks upward. In addition, the tadpole's tail is bended to the lesioned side (Fig. 4). The degree of movement and eye posture defect decreases during further development; the duration for this so-called *vestibular compensation* is correlated with the tadpole's age at which the labyrinthectomy was performed. Young tadpoles compensate the abnormal swimming and asymmetrical eye posture completely, but the rVOR depression only partially. Old labyrinthectomized tadpoles restore only their symmetrical eye posture (balance compensation) but never compensate for their locomotion defect (Rayer *et al.*, 1983) and for their rVOR gain and amplitude (gain compensation) (Rayer and Horn, 1986).

Simultaneous destruction of both labyrinths deprives the central vestibular nuclei from the input of the macular sense organ completely. By this treatment, the development is strongly retarded and a severe inactivity is induced in all *Xenopus* tadpoles, independently of their age. Their body shows significant malformations including left-, right-, upward- or downward-directed bendings of the tail (Fig. 4) which persist throughout the life, while these animals keep their eyes always symmetrically during horizontal posture. Swimming movements are irregular throughout the life, regardless of the developmental stage at which both labyrinths were destroyed. Most of the young lesioned tadpoles never exceed the developmental stages 57/58; the older ones can live very long despite of the persisting locomotory inactivity.

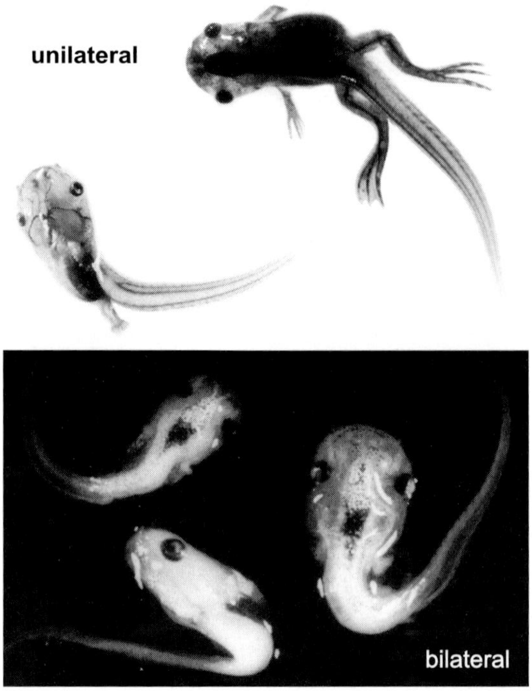

Fig. 4. Deprivation of the central vestibular nuclei from sensory input of the vestibular organ by uni- or bilateral labyrinthectomy induces malformations of the body. In hemilabyrinthectomized tadpoles of the Southern Clawed Toad (*X. laevis*), the tail is bended towards the lesioned side (unilateral; upper figures). In bilaterally labyrinthectomized tadpoles, the tails are also bended, but in different directions (lower figure).

Bilateral surgical deprivation can also be performed sequentially, i.e., the second labyrinth is destroyed after a couple of days or weeks after the first one. The technique was introduced by Bechterew to demonstrate the central nervous nature of vestibular compensation (*Bechterew compensation*; cf. Schaefer and Meyer, 1974). The basic observation after the lesion of the second gravity sense organ is the occurrence of symptoms quite similar to those following the destruction of one sense organ only. In *Xenopus*, this phenomenon occurs independently on the developmental stage at which the first operation was done, the time span between the first and second lesion, and whether the animals had compensated for the first labyrinthectomy, or not (Table 1) (Rayer *et al.*, 1983). The rotatory swimming defects following the second labyrinthectomy disappeared after some time. However, instead of becoming normal, swimming movements became irregular probably caused by a loss of the preceding central vestibular compensation. The asymmetrical eye posture, which occurred after the second lesion also, disappeared and the eyes returned to their symmetrical posture. After the second lesion, the asymmetrical eye

Table 1

Vestibular deprivation by a successive, bilateral labyrinthectomy (*Bechterew approach*)

Days between 1st and 2nd lesion	L48		L52		L56		L60	
	Swimming	Eye	Swimming	Eye	Swimming	Eye	Swimming	Eye
5	1	7	–	–	–	–	–	–
7	–	–	7	7	7	10	4	22
10	4	10	–	–	–	–	–	–
14	–	–	4	13	7	13	7	44
15	4	13	–	–	–	–	–	–
20	10	7	–	–	–	–	–	–
21	–	–	10	10	16	7	7	25
28	–	–	16	10	22	28	4	>42
35	–	–	19	19	13	28	–	–
42	–	–	16	13	4	25	–	–

Data indicate the number of days following the 2nd labyrinthectomy when symptoms induced by the 2nd lesion disappeared in 50% of the animals. Rotatory swimming towards the side of the 2nd lesion changed to irregular swimming. Asymmetrical eye posture characterized by a downward and upward directed view of vision for the eye on the side of the 2nd and 1st lesion, respectively, changed to symmetry. The destruction of both labyrinths were done successively at time intervals indicated in the left column (days between the 1st and 2nd lesion). L48, L52, L56, L60 developmental stage at the time of the 1st lesion (from Rayer *et al.*, 1983).

postures were maintained longer than rotatory swimming (Table 1). This observation indicates a compensation process of higher stability in the vestibuloocular pathway than in the vestibulospinal one (Rayer *et al.*, 1983).

Cricket: Lesion-induced deprivation from the cercal gravity sense organ was extensively studied in crickets (*G. bimaculatus*). The immediate effect of a surgical elimination of one cercal gravity sense organ was a depression of the rCHR by approximately 60–70%. Further development of the rCHR was influenced by both morphological regeneration and physiological readaptation. Four response classes were identified which correlated morphological with physiological regeneration: identical numbers of clavate sensilla on the regenerated and the normal cercus correlated with equal (Class 1a) or still depressed rCHR (Class 1b). Different numbers of clavate sensilla on the regenerated and the normally developed cercus correlated with equal (Class 2a) or different (Class 2b) rCHR amplitudes. Thus, regenerated gravity sense organs can reach full efficiency by following a basic genetic program, which needs a specific number of molts to reach normality of structure and function (cf. response classes 1a and 2b). In contrast to these apparently rigid mechanisms, complete morphological regeneration is not a prerequisite for normalization of the rCHR. Neuroplastic mechanisms are obviously involved in the processes of physiological normalization (cf. response classes 1b and 2a) (Horn and Föller, 2001). In *Gryllus*, the rCHR is also induced by stimulation of leg and antennal proprioceptors (Horn and Bischof, 1983; Horn and Föller, 1985). This multi-channel GSS might support the compensation for a decreased input from the regenerating cercal GSS, because it is continuously stimulated by dislocations of the trunk or the antennae during walking.

Sensory deprivation by microgravity

Deprivation by microgravity exerts an effect on all gravity sensitive organs and related systems. In some instances, exposure to microgravity reveals neuroplastic mechanisms, which are masked on ground due to normal gravity-related behavior. According to the otolith asymmetry theory, asymmetries between the bilateral otolithic organs develop during growth and maturation. One hypothesis for vestibular function is that under 1*g*-condition on Earth, neuronal mechanisms compensate for the morphological asymmetry between the otoliths of the left and right labyrinths and normal body posture or swimming are gained—but only on costs of an asymmetry in the activity between the left and right central vestibular nuclei. In absence of the macular input under microgravity this physiological asymmetry becomes obvious. In fact, a systematic study in two fish species, the swordtail (*Xiphophorus helleri*) and a cichlid (*O. mossambicus*), revealed that the extent and direction of rotatory swimming during microgravity were positively correlated with the otolithic asymmetry (Hilbig *et al.*, 2001).

In contrast to lesion-induced deprivation, microgravity-induced adaptive mechanisms affect intact sensory systems. Similar to lesion-induced deprivation, adaptive processes occur in each part of the pathway which connects the gravity sense organs with the effector organs. In fact, exposure to microgravity modifies behavior, physiology of neurons, gene expression or biochemistry of central nuclei in adult as well as developing animals (cf. Buckey and Homick, 2002). Systematic observations on compensatory head or eye movements in developing animals using spaceflights are rare. Most extensive studies were done during post-flight measurements after the missions STS-55 (German D-2 mission; 1993), STS-84 (Shuttle-to-Mir mission SMM-06; 1997), STS-90 (Neurolab; 1998) and a flight of a Soyuz spacecraft to the International Space Station (ISS) (Andromeda; 2001).

Fish: Effects of microgravity on the development of the rVOR were studied in young fish (*O. mossambicus*) which flew on STS-55 or on STS-84. The studies included: (1) animals which had reached stages 11–12 and had not yet developed the rVOR at μg-onset, and (2) older ones stages 14–16 which had already developed the rVOR. During the spaceflights, all fish embryos continued to develop at approximately a normal pace. Recordings of the rVOR taken between the post-flight days 1–4 revealed no modification of the sine-like shape of the rVOR characteristics irrespective of the animals' age at onset of microgravity. In contrast, both rVOR gain and amplitude were significantly

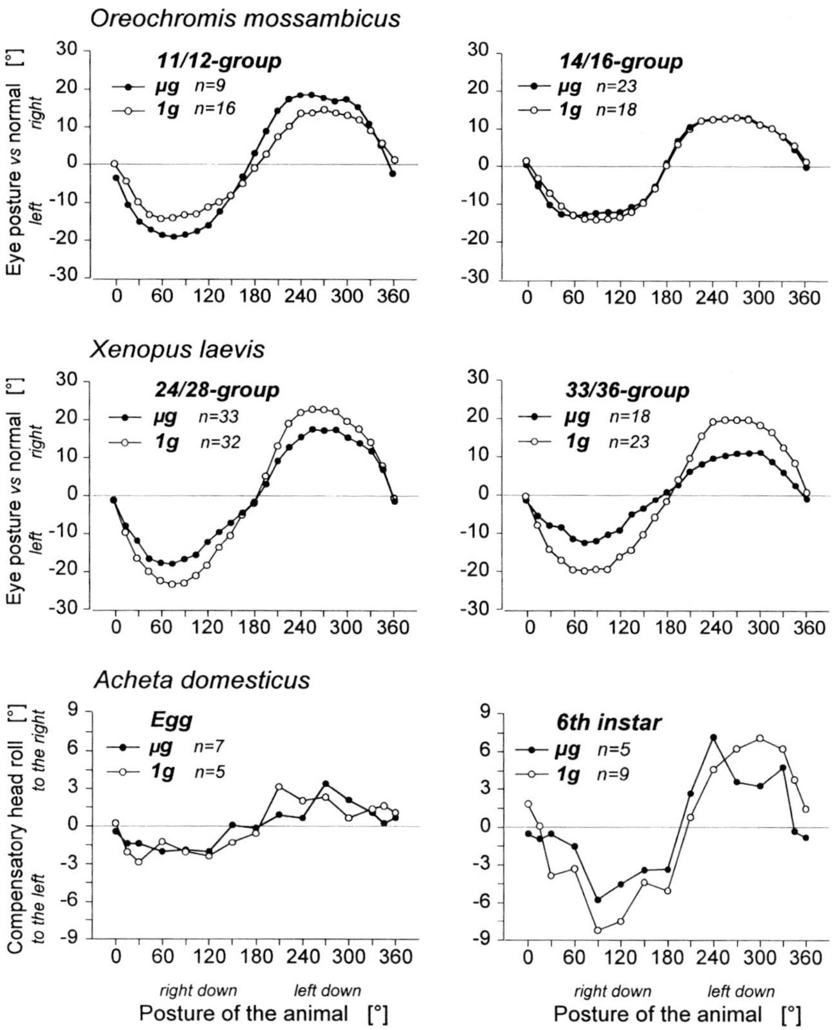

Fig. 5. Effects of microgravity (µg) on the development of compensatory eye or head responses in developing animals. Groups are defined by the animals' developmental stage at onset of µg. Note the lack of µg-effects in crickets (*Acheta*), a stage-dependent rVOR depression in fish (*Oreochromis*), and rVOR depression in both tadpole stages (*Xenopus*). *Acheta*: modified from Förster (2001) and Horn (2003); *Oreochromis*: modified from Sebastian *et al.* (2001); *Xenopus*: modified from Sebastian *et al.* (1996) and Sebastian and Horn (1998).

larger in 11/12-fish while the rVOR was not affected in the 14/16-fish (Fig. 5, upper). Post-flight readaptation was completed in the affected animals within 11 days; during the same period, 14/16-fish with µg-experience continued to develop as their 1*g*-ground control did. Thus, microgravity exposure increased the vestibular sensitivity only in those fish embryos, which had not yet developed their rVOR at onset of the spaceflight (Sebastian *et al.*, 2001).

Amphibian: Effects of microgravity on the development of the rVOR were studied in *X. laevis* tadpoles which flew either on STS-55, STS-84 or on ISS. The flights lasted 9 or 10 days. Studies included young animals (stages 25–36) which had not yet developed the rVOR at μg-onset and older ones (stages 45) which had already developed the rVOR. All studies revealed a significant depression of the rVOR in those tadpoles which had not yet developed their rVOR at μg-onset (Sebastian *et al.*, 1996; Sebastian and Horn, 1998) (Fig. 5, middle), in particular in those animals, which developed an upward bended tail during microgravity exposure (Sebastian and Horn, 2001). In contrast, stage 45 tadpoles which flew on ISS developed an augmented rVOR which persisted for 3 days after return to Earth (Horn *et al.*, 2002).

Studies following the STS-84 flight, revealed a minimum duration of microgravity necessary to induce the rVOR depression in young tadpoles. The rational for this particular rVOR experiment was to test the hypothesis whether microgravity exposure is effective if it starts before the rVOR has developed, while it is uneffective if it started thereafter. Tadpoles were exposed to μg either during the first (MC-group) or second half of the mission (CM-group), or throughout the 9-day mission (MM-group). It was expected that the results fulfill the equation $A[\mathrm{MM}] \leq A[\mathrm{MC}] \ll A[\mathrm{CM}] \leq A[\mathrm{CC}]$. In this equation, A stands for the rVOR amplitude. Inflight-1g was simulated by centrifugation (CC-group). The results, however, revealed the following rank order: $A[\mathrm{MM}] < A[\mathrm{MC}] = A[\mathrm{CM}] = A[\mathrm{CC}]$; i.e., the rVOR was only reduced in the MM-group while both the MC- and CM-groups were not affected by the 4- and 5-day μg-exposure. Thus, the original hypothesis could not be answered because the duration of gravity deprivation was to short. Despite the lack of rVOR modifications, the CM-tadpoles revealed a significant susceptibility to gravity deprivation; similar to MM-tadpoles, they developed dorsally bended tails which disappeared in both groups within 2 weeks after landing (Sebastian and Horn, 1998).

Cricket: Effects of microgravity on the development of the rCHR was studied in *A. domesticus* which flew on the 16-day mission Neurolab. Eggs, 1st, 4th and 6th stage larvae were sent into space in order to explore the possibility of an age-related sensitivity to microgravity. Additionally, lesions of the cercal GSS were done in the stage 6th larvae. Eggs were chosen because the 1st larvae, which hatched in space, had no rCHR-experience prior to launch. First instars were chosen because they possess only a single sensory cell on each cercus; i.e., in this stage, gravity experience before launch can be regarded as the minimum possible under earth-bound conditions (Horn and Föller, 1998). Older stages were chosen because of the increasing connectivity within their nervous system which might affect the extent of adaptation to microgravity. The rCHR characteristics recorded from μg-animals between post-flight days 1–8 revealed no significant differences to those recorded from 1*g*-ground controls. Morphological and functional regeneration occurred under microgravity. As in the 1*g*-ground controls, the regenerated gravity sense organ possessed less sensory units than the normal developed one and the rCHR was weaker in animals with a regenerate than in the intact controls. But neither for the lesioned nor for the intact larvae, significant behavioral differences were obtained between the cumulated data from flight versus 1*g*-ground larvae (Fig. 5, lower) except a retardation of rCHR development in slowly developing animals in the oldest larvae, and a retardation of 1*g*-readaptation in larvae which hatched in orbit (Förster, 2001; Horn *et al.*, 2001; Horn, 2003).

Stimulus augmentation by centrifugation

In lower vertebrates and insects, long-lasting gravity augmentation by centrifugation revealed effects on many systems of the body including sensory, central nervous and motor systems. Recordings during exposure to hypergravity—as induced by centrifugation—give a clear picture of adaptation processes to this new environment; however, those studies are rarely done (cf. for example, Etard *et al.*, 1999), probably due to technical complications. Most studies were performed after hg-termination, in particular those on development. Observations on compensatory head or eye movements were done in young *O. mossambicus*, *X. laevis* tadpoles and in the cricket *A. domesticus*. The threshold for the efficiency of hypergravity was determined for *Oreochromis*. The rVOR of fish exposed to hypergravity when they had reached stage 11/12 was not affected by 2.0*g* or 2.5*g* while 3*g*-reared fish revealed a significant rVOR depression (Sebastian *et al.*, 2001).

Fish: Two *Oreochromis* stages (11/12 and 15) were exposed to 3*g*. In both groups, the stage reached by 3*g*-exposed fish was significantly more advanced than that reached by the 1*g* controls. Their rVOR characteristics were still sine-like after 3*g*-exposure. In both 3*g*-animals from the 11/12- and 15-group, rVOR gains recorded during the 1st post-3*g* week were significantly smaller than those from their 1*g*-control. In addition, the rVOR amplitude was reduced by the 3*g*-exposure, in the 11/12-fish to 26.6° ($n=35$) compared to 30.8° ($n=35$) in the 1*g*-control, and in the 15-fish to 19.9° ($n=25$) compared to 25.8° ($n=40$) in the 1*g*-controls (Fig. 6, upper). During the post-3*g* days 8–11, the rVOR characteristics remained sine-like in all groups but now, rVOR gains and amplitudes from 3*g*-fish had adapted to the levels of their 1*g*-controls in both, stage 11/12 and stage 15-fish. A third recording was performed 4 weeks after 3*g*-termination when the animals had developed to stages 24 or 25. At this time, the rVOR gains were in the range of the values recorded during the 2nd week.

A stage-matched analysis for the first recording period revealed that in the pooled stages 19/20, the rVOR gain was reduced in the 3*g* animals rolled from

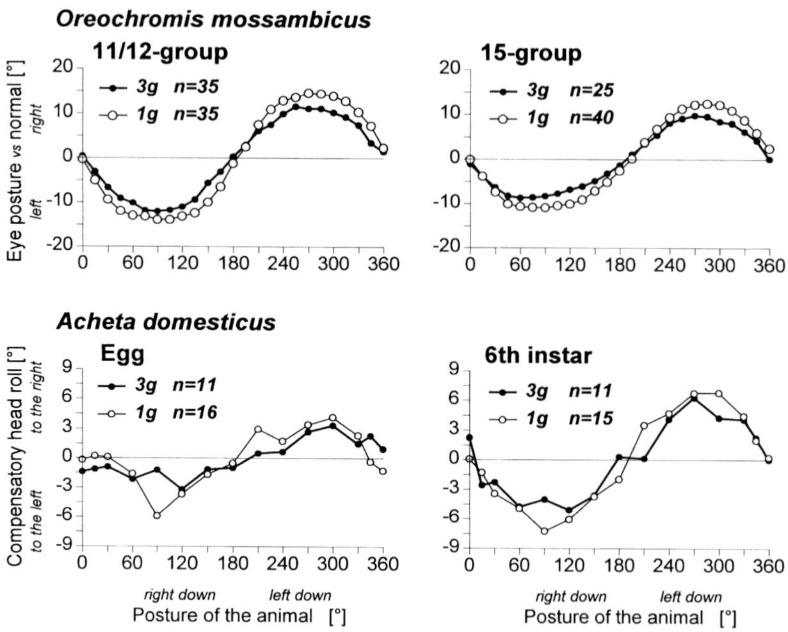

Fig. 6. Effects of 3*g*-hypergravity on the development of compensatory eye or head responses in developing animals. Groups are defined by the developmental stage the animals had reached at onset of 3*g*. Note the lack of μg-effects in crickets (*A. domesticus*), and the depression of the rVOR in fish (*O. mossambicus*). Observations in *X. laevis* (not included in the figure) resemble those in *Oreochromis*. *Acheta*: modified from Förster (2001); *Oreochromis*: modified from Sebastian *et al.* (2001).

the inclined to the horizontal position by 30° but not in the pooled stages 21 and 22. For the rVOR amplitude, the reduction was significant for the pooled 21/22-stage fish. No differences between stage-matched 3*g*- and 1*g*-samples were observed for the 2nd and 3rd recording period. Thus, the reduction of the rVOR amplitude after 3*g*-exposure is caused by both a specific gravity effect and the 3*g*-induced acceleration of morphological development (Sebastian *et al.*, 2001).

Amphibian: The main finding was that the rVOR amplitudes and gains recorded from *X. laevis* tadpoles with 3*g*-experience were depressed with respect to those recorded from their 1*g*-reared siblings, and that the subsequent development depended on the stage at onset of the 3*g*-period. In particular, in the stage 6/9-sample, no difference between the 3*g*- and 1*g*-samples was recorded for the first 2 days after 3*g*-exposure. During the following 4 days, the rVOR amplitude increased significantly only in the 1*g*-controls, so that 3*g*- and 1*g*-reared tadpoles became significantly different. In the other stages, a rVOR depression was recorded in 3*g*-exposed tadpoles with respect to their 1*g*-reared siblings during the first ten post-3*g* days. However, the extent of further development of the rVOR amplitude during the 10-day recording period was linked to the first appearance of the rVOR. Tadpoles which were exposed to 3*g* at developmental stages < 42/43 (at these stages, the rVOR appears for the first time, Horn *et al.*, 1986a) did not show a further increase of their rVOR amplitude during 1*g*-readaptation, while the stage 45 tadpoles did so even during the 1st week (Horn and Sebastian, 1996; Sebastian *et al.*, 1996). Thus, the period of the first appearance of the rVOR is obviously a critical milestone for the development of adaptive properties within the neuronal network which mediates the rVOR.

Crickets: Effects of 3*g*-hypergravity on the rCHR development of *Acheta* were studied according to the time schedule of the Neurolab mission. In particular, behavioral studies were performed with animals which had reached the late embryonal status (eggs), the 1st, 4th and 6th larval stage at 3*g*-onset. The results from the behavioral observations were similar to those obtained from the μg-exposed crickets. Under 3*g*-conditions, 1st instars hatched and larvae molted at the same ratio as the 1*g*-controls. The rCHR-characteristics as well as the rCHR amplitudes and gains were mainly unaffected by the 16-day 3*g*-exposure for all four experimental groups compared to their 1*g*-reared siblings (Fig. 6, lower). Also during further post-embryonic development, no rCHR differences between 1*g*- and 3*g*-larvae became obvious (Förster, 2001; Horn *et al.*, 2001; Horn, 2003).

Stimulus modification by body-loading or deloading

Loading or deloading of the body is a simulation method to test effects of gravity augmentation or deprivation, respectively, on sensory, neuronal and motor systems in adults and during development. Most studies were done on the development of locomotion in rats using the tail suspension technique. They revealed that motor systems which control walking or swimming need weight experience for normal development (Walton, 1998). At the beginning of the seventies, first attempts with loading and deloading were done to study mechanisms of gravity perception in insects (Jander *et al.*, 1970; Wendler, 1975). Their observations prompted the hypothesis that insects use displacements of their body for the detection of the gravity vector (*proprioceptive GSSs*, cf. Horn, 1985b), a principle which is also common in man (Massion *et al.*, 1995). Loading improved geotactic orientation, deloading impaired it (Fig. 7). In a few insect species such as honey bees, loading was uneffective (Horn, 1973) although they use gravity for orientation as revealed by the elicitation of compensatory head responses during lateral roll (Table 2). Body-loading studies were not yet extended to developing insects; but they could be used for ground based studies in the preparation of spaceflight experiments.

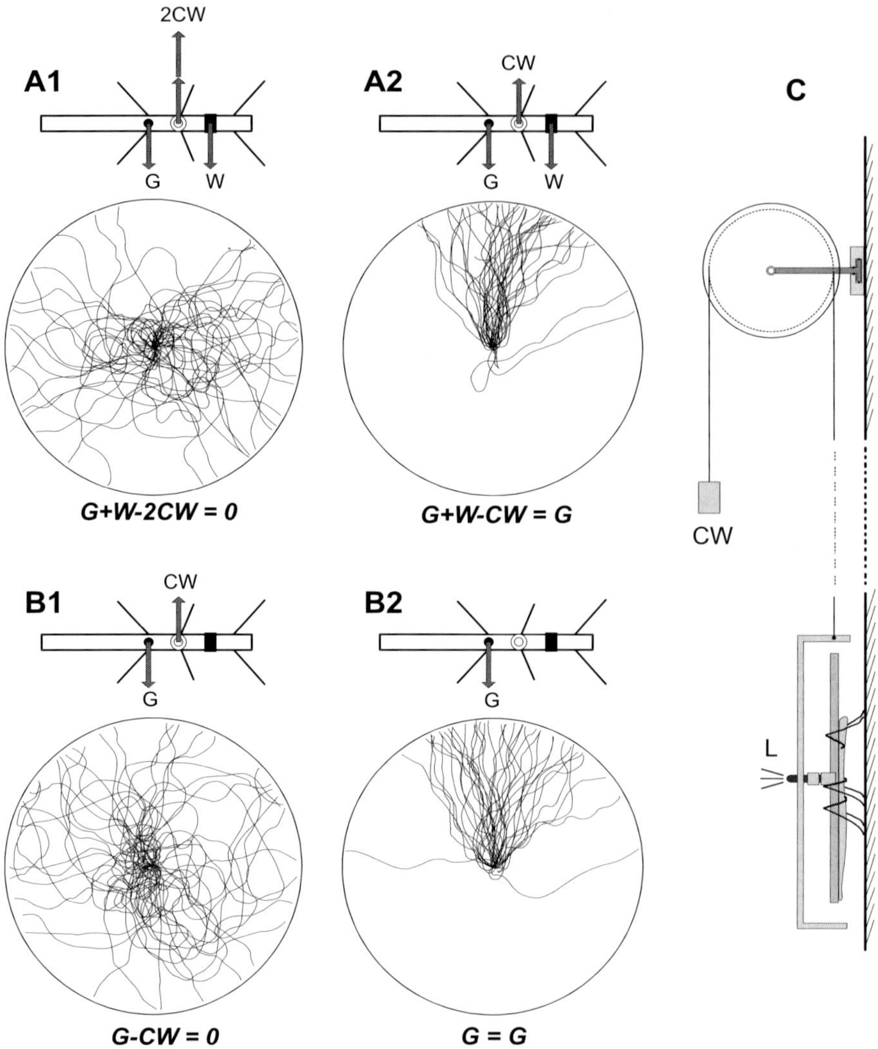

Fig. 7. Body deloading in insects. Right part: Experimental method of deloading by a counterweight (CW); L, light source necessary for recording of the animal's tracks. Left part: Effects of body deloading on the geotactic orientation in the stick insect (*Carausius morosus*). Superimposed walking traces of five animals on a vertical surface at four experimental conditions characterized by the schematic drawings of the stick insect. Each arrow represents the weight of the animal and its direction of action. Upwards directed arrows represent the counterweights (CW) acting at the axis of rotation during walking (concentric circles); the downward directed arrow G represents the body weight acting from the center of gravity (black dot); the downward directed arrow W represents a weight necessary to eliminate the torque. Note the preference of upward walking in case of dominating gravity input (A2, B2) and the loss of geotactic behavior in case of compensated gravity input by deloading (A1, B1) (modified from Wendler, 1975).

Table 2

Modification of geotactic behavior in insects by loading (L) or deloading (DL) of their body

Species	Order	L	DL	rCHR
Enallagma cyathigerum	Odonata	0	–	–
Gryllus bimaculatus	Saltatoptera	1	–	**C, P**
Acheta domesticus	Saltatoptera	1	–	**C, P**
Locusta migratoria	Saltatoptera	1	–	–
Forficula auricularia	Dermaptera	1	–	–
Carausius morosus	Phasmida	–	1	–
Blaberus craniifer	Blattaria	1	–	–
Leucophaea maderae	Blattaria	1	–	–
Capsodes cingulatus	Rhynchota	1	–	–
Abax ater	Coleoptera	1	–	–
Blaps mortisaga	Coleoptera	1	1	–
Tenebrio molitor	Coleoptera	1	1	–
Sialis flavilatera	Megaloptera	1	–	–
Osmylus chrysops	Planipennia	< 1	–	–
Chrysopa vulgaris	Planipennia	0	–	–
Panorpa communis	Mecoptera	1	–	–
Calliphora erythrocephala	Diptera	< 1	–	**P**
Pieris napi	Lepidoptera	0	–	–
Tenthredo mesomelas	Hymenoptera	0	–	–
Bombus terrestris	Hymenoptera	0	–	–
Apis mellifera	Hymenoptera	0	–	**P**

Modifications were high (1), moderate (< 1) or absent (0). Some of the insect species were positively tested for their compensatory head response (rCHR) by stimulation of cercal gravity sense organs (C) or their leg proprioceptive GSS (P); a minus sign means not investigated (modified from Jander *et al.*, 1970; Horn and Bischof, 1983; Horn and Lang, 1978; Horn, unpubl.; Wendler, 1975).

Modification of central nervous and motor activity by sensory deprivation or stimulus augmentation

Neurons within the vestibular system can be irreversibly deprived from their input from the macular organs by bilateral labyrinthectomy. Neurophysiological studies in alert guinea pigs (Ris and Godaux, 1998), anesthetized cats (Ryu and McCabe, 1976) and awake monkeys (Waespe *et al.*, 1992) had revealed that despite absence of any labyrinthine input, vestibular nuclei were able to restore the initially depressed neuronal activity. In contrast to bilateral labyrinthectomy, the macula systems remains intact in microgravity, however, the otolithic mass is unweighted so that the otolith cannot be displaced by body roll and tilt. Absence of weight-stimulation of utricular sensory hair cells makes it likely that macular afferents and central vestibular neurons decrease their resting discharges after μg-onset, but restore normal activity during further spaceflight due to their neuroplastic properties (cf. Precht and Dieringer, 1985; Smith and Curthoys, 1989; Dieringer, 1995).

There is only one neurophysiological study on a GSS during altered gravity. It shows that the resting discharge of primary neurons of bull frog statoreceptors decreased immediately after μg-onset and increased during further μg-exposure beyond 1*g*-preflight discharge levels (Bracchi *et al.*, 1975). But also recordings performed after termination of altered gravity give insight into what might happen during the exposure to AG conditions. Similar to memory formation, adaptive processes activated during μg- or hg-exposure will not be extinguished immediately with re-entrance of the organisms to 1*g*-condition. They will persist for some hours, weeks or even months. But even the number of studies in developing animals is low. Mass recordings from the 8th nerve were done post-flight in chicken exposed as embryos for 5 days to microgravity (STS-29) (Jones *et al.*, 1993). More extensive studies were done on identified gravity sensitive neurons in cricket larvae (*A. domesticus*) (Riewe, 2001; Horn *et al.*, 2001; Horn, 2003) and, concerning the motor output, in *Xenopus* tadpoles after exposure to microgravity or 3*g*-hypergravity (Böser and Horn, 2003; Böser *et al.*, 2003).

Microgravity and hypergravity

Amphibian: High speed camera recordings have shown that tadpoles (*X. laevis*) raised in microgravity exhibited a lower tailbeat frequency than 1*g*-controls (Fejtek *et al.*, 1998). The swimming pattern is generated by a central oscillator (Kahn and Roberts, 1982; Dale, 1995). It is possible to observe the rhythmical, burstlike activity of motoneurons by extracellular recordings from ventral roots (*VR*) of the spinal cord in paralyzed animals (*fictive swimming*). The characteristics of VR activity are similar to the swimming motor pattern, although in paralyzed animals there is no sensory feedback (Kahn and Roberts, 1982). One typical parameter for both fictive swimming and swimming without paralyzation is the rostrocaudal delay of activity in ipsilateral myotomes of different segments (Kahn *et al.*, 1982). Ventral root recordings in tadpoles with μg-experience were performed for the first time with *Xenopus* tadpoles following the 10-day flight during the Andromède mission to ISS in October 2001. At onset of microgravity, the embryos had reached the stage 26/27. After μg-exposure, the duration of fictive swimming episodes was increased and the rostrocaudal delay was decreased compared to ground-reared controls; in addition, the burst duration of fictive swimming was slightly decreased (Böser *et al.*, 2002).

Hypergravity exposure during early development also affects the characteristics of VR activity in *Xenopus* tadpoles. The VR bursts recorded from tadpoles with 3*g*-experience were longer but reached the level of 1*g*-tadpoles during the post-3*g* days 7 and 8. In particular, the burst duration recorded during the first 5 days after termination of the 3*g*-period from the rostral VR was 20.7 ms for 3*g*-tadpoles but only 15.9 ms in the 1*g*-control group. The respective results from the caudal VR were 17.2 and 12.9 ms, respectively.

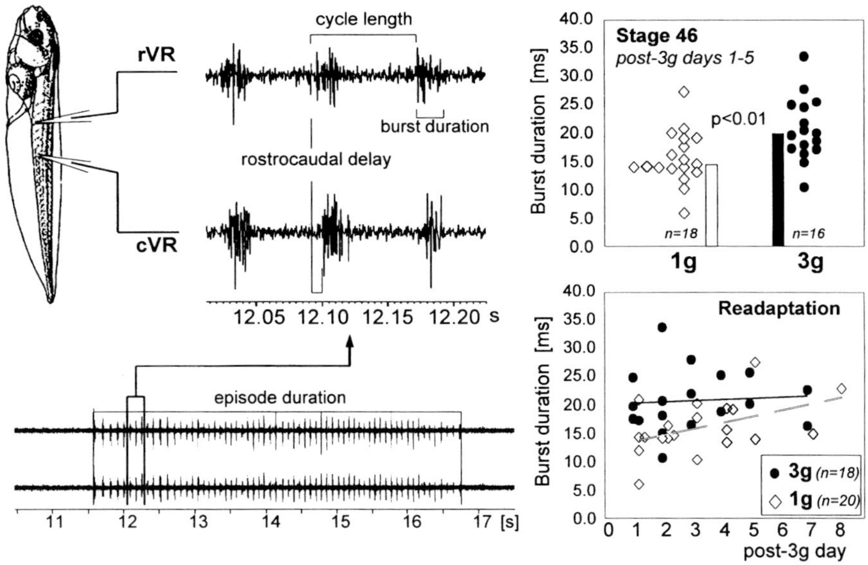

Fig. 8. Ventral root recording technique to study the motor output in paralyzed *X. laevis* tadpoles. Left: Two glass electrodes are attached to two ventral roots (rVR, rostral root; cVR, caudal root) to determine typical parameters of the swimming pattern such as cycle length, burst duration and rostrocaudal delay. Episode duration is the total time of swimming activity (lower). Right: The effect of a 10-day 3*g*-exposure on the development of the swimming movement pattern in *Xenopus* tadpoles. All animals had reached stage 46 during the first 5 days of recording. On the 7th and 8th post-3*g*-days, the animals had developed to stage 47 in both groups. Note the significantly increased burst duration in tadpoles during the first 5 days after 3*g*-exposure (upper) and the adjustment between both groups within 1 week (lower) (modified from Böser and Horn, 2002).

Recordings taken from post-3*g* days 7 and 8 revealed an adjustment between both groups; at this time the tadpoles were at stage 47. Cycle length, rostrocaudal delay and episode duration were not affected by 3*g* (Fig. 8) (Böser and Horn, 2002).

Readaptation of VR-activity to standard levels was completed within 8 days for both, tadpoles with μg- and tadpoles with 3*g*-experience. There is evidence that 3*g*- or μg-induced changes in fictive swimming have a vestibular origin. The μg-induced reduction of macular activity may affect the development of descending reticulospinal and raphespinal projections to the spinal cord in a reversible manner while hypergravity exposure may lead to an advanced development (Böser *et al.*, 2002; Böser and Horn, 2002).

Cricket: In the larval house cricket (*A. domesticus*), postural information originating in the cercal gravity sense organ is projected to a position sensitive interneuron (PSI) whose soma and main dendritic arborization are located in the terminal (=cercal) ganglion. Lateral roll is an effective stimulus to the PSI (Fig. 3, right) (Sakaguchi and Murphey, 1983). The PSI revealed a significant

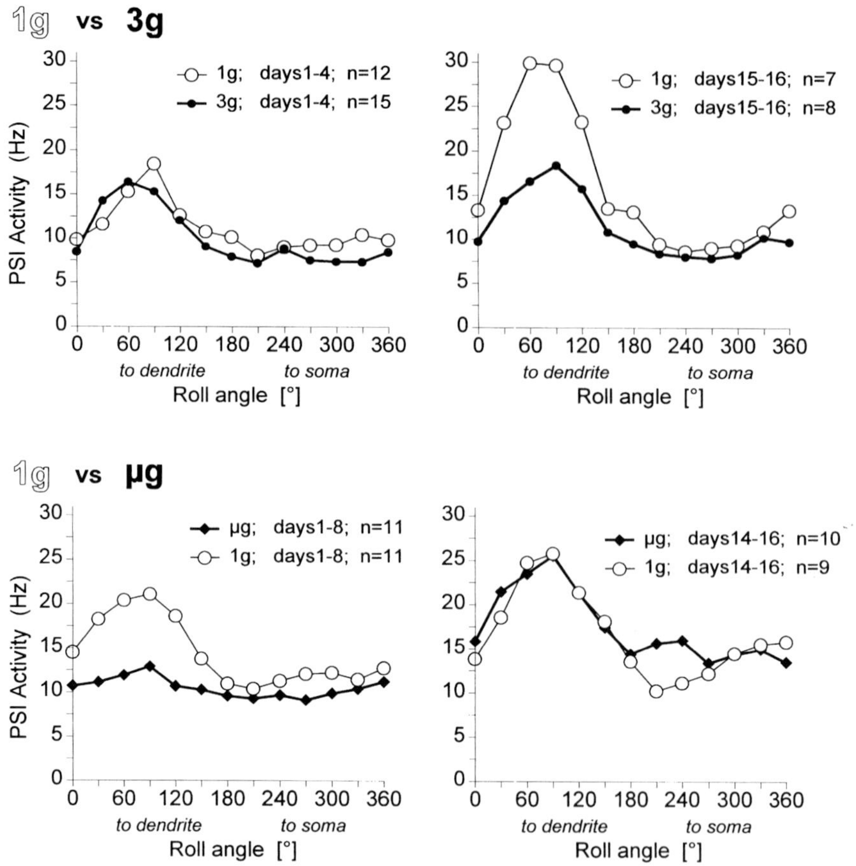

Fig. 9. Sensitivity of the PSI of the cricket *A. domesticus* after a 16-day µg-exposure during a spaceflight (STS-90 Neurolab) (lower plots) or after a 16-day 3*g*-exposure (upper plots). At onset of altered gravity, the larvae had reached stage 4. After AG-termination, recordings were taken between the 1st and 8th day (µg-animals) or between the 1st and 4th day (3*g*-animals) to determine early effects (left plots) and between the 14th and 16th day to study modifications in PSI activity during 1*g*-readaptation (right plots) (modified from Riewe, 2001).

susceptibility to microgravity and impaired readaptation properties after 3*g*-exposure. These observations contrasted the behavioral ones. In larvae with a 16-day µg-experience, the modulation of PSI activity during a roll to the dendritic side of the PSI was significantly reduced between post-flight days 1 and 8 compared to the ground controls. Normalization of the discharge modulation was observed between post-flight days 15 and 16 (Fig. 9, lower). After a 16-day hypergravity exposure, larvae revealed no significant difference to the 1*g*-controls between post-3*g* days 1 and 4. But after a 2-week 1*g*-readaptation, the response modulation in animals with 3*g*-experience was maintained without any change while their 1*g*-controls increased the extent of impulse frequency modulation according to standard development (Fig. 9, upper) (Riewe, 2001; Horn *et al.*, 2001).

Combination of gravity deprivation or augmentation with lesion-induced deprivation

The asymmetry between the left and right macular system induced by a unilateral labyrinthectomy is compensated on ground by modifications of the activity within the central vestibular system. This conclusion is derived from behavioral experiments using the Bechterew approach as well as from neurophysiological studies which revealed the normalization of neuronal activity after some time (cf. Schaefer and Meyer, 1974; Precht and Dieringer, 1985). The unequivocal proof of this conclusion, however, became possible by the use of spaceflights. In microgravity, absence of the macular input reveals the existing central compensatory action similar to that observed in the Bechterew paradigm and, therefore, the appearance of the asymmetrical behavior. In fact, recent observations in rats during parabolic flights supported these theoretical considerations. Head posture and neck muscle activity (EMGs) were examined in unilateral (UL) and bilateral (BL) vestibularly lesioned rats in hypogravity during parabolic flights. Compared with BL rats taken as control, the head and the body of UL deviated toward the lesion side at hypogravity. Recorded in head fixed condition, left and right EMGs remained symmetrical in BL while UL rats displayed no asymmetry between left and right muscles at μg (Etard *et al.*, 1999). The technique of a superposition of microgravity conditions with lesion-induced deprivation was not yet used in studies on development and aging; but it offers the possibility to study the balance between adaptive neuroplastic processes of the central nervous system induced either physiologically by modifications of the environment or by pathological conditions.

Morphological malformations by sensory deprivation

Deprivation of the GSS by lesioning the sense organ or by exposure to microgravity can induce malformations of the body (Fig. 4). They were frequently observed in *Xenopus* (Rayer *et al.*, 1983; Wassersug and Murphy, 1987; Neubert *et al.*, 1994; Snetkova *et al.*, 1995; Sebastian and Horn, 1998). The tails of bilaterally labyrinth-ectomized *X. laevis* tadpoles and the body of juveniles are strongly left-, right-, upwards- or downwards-bended. In addition, these animals are extremely inactive and strongly retarded in their development. Many of the young ones never exceed developmental stage 57. Older ones, however, can live very long despite of their extreme inactivity and their malformations which last for their whole life (Rayer *et al.*, 1983).

Gravity deprivation during spaceflights causes morphological changes. After termination of μg, they persist only for some days. In *X. laevis* tadpoles, the most obvious modification of the body is the hyperextension of the tail. It was observed in tadpoles which were exposed to microgravity when they had reached the stages 25–28 (Snetkova *et al.*, 1995; Sebastian and Horn, 1998) but not in tadpoles which developed from eggs fertilized in microgravity (Souza *et al.*, 1995). After re-exposure to 1*g*-conditions the hyperextension of the tail disappeared. Data from STS-84 in 1997 revealed that the number of tadpoles with abnormal tail development increased with increasing duration of μg-exposure. In particular, 21 out of 35 tadpoles with a 9-day μg-experience but only 13 out of 28 tadpoles with a 5-day μg-experience developed the hyperextension of the tail. At the end of the 2nd post-flight week, only one tadpole in each of the affected groups had the upward bended tail which demonstrates morphological normalization under the influence of Earth gravity (Sebastian and Horn, 1998).

The reason for these malformations is, so far, unknown. One explanation was based on the observation that μg-animals with upward bended tails show muscle degeneration, probably caused by the deloading during μg-exposure (Snetkova *et al.*, 1995). But also vestibular mechanisms have to take into consideration especially in lesioned animals. In general, activity exerts trophic effects on developing neurons and their interneuronal connectivity patterns as well as on the development of muscles (cf. Schmidt, 1988; Fawcett, 1988; Brueckner *et al.*, 1999). The macular vestibular structures do not only control static compensatory eye movements but also tonically activate the trunk muscles via descending projections to the spinal cord. Thus, reduction or absence of tonic vestibular activity might influence the development of vestibulospinal projections and body muscles which might also cause modifications of the body shape.

Critical (sensitive) periods in gravity sensory systems

A feature of developing sensory, neuronal and motor systems is their critical period (cf. Wiesel, 1982). Critical periods of GSSs can be studied on ground using modifications of the gravitational environment such as hypergravity, simulated microgravity by means of 2D- or 3D-clinostates or rotating bioreactors, or by lesion-induced gravitational deprivation. In the zebrafish, vestibular development was investigated on ground using a bioreactor developed by NASA. In a first study, eggs and hatchlings stayed in the rotating bioreactor for either 72 or 96 h after fertilization. These animals displayed a swimming behavior that was indistinguishable from control animals when illuminated from above. When incubated in the bioreactor for 72 h, some animals were missing one or more otoliths. In contrast, all control animals had two otoliths on each side. When incubated in the bioreactor for 96 h, experimental animals showed a weaker rVOR during passive head-down and -up tilt compared to the age-matched control animals for 1 day (Moorman *et al.*, 1999). In a more detailed study, eggs were placed in the bioreactor 3, 24, 30, 36, 48, or 72 h after fertilization and maintained there until they became 96 h old;

other eggs were placed in the bioreactor shortly after fertilization and stayed there until they reached the age of 24, 36, 48, 60, 66, 72, or 96 h. Based on recordings of the rVOR induced by head-down or -up tilt, the studies revealed the existence of a critical period which included the period of life from 24 to 72 h after fertilization. This postulation was confirmed by an experiment with zebrafish eggs which were rotated in the bioreactor from 3 to 96 h after the fertilization except for the period between 24 and 72 h. These larvae developed a normal VOR (Moorman *et al.*, 2002).

However, the most critical approach is gravity deprivation, i.e., exposure to long-lasting microgravity. The approach is limited by the facts that: (1) microgravity has to be started at many developmental stages, and (2) during different periods of post-embryonal life. Despite of unfavorable conditions such as the limited number of spaceflight opportunities, a first attempt was done in developing fish (*Fundulus heteroclitus*) aboard Skylab 3. Behavioral studies were performed with fish which hatched during the 59-day spaceflight. Most larvae displayed normal swimming behavior (von Baumgarten *et al.*, 1975). The post-flight tests showed that young fish launched before their labyrinth was formed exhibited a decreased positive geotaxis when compared with their 1*g*-ground controls. In contrast, fish launched when their labyrinth had developed behaved normally (Hoffman *et al.*, 1977). Sixteen to 17 months after flight, the youngest group revealed persisting physiological changes. They were less disoriented during short-term μg-periods (parabolic flights) but hypersensitive to hypergravity (Hoffman *et al.*, 1978).

The period of hatching was also critical for the development of the rVOR in *Xenopus* tadpoles concerning its adaptive properties during readaptation to 1*g*-conditions after 3*g*-exposure. Animals which hatched during 3*g*-conditions were significantly impaired in their readaptation capabilities compared to tadpoles which hatched before onset of 3*g* (Horn and Sebastian, 1996). Also crickets exposed to microgravity revealed that the time of hatching is a critical landmark for further development. The rCHR development of larvae which hatched during orbital flights was significantly delayed during 1*g*-readaptation compared to animals which hatched before the flight (Horn *et al.*, 2001). Another critical milestone in the development of the rVOR is the time of the first appearance of the rVOR (Sebastian *et al.*, 2001).

All these studies gave clear evidence for a general sensitivity of the developing GSS for modifications of the gravitational environment; but clear arguments for the existence of a critical period in the GSS development are still missing, especially concerning the irreversibility of induced behavioral, physiological or anatomical modifications. Nevertheless, persistence of μg- or hg-induced effects cannot be excluded for some reasons: (1) in all studies, the exposure time was restricted to short periods during development compared to the time necessary for the development from a fertilized egg to an adult, (2) in slowly developing *Xenopus* tadpoles, behavioral modifications persist longer during 1*g*-readaptation than in animals which develop with normal pace

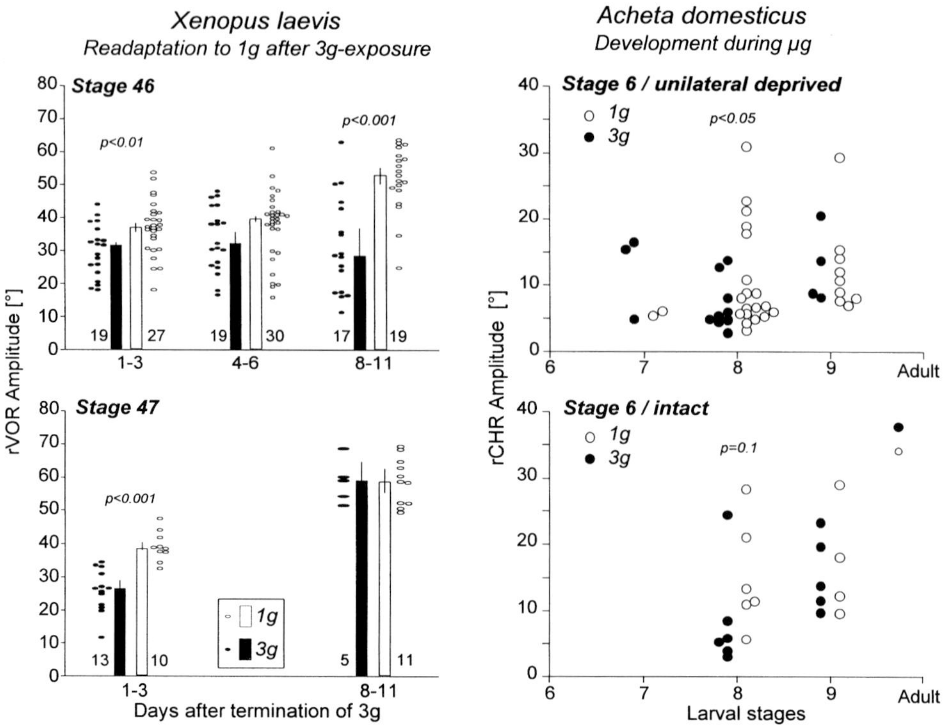

Fig. 10. Effect of developmental velocity on rVOR readaptation after hypergravity exposure in tadpoles and on rCHR development during microgravity exposure in cricket larvae. Right: A retardation of rCHR development in slowly with respect to fast developing larvae was observed for 6th instars of the cricket *A. domesticus*. During the 16-day Neurolab flight, larvae developed to 7th, 8th or 9th instars as the ground controls did. From larvae with a regenerate (upper right), the microgravity group which had molted only twice showed a significant weaker rCHR than their stage-matched ground-reared siblings. No difference existed between larvae which had molted three times in orbit and on ground. In the intact controls (lower right), there was a tendency ($P < 0.1$) for a similar stage related susceptibility (Horn *et al.*, 2001; Horn, 2003). Left: In *X. laevis* tadpoles which developed slowly and maintained at stage 46 during 1*g*-readaptation, the rVOR did not show any increase as it was observed in the age-matched 1*g*-controls; in contrast, tadpoles which developed from stages 46 to 47 during the same period normalized their rVOR depression to the 1*g*-control level (Sebastian *et al.*, 1996).

(Sebastian *et al.*, 1996) (Fig. 10, left), (3) *Xenopus* tadpoles needed longer to normalize their rVOR if stimulated by small lateral rolls compared to large rolls (Sebastian *et al.*, 1998).

General considerations

The influence of AG conditions or lesion-induced deprivation on the development of gravity induced behavior are not yet completed and need further spaceflights. These studies should be guided by aspects which are characteristic for GSSs such as mechanisms of neuronal adaptation to altered gravity, AG-induced developmental retardation or acceleration, or intersensory convergence and its influence on behavioral normalization. But also non-neuronal elements should be taken into account. It has been postulated that abnormal swimming behavior during or after microgravity exposure relates to the control of buoyancy which requires proper lung inflation (Wassersug and Souza, 1990; Pronych and Wassersug, 1994). Ground studies also indicated that hatchlings denied access to the water's surface do not inflate their lungs which normally occurs a day or two after hatching (Pronych and Wassersug, 1994); as a result they have abnormally small lungs and their overall growth is severely retarded (Wassersug and Murphy, 1987) as it was also observed in lesion-induced gravity deprivation (Rayer *et al.*, 1983). Wassersug and colleagues supposed that those tadpoles in microgravity failed to locate the air–water boundary and to inflate their lungs in a timely fashion (Souza *et al.*, 1995). Accordingly, movement control and buoyancy regulation during swimming is likely to be the less affected by gravity deprivation the older tadpoles are at onset of the AG environment.

Mechanisms which cause adaptation to altered gravity

The comparative approach reveals some common features of how altered gravity and sensory deprivation effect GSS development which might be important for human medicine. First, vestibular desensitization occurs during exposure to hypergravity in fish and amphibian, because their rVOR was always depressed after a 9-–12-day 3*g*-exposure (Sebastian *et al.*, 1996, 2001). Most obvious is the sensitization of the GSS by gravity deprivation which was observed in man as well as in adult and young animals. Stimulation of the macular organ of adult toad fish (*Opsanus tau*) by translational linear acceleration revealed a higher nervous activity of the 8th nerve in animals which were exposed to microgravity for 16 days during the Neurolab mission than their 1*g* ground controls (Boyle *et al.*, 2000). Astronauts overestimated their tilt sensation during the first days after their return to 1*g*-condition on Earth (Clément *et al.*, 2001). Developing cichlid fish (*O. mossambicus*: Sebastian *et al.*, 2001) and clawed toad tadpoles (*X. laevis*: observations by the author after the Andromeda mission to ISS in 2001) revealed an increased rVOR after a 9-–10-day spaceflight. In astronauts and all species, the μg-induced sensitization as well as the hg-induced desensitization disappeared after some days.

From the physiological point of view, adaptation is the most simple mechanism for vestibular sensitization under microgravity. Most receptor cells and neurons decrease their physiological activity after a steplike decrease of the stimulus level, but adapt to a more or less higher activity level during maintained stimulation (*phasic–tonic response pattern*). Vestibular neurons also decrease their resting activity transiently after they were deprived from labyrinthine input, but they recover it to higher levels thereafter (Precht, 1974).

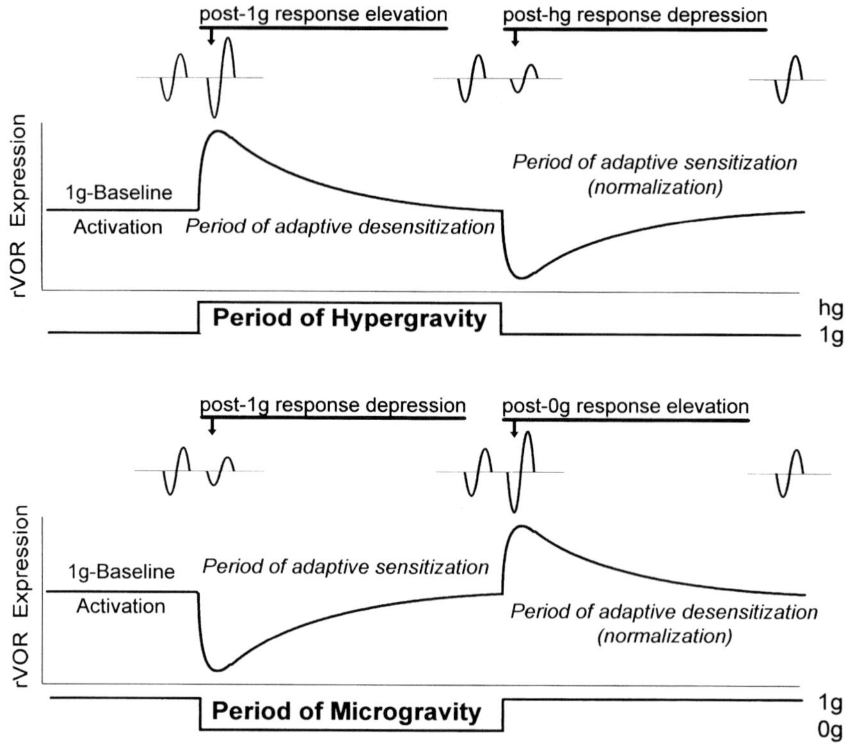

Fig. 11. Theoretical considerations about sensitization and desensitization of gravity sensitive systems during exposure to altered gravity. Upper: Exposure to hypergravity; lower: exposure to microgravity. Explanations within the text.

Similarly, macular afferents decrease their resting discharges after μg-onset (Bracchi *et al.*, 1975). During microgravity, neuroplastic properties of the vestibular system may increase its sensitivity (Fig. 11, lower). Macula activity originated by translatory accelerations during swimming or other movements (cf. Clément and Reschke, 1996) might initiate the increase of spontaneous activity, similar to the initiation of vestibular compensation following hemilabyrinthectomy (Dieringer, 1995). If neuroplastic capabilities are also used for the adaptation to hypergravity and if the underlying adaptive processes are guided by genetically pre-programmed set points within the nervous network to which gravity guided neuronal activity adjust during development—similar to those for the adaptation to microgravity—, animals with hg-experience have to show transiently lower response levels if they are tested in 1*g* (Fig. 11, upper).

Biochemical and structural studies point to possible mechanisms of the adaptive sensitization or desensitization. In fish (*F. heteroclitus*) creatine kinase activity—a biochemical marker for neuronal activity—was increased at primary vestibular projection sites of the cerebellar cortex after a spaceflight (Krasnow, 1977). In rats, the number of synapses of Type II hair cells was increased by 57% after μg-exposure (Ross, 1992), but decreased by 30% after 2*g*-exposure (Daunton *et al.*, 1991). In neonate swordtail fish *X. helleri*, the number of synapses within the vestibular N. descendens, a vestibular integration center, increased during a 16-day spaceflight (STS-90) while the vestibular N. magnocellularis which receives inputs from the lateral line and the visual N. corticalis were not affected (Ibsch *et al.*, 2000). If synapse numbers are directly correlated with the sensitivity of the vestibular system, changes such as these might explain sensitivity modifications.

In contrast to young *O. mossambicus*, depression of the rVOR was observed in zebrafish after simulated microgravity (Moorman *et al.*, 1999, 2002) and in *Xenopus* tadpoles after orbital microgravity which was explained by a retardation of rVOR development (Sebastian *et al.*, 1996). Retarded vestibular development was also postulated in fetal rats exposed to orbital microgravity (Fritzsch and Bruce, 1995). The opposite μg-effects in young fish and tadpoles do not exclude species-independent mechanisms responsible for the observed effects of gravity deprivation on rVOR development. A reasonable working hypothesis is that gravity deprivation retards rVOR development in very young stages when the development of the otic vesicle is just initiated, while neuronal adaptation might be activated in older stages when the sensory cells became sensitive to roll-stimulation.

Normalization by using input from non-gravity or accessory gravity sense organs

Modification of behavior and neurophysiological activity are mostly normalized after re-exposure to 1*g*-conditions. However, the time constants considerably differ not only between specific neuronal networks of one species but also between species. Different readaptation velocities between neuronal networks were demonstrated in *Xenopus* tadpoles. After termination of microgravity, loop-swimming (Neubert *et al.*, 1988) which is mediated by descending projections from the vestibular nuclei to the spinal cord (vestibulospinal pathway) disappeared earlier during 1*g*-readaptation than the rVOR depression which is mediated by vestibuloocular pathways (Sebastian and Horn, 1998). Balance compensation following unilateral lesion-induced labyrinthine deprivation was faster for the tadpole's swimming than for the eye posture (Rayer *et al.*, 1983). The complexity of normalization is also demonstrated by two other observations. In *Xenopus* tadpoles with 3*g*-experience, the duration of 1*g*-readaptation increased with decreasing test roll stimuli; i.e., during 3*g*-adaptation the vestibular system was modified so that the threshold for the detection of the adequate stimulus was increased. This modification persisted for many weeks (Fig. 12). Second, *Xenopus* tadpoles which developed slowly during 1*g*-readaptation maintained the depression of their rVOR for a

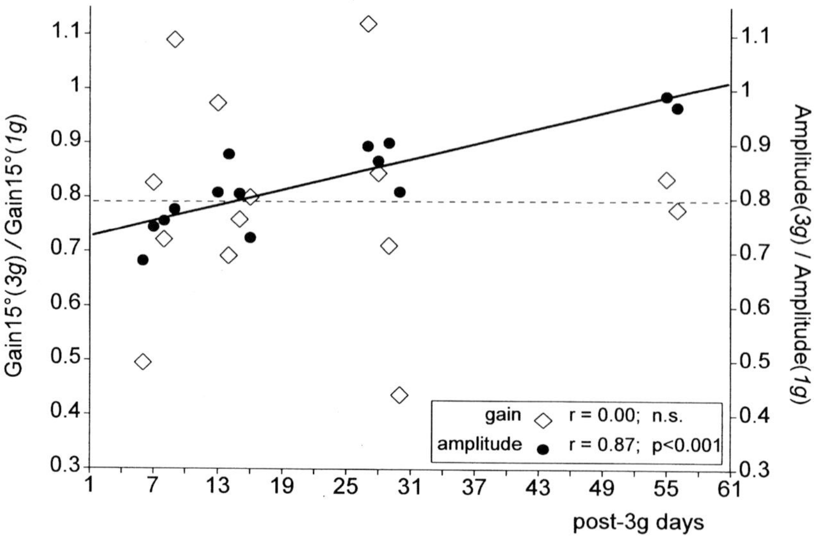

Fig. 12. Long-term observation during 1*g*-readaptation of tadpoles exposed to 3*g* for 9 days beginning approximately 2 weeks after fertilization (= developmental stage 45). Recordings started 1 day after the end of the 3*g*-period. Upper and middle plot: rVOR gain for a 15° lateral roll (right ordinate) with the corresponding angle of compensatory eye response (left ordinate), and rVOR amplitude. In both plots, dots represent values from 3*g*-tadpoles, circles those from 1*g*-controls. Levels of significances between 3*g*- and 1*g*-reared samples are indicated for each recording day. Lower plot: rVOR amplitudes (dots) and gains (rhombus) in tadpoles with 3*g*-experience as a percentage of those values recorded in their 1*g*-reared siblings (= 100%). Note that for a 15° roll stimulus, the ratio is maintained at a level of 80% throughout the experimental period, i.e., readaptation is absent during this time (modified from Sebastian and Horn, 1998).

longer time than those tadpoles which developed fast (Fig. 10, left) (Sebastian *et al.*, 1996).

Interspecific differences for the normalization of μg- or hg-induced behavioral or physiological modifications revealed an improvement of readaptation by the use of *input originating from non-gravity or accessory gravity sense organs*. This observation is known from lesion-induced effects; gravity related behavior compensates faster the more reference systems such as light, pressure and water or air flow are available (Schaefer and Meyer, 1974). The use of non-gravitational systems for the normalization of lesion-induced vestibular defects was called "adjustment" (Kornhuber, 1966). According to this concept, the faster rVOR-readaptation to 1*g* after exposure to altered gravity of the fish *O. mossambicus* compared to that in *X. laevis* is probably caused by the strong influence of the light on the orientation of fish. They use the convergence between the visual and vestibular system to stabilize posture during normal life and to normalize behavioral defects following labyrinthine lesions (von Holst, 1950; Schoen, 1950), while in *Xenopus* light plays only a minor role in these processes (Rayer *et al.*, 1983; Rayer and Horn, 1986). In the cricket *A. domesticus*, behavioral studies revealed no or only weak effects of micro- or hypergravity on the rCHR. These animals can use input from gravity receptors located in their antennae and legs (cf. Horn and Föller, 1985) because their projection pathways to the motor centers of the neck muscles are outside of the recording site used for the sensitivity studies of the PSI neuron (Förster, 2001; Riewe, 2001). Thus, from the comparative point of view, it is useful to consider no or weak effects as a very fast 1*g*-readaptation.

Perspectives

Developmental studies in general, but in particular those which analyze critical periods for gravity sensory and related motor systems need long-lasting exposures to microgravity. So far, there is a lack of Space verified animal holding facilities which fulfill this demand. Most of the existing facilities such as the ARF (Aquatic Research Facility; cf. Schatten *et al.*, 1999) developed by the Canadian Space Agency (CSA), the CEBAS incubator (Closed Equilibrated Biological Aquatic System; Blüm and Paris, 2001), the Biorack-miniaquarium (Sebastian *et al.*, 1998; Horn and Sebastian, 1999) or the BOTEX incubator (Botany Experiment Incubator; cf. Kern and Hock, 1994; Förster, 2001) all developed by the German Space Agency (DLR), or the various aquatic animal experiment facilities developed by the Japanese Space Agency (NASDA) (Sakimura *et al.*, 1999) offer the possibility for approximately 2-week exposures to altered gravity. However, most animals such as *Xenopus*, medaka fish, etc., need longer to develop from a fertilized egg to the adult animal. There is also a lack of hardware to perform multi-generation studies in species such as *Drosophila*.

The most attractive question concerning development on the systemic level of the organism is whether the all-over development of behavior and central nervous activity is insensitive to micro- or hypergravity exposure lasting many months, or not. In general, μg- and hg-induced behavioral or physiological modifications normalize during 1*g*-readaptation even if the normal development has not been completed. This observation is not necessarily expected but it points to the dominant influence of the genetic programs. Some observations revealed transient deviations from normal development in response to AG conditions which are overcome by regulatory mechanisms. No differences in development of frog eggs (*Rana pipiens*) were detected after a 45-h flight in the Biosatellite II at which microgravity started after the first cleavage (the stage most sensitive to inversion) although the orbited embryos were somewhat slower to reach certain stages of development (Edwards, 1969). In *Xenopus* embryos exposed to long- or short-term microgravity during spaceflights, a thickening of the blastocoel roof was observed (Souza *et al.*, 1995; De Maziere *et al.*, 1996; Ubbels, 1997). These modifications occurred only transiently, and animals developed to normal larvae (Souza *et al.*, 1995). Regulatory capabilities of embryos were also demonstrated in the amphibian *Pleurodeles waltl* in which neuronal precursor cells isolated in culture immediately after neural induction, differentiated in microgravity without any significant abnormalities (Husson *et al.*, 1998; cf. Duprat *et al.*, 1998). In an experiment performed on Neurolab (STS-90) it was shown that crickets regenerate a destroyed GSS even in the absence of gravity (Förster *et al.*, 2000; Horn, 2003). These observations favor the hypothesis that developmental abnormalities which are induced by gravity deprivation or augmentation are regulated to normality during altered gravity conditions. This might become possible by increasing regulatory efforts in case of developmental retardation, or vice versa, by decreasing their regulatory efforts in case of accelerated development.

Future studies in developmental system neurobiology have to be focused: (1) on the analysis of conditions that induce irreversible morphological, physiological and/or behavioral effects during exposure to modified gravity, i.e., to verify the existence or absence of a critical period during GSS development, and (2) on the search for mechanisms that allow stable life in orbit after expression of normal development or even in case of irreversible developmental modifications. Normal development from the egg to the adult is as likely as the manifestation of irreversible structural or physiological changes. It depends on the existence of regulatory mechanisms, which are based either on genetic programs or on the activation of compensatory responses within the developing system after detection of deviations from standard development. The ISS is the necessary platform of excellence if biological research is focussed on the analysis of long-term space effects on organisms.

Acknowledgements

The projects of the author were supported by grants from the German Space Agency (DLR) and the German Science Foundation (DFG). All experiments of the author comply with the "Principles of Animal Care", publication No. 86-23, revised 1985 of the National Institutes of Health, and with the "Deutsches Tierschutzgesetz", BGBl from February 17, 1993. Permissions for the experiments were given by the Regierungspräsidium of Tübingen (Germany), numbers 399/Ulm, 506/Ulm and 657/Ulm, and by the Animal Care and Use Committee (ACUC) at Kennedy Space Center/Florida. The author thanks all his former and recent co-workers, the members of the Astrium, Bionetics and OHB teams, and in particular the crew members from the missions D-2, SMM-06, Neurolab and Andromeda for making possible all the above mentioned studies.

References

Anken, R.H., Kappel, T., Slenzka, K. and Rahmann, H. (1993) The early morphogenetic development of the cichlid fish *Oreochromis mossambicus* (Perciformes, Teleostei). Zool. Anz. 231, 1–10.

Berthoz, A., Jeannerod, M., Vital-Durand, F. and Oliveras, J.L. (1975) Development of vestibulo-ocular responses in visually deprived kittens. Exp. Brain Res. 23, 425–442.

Blüm, V. and Paris, F. (2001) Aquatic modules for bioregenerative life support systems based on the C.E.B.A.S. biotechnology. Acta Astronautica 48, 287–297.

Böser, S. and Horn, E. (2002) Hypergravity exposure affects ventral root activity in tadpoles (*Xenopus laevis*). ESA SP (ESA-Symposium, Stockholm, 2002).

Böser, S., Dournon, C., Gualandris-Parisot, L. and Horn, E. (2002) Ventral root recordings during fictive swimming in tadpoles (*Xenopus laevis*) with microgravity or hypergravity experience. 18th ASGSB-Conference 2002, Cape Canaveral.

Boyle, R., Mensinger, A.F., Yoshida, K., Usui, S., Intravaia, A., Tricas, T. and Highstein, S.M. (2000) Neural readaptation to Earth's gravity following return from space. J. Neurophysiol. 86, 2118–2122.

Bracchi, F., Gualierotti, T., Morabito, A. and Rocca, E. (1975) Multiday recordings from the primary neurons of the statoreceptors of the labyrinth of the bullfrog. Acta Otolaryngol. (Stockh) Suppl 334, 3–27.

Brueckner, J.K., Ashby, L.P., Prichard, J.R. and Porter, J.D. (1999) Vestibulo-ocular pathways modulate extraocular muscle myosin expression patterns. Cell Tissue Res. 295, 477–484.

Buckey, J.C. Jr. and Homick, J.L. (Eds.) (2002) *The Neurolab Spacelab Mission: Neuroscience Research in Space*. NASA SP-535, Huston, Texas.

Budelmann, B.U. (1970) Die Arbeitsweise der Statolithenorgane von *Octopus vulgaris*. Z. Vergl. Physiol. 70, 278–312.

Clément, G. and Reschke, M.F. (1996) Neurosensory and sensory-motor functions. pp. 178–258. *In* Moore, D., Bie, P. and Oser, H. (eds.) *Biological and Medical Research in Space*. Springer Verlag, Berlin, Heidelberg, New York.

Clément, G., Moore, S.T., Raphan, T. and Cohen, B. (2001) Perception of tilt (somatogravic illusion) in response to sustained linear acceleration during space flight. Exp. Brain Res. 138, 410–418.

Cohen, B. (1974) The vestibulo-ocular reflex arc. *In* Kornhuber, H.H. (ed.)Vestibular System. Part 1: Basic Mechanisms. pp. 477–540 *In* "*Handbook of Sensory Physiology*" Vol. IV/1, Springer Verlag, Berlin, Heidelberg, New York.

Dale, N. (1995) Experimentally derived model for the locomotor pattern generator in the *Xenopus* embryo. J. Physiol. 489, 489–510.

Daunton, N.G., Ross, M.D., Fox, R.A., Corcoran, M.L., Cutler, L.K. and Wu, L.C. (1991) Effects of chronic hypergravity on the righting reflex and vestibular endorgan morphology in the rat. Soc. Neurosci. Abstr. 17, 316.

De Maziere, A., Gonzalez-Jurado, J., Reijnen, M., Narraway, J. and Ubbels, G.A. (1996) Transient effects of microgravity on early embryos of *Xenopus laevis*. Adv. Space Res. 17, 219–223.

Dews, P.B. and Wiesel, T.N. (1970) Consequences of monocular deprivation on visual behavior in kittens. J. Physiol. 206, 437–455.

Dichgans, J. and Brandt, T. (1978) Visual–vestibular interaction: effects on self-motion perception and postural control. pp. 755–804. *In* Held, R., Leibowitz, H.W. and Teuber, H.L. (eds.) *Perception*. Springer Verlag, Berlin, Heidelberg, New York.

Dieringer, N. (1995) 'Vestibular compensation': Neural plasticity and its relations to functional recovery after labyrinthine lesions in frogs and other vertebrates. Prog. Neurobiol. 46, 97–129.

Duprat, A.M., Husson, D. and Gualandris-Parisot, L. (1998) Does gravity influence the early stages of the development of the nervous system in an amphibian?. Brain Res. Rev. 28, 19–24.

Edwards, B.F. (1969) Weightlessness experiments on Biosatellite II. Life Sci. Space Res. 7, 84–92.

Etard, O., Hardy, O., Reber, A., Quarck, G., Denise, P. and Corvisier, J. (1999) Hyper- and hypogravity alter posture in rats compensated on Earth for a vestibular asymmetry. Neuroreport 10, 669–673.

Fawcett, J.W. (1988) Retinotopic maps, cell death, and electrical activity in the retinotectal and retinocollicular projections. pp. 395–416. *In* Parnavelas, J.G., Stern, C.D. and Stirling, R.V. (eds.) *The Making of the Nervous System*. Oxford University Press, Oxford, New York, Tokyo.

Fejtek, M., Souza, K., Neff, A. and Wassersug, R. (1998) Swimming kinematics and respiratory behavior of *Xenopus laevis* larvae raised in altered gravity. J. Exp. Biol. 201, 1917–1926.

Förster, S., Sebastian, C., Böser, S., Riewe, P., Horn, E., Kämper, G. and Agricola, H. (2000) Was sagen uns Weltraum-Grillen über die Entwicklung des Nervensystems?, pp. 482–492 *in* Sahm, P.R. and Keller, M.H. (Eds.) *Forschung unter Weltraumbedingungen*. Köln, WPF, DLR.

Förster, S. (2001) Einfluss der Schwerkraft auf die postembryonale Entwicklung des Schweresinns bei Grillen (*Acheta domesticus*). Dissertation, Universität Karlsruhe.

Fritzsch, B. and Bruce, L.L. (1995) Utricular and saccular projections of fetal rats raised in normal gravity and microgravity. ASGSB Bull. 9, 97.

Hilbig, R., Anken, R.H., Sonntag, G., Kretschmer, N., Bäuerle, A., Baumhauer, N., Knappenberger, M. and Rahmann, H. (2001) Otolith asymmetry and inner ear morphology of motion sick fish: a parabolic aircraft flight study. Abstract, 12th ELGRA-Symposium 2001, Banyuls-sur-mer, France.

Hoffman, R.B., Boyd, S.J., von Baumgarten, R.J. and Baky, A.A. (1977) Behavioral analyses of killifish exposed to weightlessness in the Apollo-Soyuz test project. Aviat. Space Environ. Med. 48, 712–717.

Hoffman, R.B., Boyd, S.J., von Baumgarten, R.J. and Baky, A.A. (1978) Effect of prehatching weightlessness on adult fish behavior in dynamic environments. Aviat. Space Environ. Med. 49, 576–581.

Horn, E. (1973) Die Verarbeitung des Schwerereizes bei der Geotaxis der höheren Bienen (Apidae). J. Comp. Physiol. 82, 379–406.

Horn, E. (1985) Funktionelle Entwicklung der Schweresinnesorgane bei Insekten und Vertebraten. DFVLR-Mitt. 85(16), 121–128.

Horn, E. (1985) Gravity. pp. 557–576. *In* Kerkut, G.A. and Gilbert, L.I. (eds.) *Comprehensive Insect Physiology, Biochemistry and Pharmacology, Vol. 6: Nervous System: Sensory*. Pergamon Press, Oxford, New York, Paris, Frankfurt, Sidney, Toronto.

Horn, E. (1985) Multimodal convergence. pp. 653–671. *In* Kerkut, G.A. and Gilbert, L.I. (eds.) *Comprehensive Insect Physiology, Biochemistry and Pharmacology, Vol. 6: Nervous System: Sensory*. Pergamon Press, Oxford, New York, Paris, Frankfurt, Sidney, Toronto.

Horn, E. (1987) Functional development of gravity receptors in insects and vertebrates. *ESA TT*-988, 91–98.

Horn, E. (2002) The development of an insect gravity sensory system in space—Neurolab Experiment 089 CRISP (Crickets in Space), pp. (in press) *in* Buckey, J.C. Jr. and Homick, J.L. (Eds.) *The Neurolab Spacelab Mission: Neuroscience Research in Space*. NASA SP-535, Huston, Texas.

Horn, E. and Bischof, H.J. (1983) Gravity reception in crickets: the influence of cercal and antennal afferences on the head position. J. Comp. Physiol. 150, 93–98.

Horn, E. and Föller, W. (1985) Tonic and modulatory subsystems of the complex gravity receptor system in crickets, *Gryllus bimaculatus*. J. Insect Physiol. 31, 937–946.

Horn, E. and Föller, W. (1998) Induction of a gravity-related response by a single receptor cell in an insect. Naturwissenschaften 85, 121–123.

Horn, E. and Föller, W. (2001) Functional regeneration of a gravity sensory system during development in an insect (*Gryllus bimaculatus*). NeuroReport 12, 2685–2691.

Horn, E. and Lang, H.G. (1978) Positional head reflexes and the role of the prosternal organ in the walking fly, *Calliphora erythrocephala*. J. Comp. Physiol. 126, 137–146.

Horn, E. and Sebastian, C. (1996) A hypergravity related sensitive period during the development of the roll induced vestibuloocular reflex in an amphibian (*Xenopus laevis*). Neurosci. Lett. 216, 25–28.

Horn, E. and Sebastian, C. (1999) A comparison of normal vestibulo-ocular reflex development under gravity and in the absence of gravity. *ESA SP*-1222, 127–138.

Horn, E., Lang, H.G. and Rayer, B. (1986) The development of the static vestibulo-ocular reflex in the Southern Clawed Toad, *Xenopus laevis* Daudin: I. Intact animals. J. Comp. Physiol. 159A, 869–878.

Horn, E., Mack, R. and Lang, H.G. (1986) The development of the static vestibulo-ocular reflex in the Southern Clawed Toad, *Xenopus laevis* Daudin: II. Animals with acute vestibular lesions. J. Comp. Physiol. 159A, 879–885.

Horn, E., Böser, S., Förster, S., Riewe, P., Sebastian, C. and Agricola, H. (2001) Crickets in Space. Acta Astronautica 49, 345–363.

Horn, E., Gualandris-Parisot, L., Dournon, C. and Böser, S. (2002) The static vestibulocular reflex and its relation to fictive swimming in tadpoles (*Xenopus laevis*) after microgravity exposure. 18th ASGSB-Conference 2002, Cape Canaveral.

Hubel, D.H. and Wiesel, T.N. (1970) The period of susceptibility to the physiological effects of unilateral eye closure in kittens. J. Physiol. 206, 419–436.
Husson, D., Gualandris-Parisot, L., Foulquier, F., Grinfeld, S., Kann, P. and Duprat, A.M. (1998) Differentiation in microgravity of neural and muscle cells of a vertebrate (amphibian). Adv. Space Res. 22, 303–308.
Ibsch, M., Anken, R.H. and Rahmann, H. (2000) Weightlessness during spaceflight results in enhanced synapse formation in a fish brain vestibular nucleus. Neurosci. Lett. 296, 13–16.
Jones, T.A., Fermin, C., Hester, P.Y. and Vellinger, J. (1993) Effects of microgravity on vestibular ontogeny: direct physiological and anatomical measurements following space flight (STS-29). Acta Vet. Brno 62(Suppl), S35–S42.
Jander, R., Horn, E. and Hoffmann, M. (1970) Die Bedeutung von Gelenkreceptoren in den Beinen für die Geotaxis der höheren Insekten (Pterygota). Z. Vergl. Physiol. 66, 326–354.
Kahn, J.A., Roberts, A. and Kashin, S.M. (1982) The neuromuscular basis of swimming movements in embryos of the amphibian *Xenopus laevis*. J. Exp. Biol. 99, 175–184.
Kahn, J.A. and Roberts, A. (1982) The central nervous origin of the swimming motor pattern in embryos of *Xenopus laevis*. J. Exp. Biol. 99, 185–196.
Kern, V.D. and Hock, B. (1994) Gravitropism of fungi—experiments in Space. ESA SP-366, 49–60.
Knudsen, E.I., Knudsen, P.F. and Esterley, S.D. (1984) A critical period for the recovery of sound localization accuracy following monoaural occlusion in the barn owl. J. Neurosci. 4, 1012–1020.
Kornhuber, H.H. (1966) Physiologie und Klinik des zentralvestibulären Systems (Blick- und Stützmotorik), pp. 2150–2351 *in* Berendes, J., Link, R. and Zöllner, F. (Eds.) *Hals-, Nasen- und Ohrenheilkunde, Vol III, Part 3*. Stuttgart, Georg Thieme Verlag.
Krasnow, I.B. (1977) Quantitative histochemistry of the vestibular cerebellum of the fish *Fundulus heteroclitus* flown aboard the biosatellite COSMOS-782. Aviat. Space Environ. Med. 48, 808–811.
Lichtenberger, B.K., Young, L.R. and Arrott, A.P. (1982) Human ocular counter rolling indcued by varying linear accelerations. Exp. Brain Res. 48, 127–136.
Massion, J., Fabre, J.C., Mouchino, L. and Obadia, A. (1995) Body orientation and regulation of the center of gravity during movement under water. J. Vestib. Res. 5, 211–221.
Meyer, D.L., Heiligenberg, W. and Bullock, T.H. (1976) The ventral substrate response. A new postural control mechanism in fishes. J. Comp. Physiol. 109, 59–68.
Moffett, S.B. (1996) *Nervous System Regeneration in the Invertebrates*. pp 208. Springer-Verlag, Berlin, Heidelberg, New York.
Moorman, S.J., Burress, C., Cordova, R. and Slater, J. (1999) Stimulus dependence of the development of the zebrafish (*Danio rerio*) vestibular system. J. Neurobiol. 38, 247–258.

Moorman, S.J., Cordova, R. and Davies, S.A. (2002) A critical period for functional vestibular development in zebrafish. Dev. Dyn. 223, 285–291.

Neubert, J., Briegleb, W., Schatz, A., Hertwig, I. and Kruse, B. (1988) The response of structure and function of the gravireceptor in a vertebrate to near weightlessness. Acta Astronautica 17, 257–262.

Neubert, J., Schatz, A., Bromeis, B. and Briegleb, W. (1994) The reaction of *Xenopus laevis* Daudin (South African toad) to linear accelerations. Adv. Space Res. 14, 299–303.

Nieuwkoop, P.D. and Faber, J. (1967) *Normal Table of Xenopus laevis (Daudin)*. pp 263. North Holland, Amsterdam.

Oakley, B. (1993) The gustatory competence of the lingual epithelium requires neonatal innervation. Dev. Brain Res. 72, 259–264.

Precht, W. (1974) Physiological aspects of the efferent vestibular system. pp. 98–109. *In* Kornhuber, H.H. (ed.) *Vestibular System. Part 1: Basic Mechanisms*" *In* "*Handbook of Sensory Physiology*" Vol. IV/1, Springer Verlag, Berlin, Heidelberg, New York.

Precht, W. and Dieringer, N. (1985) Neuronal events paralleling functional recovery (compensation) following peripheral vestibular lesions. pp. 251–268. *In* Berthoz, A. and Melvill Jones, G. (eds.) *Adaptive Mechanisms in Gaze Control. Facts and Theories*. Elsevier Science Publ BV, Amsterdam, New York, Oxford.

Pronych, S. and Wassersug, R.J. (1994) Lung use and development in *Xenopus laevis* tadpoles. Can. J. Zool. 72, 738–743.

Rayer, B., Cagol, E. and Horn, E. (1983) Compensation of vestibular-induced deficits in relation to the development of the Southern Clawed Toad, *Xenopus laevis* DAUDIN. J. Comp. Physiol. 151, 487–498.

Rayer, B. and Horn, E. (1986) The development of the static vestibulo-ocular reflex in the Southern Clawed Toad, *Xenopus laevis* Daudin: III. Chronic hemilabyrinthectomized tadpoles. J. Comp. Physiol. 159A, 887–895.

Riewe, P. (2001) Gravitationsphysiologie der Grille: Neuronale Änderungen in der Entwicklung des cercalen Schweresinns unter veränderten Gravitationsbedingungen. Dissertation, Universität Ulm.

Ris, L. and Godaux, E. (1998) Neuronal activity in the vestibular nuclei after contralateral or bilateral labyrinthectomy in the alert guinea pig. J. Neurophysiol. 80, 2352–2367.

Ross, M.D. (1992) Synaptic plasticity in utricular maculas of rats exposed to microgravity. ASGSB Bull. 6, 100.

Ryu, J.H. and McCabe, B.F. (1976) Central vestibular compensation. Effects of the bilateral labyrinthectomy on neural activity in the medial vestibular nucleus. Arch. Otolaryngol. 102, 71–76.

Sakaguchi, D.S. and Murphey, R.K. (1983) The equilibrium detecting system of the cricket: physiology and morphology of an identified interneuron. J. Comp. Physiol. 150, 141–152.

Sakimura, T., Suzuki, T., Matsubara, S., Uchida, S., Kato, M., Tanemura, R. and Honda, S. (1999) NASDA aquatic animal experiment facilities for space shuttle. Biol. Sci. Space 13, 314–320.

Schaefer, K.P. and Meyer, D.L. (1974) Compensation of vestibular lesions. *In* Kornhuber, H.H. (ed.) pp. 463–490. Vestibular System. Part 1: Basic Mechanisms" *In* "*Handbook of Sensory Physiology*" Vol. IV/1, Springer Verlag, Berlin, Heidelberg, New York.

Schatten, H., Chakrabarti, A., Levine, H.G. and Anderson, K. (1999) Utilization of the aquatic research facility and fertilization syringe unit to study sea urchin development in space. J. Gravit. Physiol. 6, 43–53.

Schmidt, J.T. (1988) Activity-dependent sharpening of the retinotopic projection on the tetum of goldfish in regeneration and development. pp. 380–394. *In* Parnavelas, J.G., Stern, C.D. and Stirling, R.V. (eds.) *The Making of the Nervous System*. Oxford University Press, Oxford, New York, Tokyo.

Schoen, L. (1950) Qantitative Untersuchungen über die zentrale Kompensation nach einseitiger Utriculusausschaltung bei Fischen. Z. Vergl. Physiol. 32, 121–150.

Schöne, H. (1954) Statocystenfunktion und statische Lageorientierung bei decapoden Krebsen. Z. Vergl. Physiol. 36, 241–260.

Sebastian, C. and Horn, E. (1998) The minimum duration of microgravity experience during space flight which affects the development of the roll induced vestibuloocular reflex in an amphibian. Neurosci. Lett. 253, 171–174.

Sebastian, C. and Horn, E. (1999) Light-dependent suppression of the vestibuloocular reflex during development. NeuroReport 10, 171–176.

Sebastian, C. and Horn, E. (2001) Features of vestibuloocular reflex modulations induced by altered gravitational forces in tadpoles (*Xenopus laevis*). Adv. Space Res. 28, 579–588.

Sebastian, C., Eßeling, K. and Horn, E. (1996) Altered gravitational experience during early periods of life affects the static vestibulo-ocular reflex of tadpoles of the Southern Clawed Toad, *Xenopus laevis* Daudin. Exp. Brain Res. 112, 213–222.

Sebastian, C., Pfau, K. and Horn, E. (1998) An age-dependent sensitivity of the roll-induced vestibuloocular reflex to hypergravity exposure of several days in an amphibian (*Xenopus laevis*). Acta Astronautica 42, 419–430.

Sebastian, C., Eßeling, K. and Horn, E. (2001) Altered gravitational forces affect the development of the static vestibuloocular reflex in fish (*Oreochromis mossambicus*). J. Neurobiol. 46, 59–72.

Smith, P.F. and Curthoys, I.S. (1989) Mechanisms of recovery following unilateral labyrinthectomy: a review. Brain Res. Rev. 14, 155–180.

Snetkova, E., Chelnaya, N., Serova, L., Saveliev, S., Cherdanzova, E., Pronych, S. and Wassersug, R.J. (1995) The effects of space flight on *Xenopus laevis* larval development. J. Exp. Zool. 273, 21–32.

Souza, K.A., Black, S.D. and Wassersug, R.J. (1995) Amphibian development in the virtual absence of gravity. Proc. Natl. Acad. Sci. USA 92, 1975–1978.

Ubbels, G.A. (1997) Fertilization and development of *Xenopus* eggs on sounding rockets and in a clinostat. ESA SP-1206, 25–36.

van der Loos, H. and Woolsey, T.A. (1973) Somatosensory cortex: structural alterations following early injury to sense organs. Science 179, 395–398.

Vinnikov, J.A., Gazenko, O.G., Lychakov, D.V. and Palmbach, L.R. (1983) Formation of the vestibular apparatus in weightlessness. pp. 537–560. *In* Romand, R. (ed.) *Development of Auditory and Vestibular Systems*. Academic Press, New York, London, Paris.

von Baumgarten, R.J., Simmonds, R.C., Boyd, J.F. and Garriott, O.K. (1975) Effects of prolonged weightlessness on the swimming pattern of fish aboard Skylab 3. Aviat. Space Environ. Med. 46, 902–906.

von Holst, E. (1950) Die Arbeitsweise des Statolithenapparates bei Fischen. Z. Vergl. Physiol. 32, 60–120.

Waespe, W., Schwarz, U. and Wolfensberger, M. (1992) Firing characteristics of vestibular nuclei neurons in the alert monkey after bilateral vestibular neurectomy. Exp. Brain Res. 89, 311–322.

Walton, K. (1998) Postnatal development under conditions of simulated weightlessness and space flight. Brain Res. Rev. 28, 25–34.

Wassersug, R.J. and Murphy, A.M. (1987) Aerial respiration facilitates growth in suspension-feeding anuran larvae (*Xenopus laevis*). Exp. Biol. 46, 141–147.

Wassersug, R.J. and Souza, K.A. (1990) The bronchial diverticula of *Xenopus larvae*: are they essential for hydrostatic assessment?. Naturwissenschaften 77, 443–445.

Wendler, G. (1975) Physiology and system analysis of gravity orientation in two insect species (*Caurausius morosus*, *Calandra granaria*). pp. 33–48. *In* Schöne, H. (ed.) *Mechanisms of Spatial Perception and Orientation as Related to Gravity*. Gustav Fischer Verlag, Stuttgart.

Wiesel, T. (1982) Postnatal development of the visual cortex and influence of environment. Nature 299, 583–591.

Developmental Biology Research in Space
H.-J. Marthy (editor)

Neurophysiology of Developing Fish at Altered Gravity: Background—Facts–Perspectives

Ralf H. Anken*

Zoological Institute, University of Stuttgart-Hohenheim, Garbenstr. 30, D-70593 Stuttgart, Germany

Abstract

During the entire evolution of life on Earth, the phylogenetic as well as the individual development of all organisms took place under constant gravity conditions, against which they achieved specific countermeasures for compensation and adaptation. On the one side, gravity represents a *factor of physical restriction*, which compelled the ancestors of all extant living beings to develop basic achievements to counter the gravitational force (e.g., elements of statics like any kind of skeleton—from actin to bone—to overcome gravity enforced size limits or to keep form). On the other side, already early forms of life possibly used gravity as an appropriate *cue for orientation and postural control*, since it is continuously present and has a fixed direction.

Due to such a thorough adaptation to the Earthly gravity vector, both orientation behaviour as well as the ontogenetic development of animals is impaired, when they have to experience altered gravity (Δg; i.e., hyper- or microgravity).

On this background, it is still an open question to which extent Δg affects the normal individual development, either on the systemic level of the whole organism or on the level of individual organs or even single cells. The present review provides information on these questions, focusing on developing fish as model systems. Special emphasis is being laid on the effect of Δg on the developing brain and vestibular system, comprising investigations on behaviour and plastic reactivities of the brain and inner ear. Moreover, clues and insights into the possible basic causes of space motion sickness-phenomena (SMS; a kinetosis) are provided.

Overall, the results speak in favour of the following concept: short-term altered gravity (≤ 1 day) can induce transitional aberrant behaviour due to malfunctions of the inner ear, originating from asymmetric otoliths or, generally, from a mismatch between canal and otolith afferents. The vanishing

* E-mail: anken@uni-hohenheim.de

aberrant behaviour is due to a reweighing of sensory inputs and neurovestibular compensation, probably on bioelectrical basis. During long-term altered gravity (several days and more), step by step neuroplastic reactivities on molecular basis (i.e., molecular facilitation) in the brain and inner ears obviously activate feedback mechanisms between the CNS and the vestibular organs for the regain of normal behaviour.

Mainly, the following areas of research with animals at altered gravity need to be addressed in the future:

(1) Maintenance of animals through two complete life cycles in the space environment (developmental deficiencies?).
(2) Investigation of the peripheral and central vestibular system by ground-based studies (mutants, hypergravity experiments...), focusing on plasticity in developing animals as well as in adults.
(3) Investigation of the effect of microgravity during critical developmental periods (imprinting phase for graviperception?).

Answers to these questions may be of crucial interest for basic gravitational research.

Background: The role of gravity during the phylogenetic development of animals

As a matter of fact, the non-linear self-organizing dynamics of biological systems are inherent in any physical theory that satisfies the requirements of both quantum mechanics and general relativity (Conrad, 1997). Gravity therefore has always been a challenge for biological systems to adapt or/and to cope with it. Concerning single cells, the earliest life forms, it has been stated that average cell size results, in part, from the physical equilibrium between the destructive influence of the force of gravity and the protective role of diffusion and the cytoskeleton (Moroz, 1984). At increased forces of gravity the cell size would thus be decreased, whereas at lower gravitational forces and weightlessness cell size would be expected to increase. Mechanisms of protection of giant cells against internal sedimentation are based on protoplasmic motion, thin and elongated shape of the cell body, increased cytoplasmic viscosity, and a reduced range of specific gravity of cell components, relative to the groundplasm. The nucleolus, due to its higher density, is considered as a possible trigger of mitosis. Although gravity limits the size even of single cells, its impact became especially apparent with the evolution of multicellular animals. There is not much known about the first multicellular animals inhabiting our planet in the late Precambrium (earlier than 570 millions of years—Myr—from now) prior to the so-called "Cambrian explosion", which showed an extremely rapid evolutionary radiation with the development of almost all nowadays phyla of invertebrate animals. With only a few exceptions, the Precambrian animals (e.g., the forms of the Ediacara-fauna, named after the Ediacara Hills north of Adelaide, Australia) did not have any sort of inner or outer skeleton. All of them were small and many species had a worm- or jellyfish-like appearance. In most cases, their relationships to the nowadays present invertebrate groups is unclear.

Obviously, elaborate anti-gravity systems had not yet been fully evolved, which would have allowed these animals to grow larger, to develop a directed locomotion and even to cope with the terrestrial impact of gravity at the stage of their exit from water to land. Development of a directed locomotion might have been one of the most important evolutionary inventions. When heterotrophic animal life decreased the nutrients (e.g., autotrophic plants) in the oceans, animals were forced to cope with this evolutionary pressure; directed locomotion was therefore developed to predate other animals or, vice versa, to escape from predators. For exercising directed locomotion especially gravity was—besides other environmental factors such as radiation (especially light), atmospheric conditions/composition, sound and electromagnetic as well as mechanical impacts—one of the most important morphogenetic factors of animal evolution which pushed the gene to elaborate adequate mechanisms for surmounting it (Vinnikov, 1995). Directed locomotion generally requires any sort of skeleton to allow the insertion of muscles; such a skeleton then could (pre-adaptively) act as a prerequisite for animals to turn from their aquatic habitat to a terrestrial life some 440 Myr ago, then following the green plants as a further source for their heterotrophic lifestyle (the first terrestrial animals were early ancestors of our nowadays spiders, belonging to the arthropods).

The first vertebrate animals which were able to cope with the terrestrial impact of Earth's gravity were early ancestors of fish some 350–400 Myr ago. Concerning modern bluefish, it has been found that these animals can accelerate at 3*g* during swimming and that the vertebral column is strong enough to withstand this force (DuBois *et al.*, 1976). This finding strongly indicates that the muscles and body structure of a bluefish would be able to withstand the force of gravity if the fish were otherwise equipped for terrestrial life (hypergravity experiments revealed in this context that development of larval fish is not impaired by 3*g*). One can therefore speculate that early fish may also have evolved some degree of strength to overcome gravity-based inertia and drag during aquatic locomotion, and this evolution may have been a prelude to vertebrate terrestrial locomotion.

Summarized, it was most probably the evolutionary pressure to develop directed locomotion which then, as a sort of prelude, allowed animals to cope with terrestrial gravity. Active, directed locomotion, and especially active maintenance of equilibrium during bodily movement (e.g., in locomotion) requires, however, appropriate sensory functions. Although many animals usually maintain their bodies with the long axis horizontal (backside up), humans being a notable exception, there are frequent departures from the usual position. A fish may dive steeply downward and a man may alter his normal orientation by lying down at full length. In no case, however, need there be any loss of equilibrium. Every deviation means an equilibrium disturbance and evokes compensatory reflex movements.

Gravity therefore did not only act as a factor of physical restriction, but was an environmental factor readily being available as an appropriate cue for orientation and postural control.

Facts: Gravity as a cue for orientation and postural control

Maintenance of equilibrium is based upon contact of the animal with the external world; several sensory systems may play a role in this context. When an animal moves over a solid surface, tactile stimuli usually predominate as cues. It has to be noted that also proprioceptors (i.e., sense organs allowing the perception of stimuli relating to the animal's own position, posture, equilibrium, or internal condition) in vertebrates and arthropods can also contribute to spatial orientation; bodily tissues like the club-shaped sensilla of arthropods (Fig. 1) under gravity weigh vertically down and stimulate internal mechanoreceptors in a way that depends on, and varies with, the animal's spatial position. When they are out of contact with the ground, many animals orient themselves in space by keeping their back (dorsal) side turned up toward the light, e.g., in the course of the dorsal light response (DLR) of fish. Visual cues also can serve equilibration; for example, through compensatory body movements

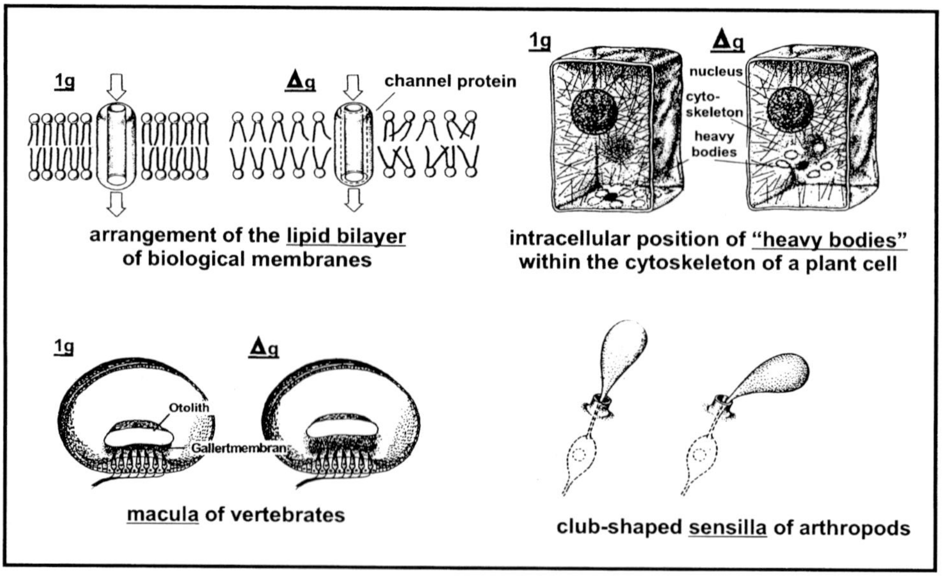

Fig. 1. Possibilities of graviperception in biological systems. Already (simulated) biological membranes can respond to the gravity vector. Plants and most animals use "heavy bodies" (called statoliths in plants, invertebrates and most vertebrates; the compact statoliths of fish are called otoliths), whereas arthropods use body extensions to perceive gravity.

(optomotor reflexes like the vestibulo-ocular reflex in fish and amphibians (Sebastian *et al.*, 1996; Horn, 2003) brought about by the shifts of the image of the environment over the retina of the eye. For the receptors mentioned thus far, however, equilibration is not the unique function. There are other sensory structures that are genuine organs of equilibrium in that they primarily and exclusively serve orientation of posture and movement in space: gravity receptors.

In general, gravity—or any sort of acceleration—can be transformed to a biological signal in several different ways (Fig. 1): already synthetic membrane bilayers containing incorporated ion-channels respond to gravity (Schatz *et al.*, 1994; Hanke, 1996). The exact signal-transduction chain remains, however, hitherto unresolved. Unicellular protozoans (e.g., *Paramecium*) perceive gravity obviously by their membrane or have already developed minute, intracellular crystals that function as "heavy bodies" (*Loxodes*).

Graviperception in unicellular animals

Already archaic extantly living eukaryotic organisms (particularly phyto-flagellates and ciliates) were shown to be gravitactic (cell orientation) or even gravikinetic (adjustment of swimming speed) (for review: Machemer, 1999). In the ciliate *Paramecium*, gravikinesis is obviously regulated by a pressure-gradient between the membrane and the surrounding aqueous medium. This gradient is only at equilibrium when the animal swims horizontally. Vertical tilts result in the opening/closing of mechano-sensitive ion-channels, which effect a local de- or hyperpolarisation which in turn selectively activates cilia for propulsion. In contrast to *Paramecium*, another ciliate—*Loxodes*—seems to perceive gravity via particular intracellular organelles, called "Müller-bodies", which consist of a membraneous pouch, containing a "heavy body" or "statolith" of $BaSO_4$ (Neugebauer and Machemer, 1997; Fig. 2).

Graviperception in multicellular animals

All multicellular animals hitherto analysed, which exhibit active (not necessarily also directed) locomotion, use so-called stato- or otoliths, being localized in differently specialized organs (e.g., the vestibular organs within the inner ear of vertebrates) or body extensions (e.g., club-shaped halters and sensilla of insects; i.e., proprioceptors, see above) for the transformation of an acceleration into a body-own signal.

Thus, in accomplishment of the "heavy body"-strategy of the unicellular protozoans, some fungi, some lower but most of the higher plants, almost all invertebrate animals and virtually all vertebrates have different types of heavy bodies to orientate themselves towards the direction of earth gravity.

Depending on the function–morphologic characteristics of the body, the respective sense organs (sensilla, inner ear maculae etc.) have been positioned in

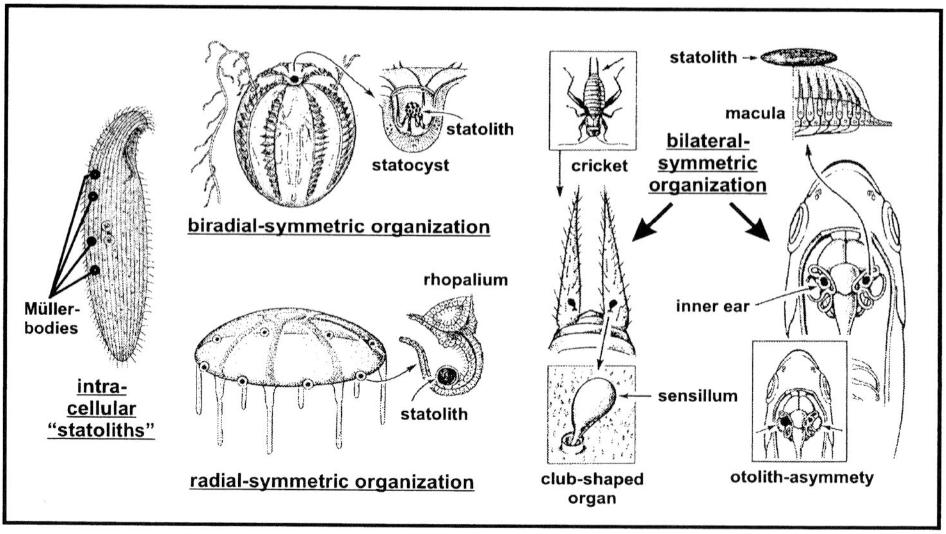

Fig. 2. Positioning of gravireceptors (for details, see text).

different ways during the phylogenetic development. Gravity receptors already appear in jellyfish. Ctenophoran jellyfish, e.g., are biradial–symmetric (two axes of symmetry) and possess one statocyst at the top of the body (Fig. 2). Radial–symmetric cnidarian jellyfish with multiple axes of symmetry have many statocysts located around the mantle (Fig. 2).

A bilateral-symmetric organization (one axis of symmetry) is found in most of the other animals. In both inner ears of vertebrates (Figs. 2 and 3), three semicircular canals (or cupular organs), located perpendicularly to each other for the detection of angular acceleration, are completed by three pouches containing either three oto-/statolith organs in lower vertebrates (utricle, saccule and lagena for the perception of linear acceleration/gravity as well as for sound with otoliths consisting mainly of calcium carbonate) or two statolith organs and a lagena/cochlea without a stato-/otolith for sound perception in higher vertebrate animals. Sound or the movement of a statolith results in the bending of hair cell-cilia, which mechanically opens or closes ion channels (Hudspeth and Gillespie, 1994) altering the electrical current of the respective sensory cell in the inner ear. Here, the transformation to computable action potentials takes place. Finally, a signal transduction on the level of the brain causes a motor response (Fig. 4). Generally, the brain integrates informations from the inner ear vestibular organs together with tactile, proprioceptive and visual cues for spatial orientation and postural control in the environment (Clément and Berthoz, 1994) (compare Fig. 5).

Not only in vertebrates but in all animals exhibiting a central nervous system, there is always a close relationship between the presence of highly developed sense organs and a specialized region of the brain, since

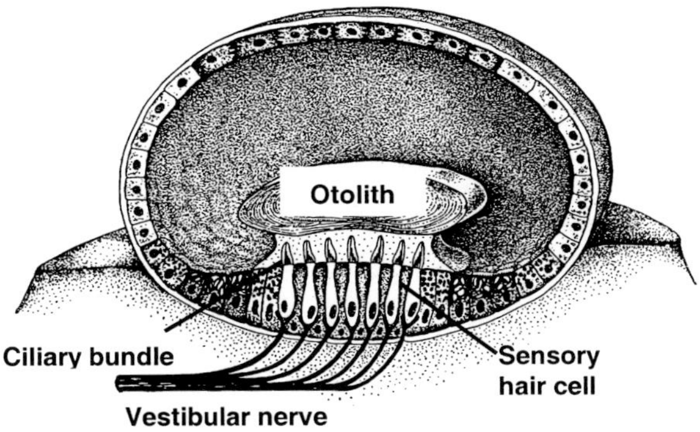

Fig. 3. Macula of a bony fish. Note the compact otolith. Amphibians, reptiles, birds and mammals reveal so-called statoliths, which are composed of hundreds of thousands of minute calcium carbonate crystals.

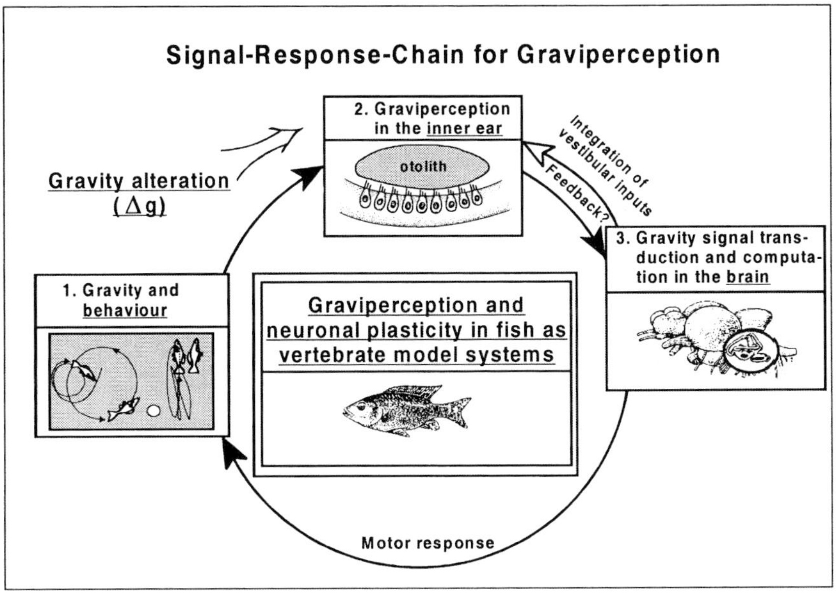

Fig. 4. The signal-response-chain for graviperception in vertebrate animals (fish). The basic principles of this response-chain (i.e., graviperception followed by a gravity signal transduction and computation, resulting in a motor response/behaviour) are found in all animals who exhibit a central nervous system.

the latter is needed to process the incoming information in order to abstract the cues of importance to a given animal, which lastly results in behaviour (compare Fig. 4). An example in vertebrates is their cerebellum, which is responsible for the control of postural control; it is especially large and efficient

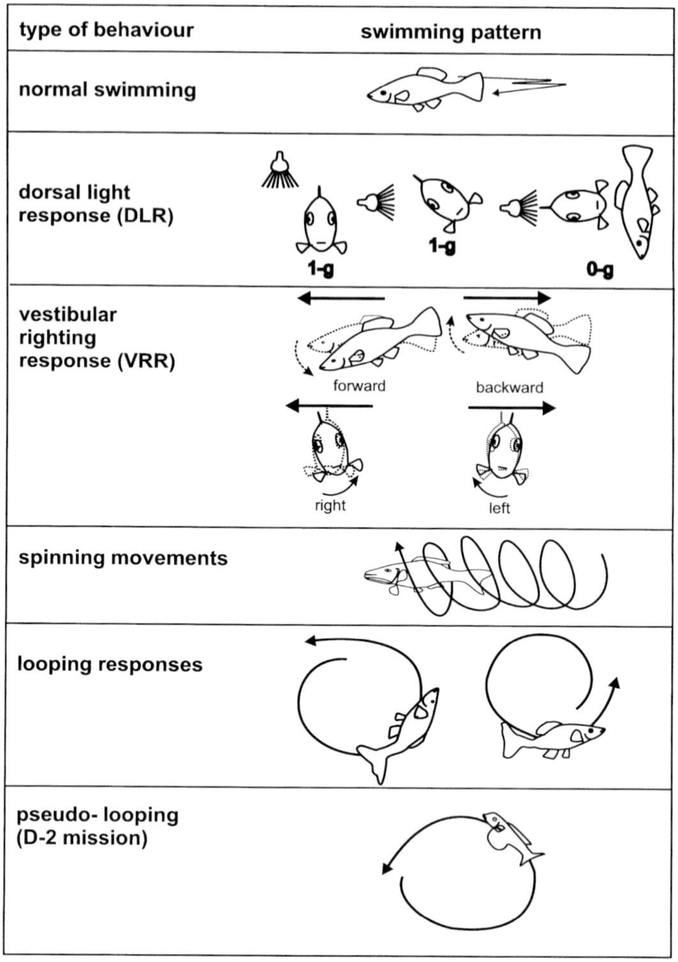

Fig. 5. Vestibular behaviour of fish. The DLR illustrates how a fish integrates visual and vestibular cues for spatial orientation and postural control. The VRR is the vestibular component of the DLR. Spinning movements and looping responses are kinetoses which frequently appear when the environmental gravity vector is diminished (e.g., at the transition from 1*g* to microgravity or from hypergravity to 1*g*). The pseudo-looping has not a vestibular origin but directly results from a microgravity environment: Without weight, yolk-sac larvae beating with their pectoral fins will inevitably move around in circles.

in such animals which need to orient themselves in all three directions of space, like birds. (The fact that such elaborate systems exist does not exclude the possibility of much shorter and simpler pathways, which provide for more localized and quicker reactions; for instance, gravireceptor-mediation of postural control in Comb Jellies seems not even to require neurons.)

Human space motion sickness

The term "motion sickness" is commonly used for the sensation of malaise (or "nausea", often including vomiting) occurring at the change of direction and/or amplitude of the acceleration vector (constant linear motion is not nauseogenic) in human subjects. Individuals frequently suffer such motion sicknesses (kinetoses) when moving or turning the head aboard boats, in cars and in aircrafts.

The respective illness experienced by many astronauts when moving at weightlessness conditions was termed "space motion sickness" (SMS). Some astronauts, however, reported to become "motion sick" during periods of rest. The terms "static SMS" and "dynamic SMS" account on these two different types of space sickness. Since "static motion" sounds somewhat contradictionary, some authors prefer to term static SMS as "vestibular space sickness" (e.g., v. Baumgarten, 1986).

SMS is an important operational problem, since the sensorymotor performance of an affected astronaut is severely impaired especially in the first 2–3 days of a space mission. Later, the astronauts adapt themselves towards the altered gravitational environment ("space adaptation syndrome"). Back at 1g normal earth gravity, a readaptation takes place, also accompanied by sensorymotor disorders. The readaptation period normally lasts a few days, depending on the duration of the space mission.

Of course human subjects are not amenable to invasive research techniques for studying the neurobiological basis of SMS and related kinetoses. Other vertebrates are therefore used as model systems since their peripheral and central vestibular system—at least at the level of primary vestibular brain nuclei—is largely homologous to that of humans.

Summarized, animals did not only acquire systems to cope with gravity during evolution, but they also evolved receptors (and respective brain parts) in order to use gravity as the only virtually constant environmental factor for orientation, maintenance of equilibrium and postural control. On the one hand side, a gravitational environment unexperienced by humans and animals therefore disturbs the neuronal processing of incoming information (especially when—at altered gravitational sensations such as hyper- or microgravity—the stato- or otolithic organs in some individuals transmit informations to the brain, which do not anymore necessarily completely agree with, e.g., the visual cues needed for a correct postural control), and "normal behaviour" is inevitably affected. In human subjects, such an "intersensory conflict" can result in orientation problems, often accompanied by motion sickness and vomiting (kinetosis). On the other hand side, Δg may have a strong influence on developmental aspects, since all systems (from sub-cellular organelles to complete organs) should be fully adapted to normal Earth gravity.

In order to understand the role of gravity on the (ontogenetic) development of animals as well as on their behaviour, experiments at Δg yielded valuable clues and insights especially using fish as vertebrate model systems (compare also Rahmann and Anken, 2000, 2002 on the effect of altered gravity on fish).

Facts: Developmental neurophysiology in fish

Fish have been proved to be the most suited vertebrate animals for basic gravity-research (Rahmann and Slenzka, 1994): they can be characterized by an absence of body weight related proprio-perception (in comparison with surface-bound terrestrial vertebrates), a reduced influence of gravity on supporting tissue, muscles, vascular tonus system etc., a relative higher sensitivity for gravity due to larger otoliths etc., a high reproduction rate (combined with higher genetic homogeneity of individuals), and mostly an external development, thus enabling better access to defined developmental stages. Moreover, there is rich information about genetics and developmental mutations available (Zebrafish *Danio*, Medaka *Oryzias*, Swordtail *Xiphophorus*), the developmental pattern of large numbers of genes are known and there exists an extensive homology to mammals not only at the molecular level but also concerning the central and peripheral vestibular system. Last not least, fertilization and development is not significantly impaired by Δg (Ijiri, 1995).

Vestibular behaviour

Dorsal light response (DLR)

Basically, fish use visual and vestibular cues for postural equilibrium maintenance and orientation. Already in 1935, the so-called DLR (Fig. 5) was described (v. Holst, 1935): illuminated from the side (light from above is the normal case) at normal 1*g* earth gravity, a fish tilts its back towards the light source. The angle of the tilt increases when raising the light intensity. Increasing the *g*-force in a centrifuge at constant illumination decreases the tilt angle (v. Holst, 1950). After bilateral labyrinthectomy or under microgravity conditions, the tilt is consequently guided by light alone. The DLR thus expresses a balance between the tilting force induced by visual information and the vestibular righting response (VRR) (Watanabe *et al.*, 1991) induced by tonic vestibular information. Possibly, the DLR is mainly integrated in the central nervous Valvula cerebelli (Watanabe *et al.*, 1991).

Vestibular righting response (VRR)

Since the DLR as a result of visuo-vestibulo-motoric integration is a behavioural response only found in fish in such a pronounced way, the VRR (Fig. 5) as the main vestibular component of the visuo-vestibular integration in fish (consult Watanabe *et al.*, 1991, for details) was analysed in blinded fish, thus excluding the visual component of the DLR. At normal earth gravity, the VRR in blinded fish results in a head-down movement (downward pitching), when a linear acceleration is applied in the headward direction of their longitudinal axis (Watanabe *et al.*, 1991). An acceleration towards the opposite direction results in a head-up movement. A respective rolling can be observed, when the acceleration is applied towards the lateral axis of the specimen (Watanabe *et al.*, 1991). These findings show that the VRR (sometimes called "gravity reference response"; v. Baumgarten *et al.*, 1970) uses the resulting vector (i.e., the gravito-inertial force) of the experimental acceleration applied and the 1*g* earth gravity causing a counterreacting movement of the fish to keep the acceleration on the inner ear otoliths relative to the fish (i.e., the effective acceleration) at a minimum.

All these findings were reproduced using unblinded specimens. The experiments yielded the general conclusion that a respective behaviour predominantly occurs in blinded fish, but that it can as well be observed in normal fish, however, not in all individuals of a given batch tested. Subsequently, it was found that the intensity of the DLR is species-specific, and that particular genetical strains of medakas (Japanese ricefish *Oryzias latipes*) differ in their DLR performance (Ijiri, 1995). Own investigations (Hilbig *et al.*, 1996, 2002b) clearly demonstrated that the performance of the DLR depends on the respective ability of an individual, thus suggesting that there are more "vestibular" and more "visual" individuals, as it is also the case in humans (e.g., Isableu *et al.*, 1997; Vaitl *et al.*, 1997).

Behaviour of fish at reduced gravity—basic findings

Already some three decades ago, the first microgravity experiment was undertaken using fish as test subjects (v. Baumgarten *et al.*, 1969). Under improved conditions, the experiment was repeated in 1972 by v. Baumgarten and co-workers. Hitherto, numerous subsequent investigators employed microgravity for investigating the behaviour of fish in the course of various experiments, only some of which can be mentioned here. Blinded goldfish—the experimenters wanted to exclude visual cues, so that they were focusing on the VRR alone—were observed during parabolic aircraft flights, in the course of which up to around 30 s of microgravity can be obtained. After the pull-up hypergravity phase of the flight, the push-over of the aircraft causes gravity beginning to get diminished. Then, the fish showed diving responses (downward pitching), until their noses faced the bottom of their aquarium. During the microgravity phase, the animals exhibited a continuing diving response, i.e., swimming inward loopings. This behaviour was called "*looping response*" (Fig. 5). At parts of the parabola, when the gravito-inertial vector became negative, the fish showed an inverted diving response towards the surface of the aquarium (upward pitching relative to the experimental setup; relative to the gravity vector, this behaviour is a downward pitching). Concomitantly with the diving responses and the loop-swimming, some—but not all—individuals performed *spinning movements* (Fig. 5) around their longitudinal axis (for more details see, e.g., v. Baumgarten *et al.*, 1972).

These early experiments clearly showed that fish face severe orientation problems at microgravity. In the following decades, many subsequent experiments at microgravity using fish were undertaken. They provided clues and insights into the understanding of the neurobiological basis of vertebrate gravity sensation and SMS. (Regarding fish, the term "kinetosis" is more appropriate than "motion sickness", since they seem not to suffer malaise at nauseogenic environments.)

Possible sources of space motion sickness

As has been argued above, the medical term "motion sickness" normally requires a dynamic component, since constant linear movements at 1g earth gravity are not nauseogenic. At lowered gravity conditions, however, motion sickness may occur with or without moving, thus being based either on a dynamic or on a static source.

Dynamic SMS

The VRR adjusts an accelerated fish towards the resulting vector of this acceleration and of the general gravitational environment (i.e., 1g earth gravity; the resulting vector is the gravito-inertial force). Respective experiments performed at microgravity showed that a headward acceleration of only 0.1g results in strong forward darting movements, while the fish did not react when similarly accelerated for the same magnitude horizontally on the ground (v. Baumgarten *et al.*, 1972). It has therefore been argued that, without weight, the otoliths and the otolithic membranes might freely float over the sensory epithelium, and their response towards accelerations will result in the VRR inducing darting movements. If the sensitivity of the otolith system of man in space would also be increased as compared to the ground, any brisk translation of the head would be interpreted erroneously. Such illusory tilts in turn would conflict with visual and other information and possibly generate SMS (v. Baumgarten, 1986). Support for this tilt translation hypothesis, however lacks. The following experiment speaks against the tilt translation hypothesis in fish: if the VRR would be the only reflex in response to acceleration, labyrinthectomized fish should not react at all upon a linear acceleration, but in fact they do. According to Watanabe *et al.* (1991), this implies the existence of some sensors additional to the labyrinth. It could be assumed that the lateral line organs may be sensitive to lateral acceleration and the swim bladder may sense it by acting as a tension receptor. Concerning humans, Mittelstaedt (1983) found out in the course of experiments on a person's subjective vertical that information about orientation needs not always to be provided only by the inner ears and the eyes. Somatic gravity information appears to originate from two distinctly different sources in man, i.e., the vascular system and the kidneys (Mittelstaedt, 1992; Vaitl *et al.*, 1997). Moreover, it has been hypothesized by several authors that fast changes in linear acceleration are interpreted by the subject as translatory motion while only slow changes are regarded as changes in the gravito-inertial force direction. Slow translational and continuously changing linear (vertical) acceleration, however, does not induce kinetotic behaviour in fish (Hilbig *et al.*, 2002b). Nevertheless, the darting of fish described needs not necessarily be based on the basis of the VRR alone, and the basis of dynamic SMS can hardly be explained simply by an increased sensitivity of the vestibular system. More likely, mismatching data from visual, vestibular, and somatic sources may contribute to the intersensory conflict resulting in static SMS.

Static SMS

The observation that motion triggers vestibular malaise in space does not necessarily prove that space sickness is only a special form of motion sickness: The lack of SMS while moving in some human astronauts is not a prerequisite of SMS-insusceptibility, since some of the same subjects report space sickness during periods of rest.

The microgravity induced behaviour of fish at rest clearly outlines the static component: their forward *looping responses* (i.e., continuously performed pitching down movements) can be initiated under reduced *g*-forces without providing an additional linear acceleration. The observed pitching movements or loopings most likely may resemble counter-movements in order to correct the (illusionary wrong) posture of the body. In 1972, v. Baumgarten and co-workers tried to explain this behaviour as follows: they argued that weightlessness may possibly allow the cilia of sensory hair cells to stretch, since the overlying otolith has no weight anymore. Such a presumed stretching would then cause a bending of cilia contrarily to their resting position at 1*g* earth gravity, which would mimic a dynamic process, in the course of which the VRR at rest would induce the looping response. Fernández and Goldberg (1976) were able to show in the course of an electrophysiological experiment carried out on squirrel monkeys that neither compressional forces (statoliths accelerated towards the sensory epithelium) nor exerting traction (statoliths accelerated away from the epithelium; i.e., "stretching" of the cilia) yielded significant responses. With this, the aforementioned hypothesis to explain the loop swimming must be disregarded, since the principal circumstances in fish most likely will not be different from that in squirrel monkeys. The looping response can as well not be explained by extrapolation of the VRR, since—with pitching down movements and forward loops—the centrifugal forces during forward loopings will cause an otolith displacement which has the same direction as if a fish at 1*g* would swim upside down. Swimming upside down will certainly not meet the task of the VRR. At times, astronauts reported at the beginning of microgravity on a phenomenon called "inversion illusion" where the subjects perceived themselves as being upside down (Graybiel and Kellog, 1966) which has been explained by a bias of saccular discharge in humans (Glasauer and Mittelstaedt, 1992). Given that fish also experience such an inversion illusion, they will naturally try to turn upside to downside, which then might possibly result in loopings. Whether this theoretical approach meets the facts in fish has still to be tested. Anken and Rahmann (1998a) suggested that erroneous inner ear afferents might make the eyes turn in the course of a reflex arc (vestibulo-ocular reflex, compare Sebastian *et al.*, 1996; Horn, 2003), and a second reflex would then make the fish turn so that the retinae of the eyes would be always positioned constantly relative to the longitudinal axis of the fish (Demski, 1983, discovered this unnamed response during electrical stimulation). This would in fact end up in the fish swimming loopings. However, blinded fish loop as well and the hypothesis forwarded requires a working visual system.

It has been observed during the spaceshuttle mission STS-89 that fish suddenly began to perform looping responses (Anken *et al.*, 1998a); this behaviour, not seen before or after this occasion, lasted for about 1 min. The NASA mission protocol shows that the aquarium was suddenly shocked, initiating the Mauthner cell escape response in the fish and resulting in the looping behaviour. Since the first phase of the escape response (fast body bend, FBB) was not followed by a return flip (second phase), the resulting continuous FBB may well represent loopings.

On the basis of the current knowledge, it is most likely that the looping responses of fish observed in various experiments do not have a common origin. An escape response based looping should only last for minutes at maximum like the fish did aboard STS-89. Aboard the 15-day IML-2 spaceshuttle mission, however, goldfish exhibited a looping throughout the mission. Moreover, these goldfish looped backwards (like blindfolded pigeons, Oosterveld and Greven, 1975), whereas the swordtails observed during STS-89 showed a forward looping as it has been earlier seen using goldfish during parabolic aircraft flights. The forward looping during parabolic aircraft flights and the short-lasting escape-response initiated forward looping aboard STS-89 therefore seem to belong to dynamically changing responses to be defined as transitional effects caused by changes in acceleration (Mori, 1995). For the initiation of a rotational motion the semicircular canals are possibly involved as well. Since parabolic flights involve short periods of rapidly changing micro- and hypergravity, the cupular organs may well contribute to the behavioural responses observed. At least, the continuous looping as often seen during space missions cannot be simply discussed as microgravity effects but may indicate unidentified gravireceptor(s) other than the otolith organs (Mori, 1995).

For completeness of this section, the "loop swimming" observed in larval cichlid fish during the D-2 mission (Rahmann *et al.*, 1995) (Fig. 5) needs to be mentioned. During the first days of the missions, the larvae showed backward, forward and sideward looping, which was, however, not a true loop-"swimming" at all. In fact, the 1*g* controls were crowded together at the ground of their aquarium due to their huge yolk-sacs, beating rapidly (and randomly) with their pectoral fins. Without weight, yolk-sac larvae beating with their pectoral fins will inevitably move around in circles.

The *spinning movements*, which are induced by microgravity in some individual fish at rest (i.e., without applying an additional acceleration) indicate another possible source of illusionary tilts: individually asymmetric vestibular maculae (caused by an asymmetric degeneration of the sensory epithelium or a respective loss of nerve fibers, or, based on side to side differences in the weight of otoliths) will, at rest and under 1*g* earth gravity, possibly cause slightly asymmetric shearing forces on the utricles, since the respective maculae are not exactly horizontally oriented. A normal posture then would require a (small amount of) "offset-compensation", which would compensate for the asymmetrical discharge rates. At microgravity, weight differences in the otoliths from one side of the body to the other side would then not anymore be given and the (primary) discharge rates should not be asymmetrical now. A continuing, but now unnecessary offset compensation, however, would then cause erroneously (secondary) asymmetric discharge rates.

This "asymmetry-hypothesis" (Anken *et al.*, 1998b,c; based on v. Baumgarten and Thümler, 1979) especially could explain the great interindividual differences in SMS susceptibility (man) or kinetotic behaviour (fish).

Several experiments speak in favour of this hypothesis: unilaterally labyrinthectomized fish, which were already compensated at 1*g* (mainly "gain-compensation"), performed spinning movements at weightlessness (erroneous offset-compensation; Watanabe *et al.*, 1991). In man, the grade of disconjugation of the ocular torsion of both eyes (i.e., rotation around the visual axis) is regarded as an otolith test at microgravity (Markham and Diamond, 1996).

Influence of reduced gravity on the swimming behaviour—the otolith asymmetry hypothesis

For technical reasons, asymmetric inner ears (especially concerning otoliths) have so far not directly been demonstrated in the course of research with human subjects. Last not least, direct proof for the assumption that kinetotic behaviour at decreased environmental gravity is based on a pronounced (utricular) otolith asymmetry comes from recent experiments using fish as model systems: in fact, asymmetrically weighed otoliths are a common feature in fish (Fig. 6) (for review: Watanabe *et al.*, 1991; Anken *et al.*, 1995).

Larval cichlid fish (*Oreochromis mossambicus*), which were reared at 3*g* hypergravity, morphogenetically developed normally and showed a normal swimming behaviour.

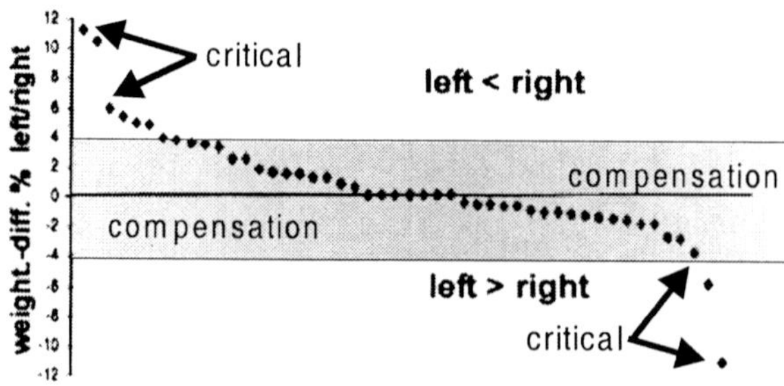

Fig. 6. Otolith asymmetry (weight differences between the left and right saccular otoliths) in Swordtail Fish, which were maintained in normal aquaria. According to the "asymmetry hypothesis", a moderate range of asymmetry can be compensated by the brain (*compensation*). Asymmetries on a level beyond the respective compensation capabilities (*critical*), however, might induce kinetotic behaviour at altered gravity.

However, as soon as the centrifuge was stopped, many individuals revealed looping responses and spinning movements (these movements represent a kinetotic behaviour, compare Fig. 5). This had also been frequently observed during spaceflight—and parabolic aircraft flight—experiments after the transfer from 1*g* earth gravity to microgravity conditions (e.g., v. Baumgarten *et al.*, 1972; DeJong *et al.*, 1996; Rahmann *et al.*, 1995, 1996; Hilbig *et al.*, 2002a,b). This kinetotic behaviour normally disappears within several hours or days (compare also the time-course of the space adaptation syndrome of astronauts; Clément and Berthoz, 1994; Rahmann *et al.*, 1995, 1996; Hilbig *et al.*, 2002b).

As predicted by the asymmetry-hypothesis, kinetotically behaving individual fish larvae after hyper-*g* and at microgravity in the course of parabolic aircraft flights indeed revealed a larger otolith asymmetry than normally behaving ones (Anken *et al.*, 1998b; Hilbig *et al.*, 2002a) (Fig. 7).

Moreover, it has been found that the asymmetry of otolithic Ca-incorporation (visualized with fluorescent tracers) is considerably higher in kinetotically swimming fish as compared to normally behaving control animals (Beier *et al.*, 2002a). These findings clearly support the otolith asymmetry hypothesis as the main source for kinetosis (especially for the static SMS).

Gravistimulated effects on inner ear otoliths

On the background of behavioural studies, we have recently provided evidence for the existence of a feedback-mechanism adjusting size, asymmetry and Ca-content of fish inner ear otoliths towards Δg (Anken *et al.*, 1998b, 2001): both in the hypergravity- (h*g*-) animals and in the 1*g*-controls, the morphogenetic development was identical. Evaluating the otoliths' growth during h*g* in comparison to the 1*g* controls, it was found that the growth both of utricular and saccular stones (lapilli and sagittae, respectively) was slowed down by h*g* (Fig. 8a) and that the development of bilateral asymmetry was considerably

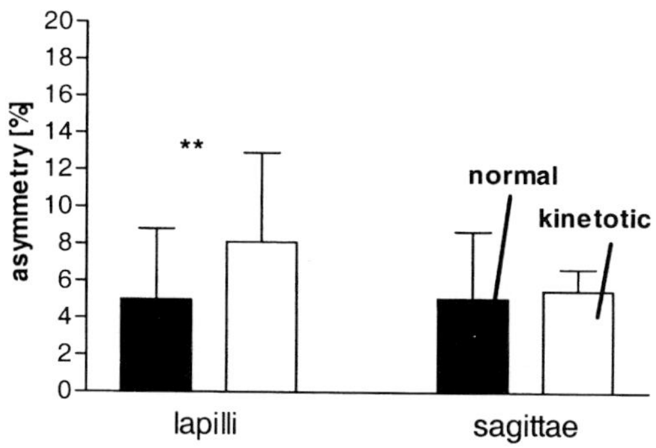

Fig. 7. Otolith asymmetry (surface area) of cichlid fish larvae in correlation to the behaviour shown during parabolic aircraft flights. Number of otolith-pairs analysed (each utricular otoliths—lapilli—and saccular otoliths—sagittae) concerning normal/kinetotic behaviour: $n=26/n=26$. $**P<0.01$. Vertical bars indicate standard deviations.

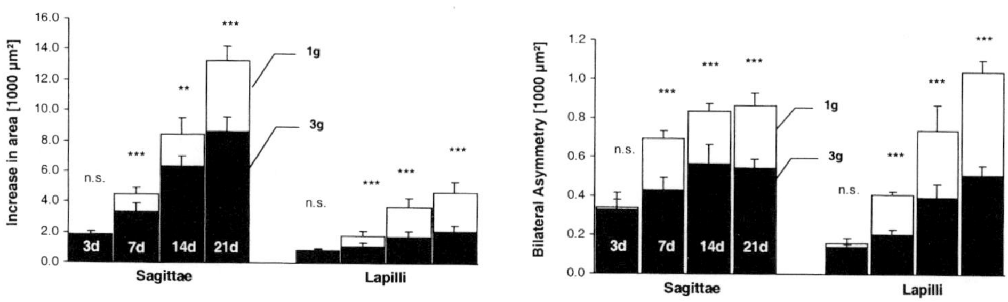

Fig. 8. Effects of hypergravity (h*g*) on the growth increase of inner ear otoliths of larval cichlid fish, which had been raised under 3*g* within a centrifuge for 3, 7, 14 or 21 days. (a) Size of otolith area grown during h*g*. $**P<0.001$; $***P<0.0001$. (b) Same otoliths as in (a). Absolute bilateral asymmetry (i.e., differences in the otoliths' area grown during hypergravity (h*g*) between left and right stones). $***P<0.0001$. n.s., not significant. Vertical bars indicate standard deviations.

decreased (Fig. 8b). Corresponding findings were obtained when measuring the otolithic calcium content (Anken *et al.*, 2001), which can serve as a proxy to determine an otoliths weight (Farrell and Campana, 1996).

This can be explained from a functional point of view as follows: under increased gravity, given otoliths will be heavier and thus cause increased shearing forces during tilts. However, any tilts need to be computed by the brain and asymmetric shearing forces need to be compensated (neuro-vestibular compensation). A feedback mechanism thus might slow down otolith growth at h*g*, so that an otolith formed at h*g* will possibly cause about the same shearing forces during tilts as a respective normally sized otolith at 1*g*. Without a feedback mechanism, a given bilateral asymmetry between two otoliths would be increased at h*g*, but for a correct interpretation of the afferent inputs to the brain for postural control and spatial orientation, any asymmetry ought not to be too pronounced in order to stay in the range of the compensation.

Therefore, it may be the task of this feedback mechanism to reduce a hg-based increased functional asymmetry in order to allow for a sufficient compensation. Such a feedback should then be active continuously during Δg.

It has been shown that the feedback is also active at normal 1*g* earth gravity (Anken *et al.*, 1998c): during ontogeny, one group of larval cichlids was constantly shifted around by a waterstream, whereas another group was kept without shifting ("shifted" vs. "stationary" specimens). In nature, cichlids are shifted around in the mother's mouth during the first weeks of their development until they can swim freely. Since the amplitude of the *g*-vector had not been altered, the otoliths (concerning their size) of both the shifted and the stationary larvae did not differ during development. During the first phase of the development, the stationary larvae were forced to lay always on the same side of their body due to their huge yolk-sacs. Therefore, in this phase, the direction of the gravity vector relative to the body axis was different between the stationary and the shifted animals. In the stationary samples, the asymmetry of utricular otoliths was highly significantly increased. During this phase, the shearing forces in the utricles of both sides of the body will have been highly asymmetric irrespective of the size of the stones. The feedback-mechanism will then have adjusted the growth of the utricular otoliths so that the shearing forces would be diminished to the possible minimum. As soon as the yolk-sacs were mostly resorbed and the animals could turn around deliberately or even swim upright, the shearing forces in the utricles would now be again asymmetric, caused by the utricular otoliths which had earlier grown asymmetric. As a consequence, otolith asymmetry rapidly decreased in the course of the further development. Significant differences in the asymmetry of the saccular otoliths (in contrast to utricular otoliths, the saccular ones are rather responsible for hearing than for graviperception) between the two stationary and shifted larvae were not obtained (Anken *et al.*, 1998c).

Indeed, our recent investigations showed that the regulation of otolith growth is obviously performed by the brain: otolithic Ca-incorporation (and thus growth) stopped after transection of the vestibular nerve (Anken *et al.*, 2000) (Fig. 9).

Summarized, the basis of a disorientated (kinetotic) behaviour in individual fish at a sudden reduction of the *g*-force is relatively well understood concerning the level of swimming behaviour and concerning gravitational effects on otolith growth and calcification.

The neuronal mechanisms, that possibly underly the space adaptation syndrome or the adaptation to hg by means of a feedback mechanism, remain so far poorly understood.

Gravistimulated neuroplastic reactivities in the brain and the inner ear

In order to ask for the possible location of the aforementioned feedback mechanism for the compensation of otolith growth following altered

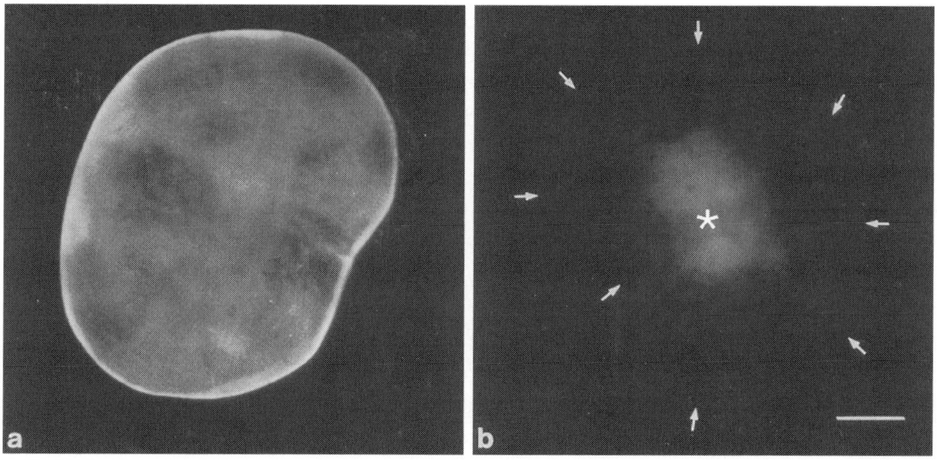

Fig. 9. Alizarin-complexone (a fluorescent calcium tracer) deposition in saccular otoliths after unilateral transection of the right vestibular nerve in a larval cichlid fish. The left stone (a) exhibits considerably more fluorescence than the right one (b). Arrows indicate the periphery of the right otolith. * artifact. Bar scale: 70 μm.

gravitational forces, different levels of neurochemical reactivities within the inner ear sensory epithelia and the brain were investigated.

General neurochemical effects

Using entire brains, gross-biochemical analyses of the neuronal energy—and plasma membrane—metabolism of fish larvae reared at hypergravity revealed an increase in energy availability (glucose 6-phosphate dehydrogenase and succinate dehydrogenase (SDH)), a decrease of plasma membrane energy transformation (plasma membrane bound creatine kinase) but an increase of mitochondrial energy transformation (mitochondrial creatine kinase), a change of the concentrations of the phosphagen creatine phosphate and of adenine nucleotides, no changes in energy consumptive processes (ATPases), but pronounced alterations in the plasma membrane configuration (ganglioside polarity, sialidase) (for review: Rahmann *et al.*, 1995). Fish larvae reared at simulated microgravity (approximately 0.01g in the fast rotating clinostat) showed a slight increase in plasma membrane related energy transformation combined with a decrease in energy availability (glucose 6-phosphate dehydrogenase). In contrast to larval fish, Δg had not such pronounced effects on adult fish (Slenzka *et al.*, 1994). This general finding, according to which larval fish may react gross-biochemically to a larger extent to altered gravitational forces than adult animals may be due to their wider range of neuronal plasticity (e.g., Hockfield and Kalb, 1993).

However, there is as yet no convincing explanation, what exactly the environmental pressures of gravity are and how they particularly affect the activity of some enzymes in larval fish, whereas others are not affected at all.

Based on earlier investigations on the effect of Δg on unicellular organisms (Machemer and Bräucker, 1992) and on cultured cells (e.g., human lymphocytes, Cogoli *et al.*, 1993), it had been argued (Slenzka *et al.*, 1993) that the mentioned gross-biochemical effects might be due to a direct influence of gravity on each neuron of a given brain, acting on the cell membrane itself (Schatz *et al.*, 1994; Hanke, 1996, review the effect of gravity on artificial bilayers as model membranes) or on cellular heavy bodies such as mitochondria or on the cytoskeleton, and that each neuron has its own gravireceptor (Slenzka *et al.*, 1993). From the gross-biochemical data (entire brains, pooled from dozens of individual fish, were homogenized to obtain the results), it cannot be derived that there acts any cellular graviperception but that the respective gross-biochemical findings may rather be based on inner ear related effects. Gravity, however, certainly acts on each cell and organ, especially if it has another density than the surrounding material. It is moreover possible that the mentioned gross-biochemical findings resemble a kind of general, "stress-related" effect of Δg. Spaceflight, e.g., increases the concentration of the blood-tension related protein ANF—atrial natriuretic factor—in some brain nuclei of frogs (Feuilloley *et al.*, 1993). Such effects may naturally change enzyme activities well in nuclei, which are not directly affected by Δg via the inner ear.

Effects on the central vestibular system (brain)

Histochemical and electronmicroscopical investigations focused on single brain centers speak in favour of the explanation that especially inner ear related effects of Δg play their role in brain enzymatics. Evaluating, for instance, the histochemically demonstrated reactivity of SDH in larval fish (Anken *et al.*, 1994, 1996) at 1g versus Δg resulted—besides of a weakly altered enzyme reactivity within entire brains—in highly significant effects only in vestibular nuclei as well as in macular ganglia, whereas non-vestibular nuclei were not affected. Similar, but less pronounced results (hypergravity only) were obtained in the case of adult fish (Anken and Rahmann, 1998b). These histochemical data strongly suggest different ranges of neuronal plasticity between juvenile and adult animals, which is in concordance with the gross-biochemical findings reviewed above.

In order to gain even deeper insights into this phenomenon, the metabolic activity of a vestibular brain nucleus was determined also on electronmicroscopical level, focused on cytochrome oxidase (CO): hypergravity yielded an increase in CO-reactivity within the nucleus, whereas microgravity resulted in a decrease (Paulus *et al.*, 1993, 1996). These findings confirm the histochemically gained results.

Moreover, hyper-*g* fish showed a tendency towards an increased formation of curved versus flat synapses as compared to 1*g* controls (Rahmann *et al.*, 1992, 1994). This has to be taken as evidence for a faster functional differentiation of the synaptic contacts in the hyper-*g* animals. Another study (Ibsch *et al.*, 2000) focused on synapse countings in primary vestibular integration centers and in a visual nucleus of spaceflown fish. Microgravity yielded an increase in synaptic contacts only within a vestibular nucleus indicating that a lack of input resulted in compensation processes.

The results of these investigations suggest that the excitatory inputs from the inner ear are reflected in the vestibular system of the brain by its neuronal activity.

Effects on the peripheral vestibular system (inner ear)

On the background of the aforementioned findings concerning neuronal plasticity within the brain it was the next step to investigate any influences of Δg on the inner ear receptor organs: carbonic anhydrase (CA) is an enzyme being functionally involved in otolithic calcium carbonate deposition, and hypergravity (3*g*, 6 h) effected a decrease in macular CA reactivity in larval cichlids (Beier *et al.*, 2002b), which is in complete agreement with the corresponding findings on otolith growth. Further investigations (Hilbig *et al.*, 2002a) revealed that kinetotically swimming fish are characterized by malformed macular epithelial cells. It has been suggested that malformed epithelia are not necessarily the direct cause for kinetosis susceptibility, but that their occurrence may well coincide with the existence of asymmetric otoliths, since an inhibition of CA coincides with the occurrence of malformed receptor cells and alters otolith growth (see above).

Another study showed that microgravity resulted in a decrease of ultramicroscopically detectable CO-activity within the utricular sensory epithelium of developing cichlid fish, but not in the saccule as compared to 1*g* controls (Nindl *et al.*, 1996). The utricles—in contrast to the saccules—when swimming upright at normal gravity are most sensitive to tilts, whereas the saccules are not working in push-pull configuration. At weightlessness, any tilt will not lead to a static displacement of the utricular otoliths since they do not have any weight anymore. Thus, enzyme activities triggered by otolith displacements will not occur in the saccules.

Hypergravity, however, did not have any effect on CO-activity within the utricles: at hypergravity, the shearing forces effected during tilts will be increased in the utricle, thus causing the development of smaller otoliths (see above), which then in turn will mimic shearing forces of normal sized otoliths at normal gravity, thus resulting in a CO-activity similar to that as seen in 1*g* controls. (Since hypergravity, however, increased the metabolic activity of a vestibular brain nucleus, these findings therefore reflect the work of the proposed feedback-mechanism between the brain and the inner ear epithelia, which alters the size and asymmetry of otoliths at hypergravity.)

Summarized, investigations of fish, which had developed at Δg, reveal a variety of short lasting, adaptive effects following Δg. The behavioural adaptation to Δg can be correlated with a compensatory otolith growth, based on a (negative) feedback mechanism between the inner ear and the brain. The neuronal control of this feedback mechanism seems to be effected by stimulation dependent enzyme activities as well as synaptic plasticity, especially in vestibular brain centers.

Conclusion

During evolution of animals, they have managed both to cope with and use the environmental gravity vector for orientation. Mechanisms concerning the latter (function of gravireceptors and the central nervous computation of respective inputs) is comparatively well investigated. Yet, however, it is not completely clear, to which extent gravity is necessary for a normal development and in which range adaptive mechanisms are efficient enough at Δg to guarantee a normal lifestyle.

The various data being obtained using especially fish as vertebrate model systems concerning gravistimulated effects reveal first of all that the normal development seems not to be significantly influenced by Δg. The studies undertaken so far, however, do not yet answer the question of whether a complete life cycle can be completed for instance under weightlessness. Secondly, the results regarding gravity-effects, particularly on fish, speak in favour of the following concept of possible interactions: short-term Δg (up to around 1 day) can induce transitional aberrant behaviour due to malfunctions of the inner ear, originating from asymmetric otoliths or, generally, from a mismatch between canal and otolith afferents. The vanishing aberrant behaviour is due to a reweighing of sensory inputs and neurovestibular compensation, probably on bioelectrical basis. During long-term Δg (several days and more), step by step neuroplastic reactivities on molecular basis (i.e., molecular facilitation) in the brain and inner ears possibly activate feedback mechanisms between the CNS and the vestibular organs for the regain of normal behaviour. It was shown that such a mechanism is involved in adjusting (assimilating) the otolith weights in fish.

The experimental data on the effect of Δg on developing and adult animals help understanding the role of gravity during evolution. The consequences for animal life at Δg are well in concordance with a concept about an evolution of the gravi-resistance and -reactivity of animals of the known Ukrainian space physiologist N. Sirotonin (1896–1977): there is a high adaptability/tolerance to Δg in arthropods (especially insects) since Δg does not seem to provoke severe effects, fish and amphibians respond to Δg mainly in most early developmental stages, and there is a wide range of responses in birds and mammals

(Serova, 1993). The reasons for this varying range of gravity-tolerance have hitherto not been disclosed.

Perspectives

As perspectives, the following areas of research with animals at Δg, exemplarily focused on aspects of normal development, need to be addressed in the future:

(1) Maintenance of—preferably vertebrate—animals through two complete life cycles in the space environment to determine whether or not developmental deficiencies are routinely produced.
(2) Investigation of the peripheral and central vestibular system by ground-based studies, involving genetic mutations (mice and fish) that eliminate the peripheral organs and by using hypergravity as a research tool, focusing both on plasticity in developing animals as well as in adults.
(3) Investigation of the effect of microgravity during critical developmental periods (imprinting phase for graviperception?) when ISS facilities are available.
(4) Studies on the effects of the vestibular system on the development of other neuronal systems such as formation and maintenance of neuronal maps and the autonomic nervous system and vice versa (e.g., effects of efferent systems on the development, maintenance and function of the vestibular apparatus) including research on how these interacting systems allow for vestibular compensation.
(5) Investigations on the effects of the vestibular system on the development of the postural and locomotive control systems to gain insights into its role in complex, navigational behaviours including studies to clarify the developmental events that lead to the differential susceptibility to sensorymotor disorders.

Answers to these questions may be of crucial interest for basic gravitational research. They might further be of importance concerning the topic "man in space", especially on the background of the ISS to be permanently manned and NASA's plans to perform a manned mission to Mars within around 20 years from now, which probably will force astronauts to live at weightlessness conditions longer than humans have ever experienced before.

Acknowledgements

I am greatly indebted to my academic instructor, Professor em. Dr. H. Rahmann, who established the field of Gravitational Zoology at the University of Hohenheim and continuously supported my scientific career. The present study was financially supported by a grant from the German Aerospace Center DLR e.V. (FKZ: 50 WB 9997).

References

Anken, R., Edelmann, E. and Rahmann, H. (2000) Fish inner ear otoliths stop calcium incorporation after vestibular nerve transection. NeuroReport 11, 2981–2983.

Anken, R., Hilbig, R., Ibsch, M., Vöhringer, P. and Rahmann, H. (1998a) Readaptation of fish to 1G after long-term microgravity: behavioural and electronmicroscopical results from the STS 89 mission. Adv. Space Res. 25, 2019–2023.

Anken, R., Ibsch, M., Bremen, D., Hilbig, R. and Rahmann, H. (1995) Mass-analytical and morphological investigations on teleost otoliths and on their relevance for the internalisation of graviperception. Proc. C.E.B.A.S. Workshops 11, 73–82.

Anken, R., Ibsch, M., Breuer, J. and Rahmann, H. (2001) Effect of hypergravity on the Ca/Sr-composition of developing otoliths of larval cichlid fish (*Oreochromis mossambicus*). Comp. Biochem. Physiol. A 128, 369–377.

Anken, R., Kappel, T. and Rahmann, H. (1998b) Morphometry of fish inner ear otoliths after development at 3g hypergravity. Acta Otolaryngol. (Stockh.) 118, 534–539.

Anken, R. and Rahmann, H. (1998a) On inappropriately used neuronal circuits as a possible basis of the "loop-swimming" behaviour of fishes under reduced gravity: a theoretical study. Adv. Space Res. 22, 277–280.

Anken, R. and Rahmann, H. (1998b) Influence of long-term hyper-gravity on the reactivity of succinic acid dehydrogenase and NADPH-diaphorase in the central nervous system of fish: a histochemical study. Adv. Space Res. 22, 281–284.

Anken, R., Slenzka, K., Neubert, J. and Rahmann, H. (1994) Altered gravity affects succinate dehydrogenase reactivity in specific nuclei of fish brain. NeuroReport 5, 1313–1316.

Anken, R., Slenzka, K., Neubert, J. and Rahmann, H. (1996) Histochemical investigations on the influence of long-term altered gravity on the CNS of developing cichlid fish: results from the 2nd German Spacelab Mission D-2. Adv. Space Res. 17, 281–284.

Anken, R., Werner, K., Ibsch, M. and Rahmann, H. (1998c) Fish inner ear otolith size and bilateral asymmetry during development. Hearing Res. 121, 77–83.

Baumgarten, R.v. (1986) European vestibular experiments on the Spacelab-1 mission: 1. Overview. Exp. Brain Res. 64, 239–246.

Baumgarten, R.v., Atema, I., Hukuhara, M. and Rocker, R. (1969) Behavioral responses to short periods of lowered gravitational force in blind goldfish. Space Life Sci. 1, 544–564.

Baumgarten, R.v., Baldrighi, G., Atema, J. and Shillinger, G. (1970) Behavioral responses to linear accelerations in blind goldfish. Space Life Sci. 3, 25–33.

Baumgarten, R.v., Baldrighi, G. and Shillinger, G. (1972) Vestibular behavior of fish during diminished G-force and weightlessness. Aerospace Med. 43, 626–632.

Baumgarten, R.v. and Thümler, R. (1979) A model for vestibular function in altered gravitational states. Life Sci. Space Res. 17, 161–170.

Beier, M., Anken, R. and Rahmann, H. (2002a) Influence of hypergravity on fish inner ear otoliths: II. Incorporation of calcium and kinetotic behaviour. Adv. Space. Res., 30, 727–731.

Beier, M., Anken, R. and Rahmann, H. (2002b) Effect of hypergravity on carboanhydrase reactivity in inner ear ionocytes of developing cichlid fish. 34th Committee on Space Res. (COSPAR) Sci. Assembly, 222.

Clément, G. and Berthoz, A. (1994) Influence of gravity on the processing of information by the central nervous system. Proc. Eur. Symp. Life Sci. Res. Space 5, 333–338.

Cogoli, A., Bechler, B., Cogoli-Greuter, M., Criswell, S., Joller, H., Joller, P., Hunzinger, E. and Müller, O. (1993) Mitogenic signal transduction in T-lymphocytes in microgravity. J. Leukocyte Biol. 53, 569–575.

Conrad, M. (1997) Origin of life and the underlying physics of the universe. BioSystems 42, 177–190.

DeJong, H., Sondag, E., Kuipers, A. and Oosterveld, W. (1996) Swimming behavior of fish during short periods of weightlessness. Aviat. Space Environ. Med. 67, 463–466.

Demski, L. (1983) Behavioral effects of electrical stimulation of the brain. *In* Northcutt, R. and Davis, R. (eds.), *Fish Neurobiology Volume 2*, University of Michigan Press, Ann Arbor, 317–359.

DuBois, A., Cavagna, G. and Fox, R. (1976) Locomotion of bluefish. J. Exp. Zool. 195, 223–235.

Farrell, J. and Campana, S. (1996) Regulation of calcium and strontium deposition on the otoliths of juvenile tilapia, *Oreochromis niloticus*. Comp. Biochem. Pysiol. 115A, 103–110.

Fernández, C. and Goldberg, J. (1976) Physiology of peripheral neurons innervating otolith organs of the squirrel monkey. II. Directional selectivity and force–response relations. J. Neurophysiol. 5, 985–995.

Feuilloley, M., Yon, L., Kawamura, K., Kikuyama, S., Gutkowska, J. and Vaudry, H. (1993) Immunocytochemical localization of atrial natriuretic factor (ANF)-like peptides in brain and heart of the treefrog *Hyla japonica*: effect of weightlessness on the distribution of immunoreactive neurons and cardyocytes. J. Comp. Neurol. 330, 32–47.

Glasauer, S. and Mittelstaedt, H. (1992) Determinants of orientation in microgravity. Acta Astronautica 27, 1–9.

Graybiel, A. and Kellog, R. (1966) The inversion illusion in parabola flight: its probable dependence on otolith function. Second Symposium on the Role of the Vestibular Organs in Space Exploration, NASA Publication SP-115.

Hanke, W. (1996) Studies of the interaction of gravity with biological membranes using alamethicin doped planar lipid bilayers as a model system. Adv. Space Res. 17, 143–150.

Hilbig, R., Anken, R., Bäuerle, A. and Rahmann, H. (2002a) Susceptibility to motion sickness in fish: a parabolic aircraft flight study. J. Gravitat. Physiol., 8, 29–30.

Hilbig, R., Anken, R., Sonntag, G., Höhne, S., Henneberg, J., Kretschmer, N. and Rahmann, H. (2002b) Effects of altered gravity on the swimming behaviour of fish. Adv. Space Res., 30, 835–841.

Hilbig, R., Schüle, T., Ibsch, M., Anken, R. and Rahmann, H. (1996) New approaches to the gravity relevant behaviour of the swordtail fish (*Xiphophorus helleri*). Proc. C.E.B.A.S. Workshop Conf. 12, 51–57.

Hockfield, S. and Kalb, R. (1993) Activity-dependent structural changes during neuronal development. Curr. Opin. Neurobiol. 3, 87–92.

Holst, E.v. (1935) Über den Lichtrückenreflex bei Fischen. Pubbl. Staz. Zool. Napoli 15, 143–158.

Holst, E.v. (1950) Die Arbeitsweise des Statolithenapparates bei Fischen. Z. vergl. Physiol. 33, 60–120.

Horn, E. (2003) The development of gravity sensory systems during periods of altered gravity dependent sensory input. *In* Marthy, H.-J. (ed.), *Developmental Biology Research in Space*, Elsevier Science, 133–171.

Hudspeth, A. and Gillespie, P. (1994) Pulling springs to tune transduction: adaptation by hair cells. Neuron 12, 1–9.

Ibsch, R., Anken, R. and Rahmann, H. (2000) Weightlessness during spaceflight results in enhanced synapse formation in a fishbrain vestibular nucleus. Neurosci. Lett. 296, 13–16.

Ijiri, K. (1995) The First Vertebrate Mating in Space. RICUT, Tokyo.

Isableu, B., Ohlmann, T., Crémieux, J. and Amblard, B. (1997) Selection of spatial frame of reference and postural control variability. Exp. Brain Res. 114, 584–589.

Machemer, H. (1999) The identification of gravikinesis from ciliates: methods and experience. Adv. Space Res. 24, 867–876.

Machemer, H. and Bräucker, R. (1992) Gravireception and graviresponses in ciliates. Acta Protozool. 31, 185–214.

Markham, C. and Diamond, S. (1996) Eye torsion in space and during static tilt pre and post-spaceflight. Proc. Eur. Symp. Life Sci. Res. Space 6, 89–93.

Mittelstaedt, H. (1983) A new solution to the problem of the subjective vertical. Naturwissenschaften 70, 272–281.

Mittelstaedt, H. (1992) Somatic versus vestibular gravity reception in man. Ann. N. Y. Acad. Sci. 656, 124–139.

Mori, S. (1995) Disorientation of animals in microgravity. Nagoya J. Med. Sci. 58, 71–81.

Moroz, P. (1984) The cell in the field of gravity and the centrifugal field. J. Theor. Biol. 107, 303–320.

Neugebauer, D. and Machemer, H. (1997) Is there an orientation-dependent excursion of the Müller-body in the "statocystoid" of Loxodes?. Cell Tissue Res. 287, 577–582.

Nindl, G., Körtje, K., Slenzka, K. and Rahmann, H. (1996) Comparative electronmicroscopical investigations on the influence of altered gravity on cytochrome oxidase in the inner ear of fish: a spaceflight study. J. Brain Res. 37, 291–300.

Oosterveld, W. and Greven, A. (1975) Flight behavior of pigeons in the weightless phase of parabolic flight. Aviat. Space Environ. Med. 46, 713–716.

Paulus, U., Körtje, K. and Rahmann, H. (1993) Effects of development and altered gravity conditions on cytochrome oxidase activity in a vestibular nucleus of the larval teleost brain: a quantitative electronmicroscopical study. J. Neurobiol. 24, 1131–1141.

Paulus, U., Körtje, K., Slenzka, K. and Rahmann, H. (1996) Correlation of altered gravity and cytochrome oxidase activity in the developing fish brain. J. Brain Res. 37, 103–107.

Rahmann, H. and Anken, R. (2000) Gravitational neurobiology of fish. Adv. Space Res. 25, 1985–1995.

Rahmann, H. and Anken, R. (2002) Gravity related research with fishes—perspectives in regard to the upcoming international space station. ISS. Adv. Space Res., 30, 697–710.

Rahmann, H., Hilbig, R., Flemming, J. and Slenzka, K. (1996) Influence of long-term altered gravity on the swimming performance of developing cichlid fish: including results from the 2nd German Spacelab Mission D-2. Adv. Space Res. 17, 121–124.

Rahmann, H. and Slenzka, K. (1994) Influence of gravity on early development of lower aquatic vertebrates. Proc. Eur. Symp. Life Sci. Res. Space 5, 147–152.

Rahmann, H., Slenzka, K., Anken, R., Appel, R., Flemming, J., Hilbig, R., Kappel, T., Körtje, K., Nindl, G. and Paulus, U. (1995) Structure- and function-related neuronal plasticity of the CNS of aquatic vertebrates during early ontogenetic development under microgravity conditions. *in* Sahm, P. (Ed.) *Scientific Results of the German Spacelab Mission D-2*. Wissenschaftliche Projektführung D-2, Aachen, 621–637.

Rahmann, H., Slenzka, K., Hilbig, R., Flemming, J., Paulus, U., Körtje, K., Bäuerle, A., Appel, R., Neubert, J., Briegleb, W., Schatz, A., and Bromeis, B. (1994) Influence of hyper- and hypo-gravity on the early ontogenetic development of cichlid fish. Behavioral and ultrastructural investigations including first results of the 2nd German Spacelab Mission D-2. Proc. Eur. Symp. Life Sci. Res. Space 5, 165–169.

Rahmann, H., Slenzka, K., Körtje, K. and Hilbig, R. (1992) Synaptic plasticity and gravity: ultrastructural, biochemical and physico-chemical fundamentals. Adv. Space Res. 12, 63–72.

Schatz, A., Reitstetter, R., Linke-Hommes, A., Briegleb, W., Slenzka, K. and Rahmann, H. (1994) Gravity effects on membrane processes. Adv. Space Res. 14, 35–43.

Sebastian, C., Eßeling, K. and Horn, E. (1996) Altered gravitational experience during early periods of life affects the static vestibulo-ocular reflex of tadpoles of the southern clawed toad. *Xenopus laevis* Daudin. Exp. Brain Res. 112, 213–222.

Serova, L. (1993) Gravity and animal development. Aviakosm. Ekolog. Med. 27, 15–18.

Slenzka, K., Appel, R. and Rahmann, H. (1993) Brain creatine kinase activity during ontogeny of the cichlid fish *Oreochromis mossambicus* and the clawed toad *Xenopus laevis*, influence of gravity?. Neurochem. Int. 22, 405–411.

Slenzka, K., Appel, R., Seibt, V., Anken, R. and Rahmann, H. (1994) Biochemical investigations on the influence of altered gravity on brain energy metabolism in developing and adult fishes. Proc. C.E.B.A.S. Workshop Conf. 10, 42–47.

Vaitl, D., Mittelstaedt, H. and Baisch, F. (1997) Shifts in blood volume alter the perception of posture. Int. J. Psychophysiol. 27, 99–105.

Vinnikov, I. (1995) Gravitational mechanisms of interactions of sensory systems in invertebrates in the evolutionary aspect. Aviakosm. Ekolog. Med. 29, 4–19.

Watanabe, S., Takabayashi, A., Tanaka, M. and Yanagihara, D. (1991) Neurovestibular physiology in fish. *In* Bonting, S. (ed.), *Advances in Space Biology and Medicine Volume 1*, JAI Press, Tokyo, 99–128.

Developmental Biology Research in Space
H.-J. Marthy (editor)

Life-Cycle Experiments of Medaka Fish Aboard the International Space Station

Kenichi Ijiri*

Radioisotope Center, University of Tokyo, Yayoi, Bunkyo-ku, Tokyo 113-0032, Japan

Abstract

Fish are the most likely candidates to be the first vertebrate to live their life cycle aboard the International Space Station (ISS). In the space-shuttle experiment using medaka, the fry born in space had the same number of germ cells as the ground control fish, and these germ cells later developed to produce the offspring on the ground. Fry hatched in space did not exhibit any looping behavior regardless of their strain, visual acuity, etc. The aquatic habitat (AQH) is a space habitat designed for long-term breeding of medaka, zebrafish and *Xenopus*, and recent advancements in this hardware also support fish life-cycle experiments. From the crosses between two strains, fish having good eyesight and less sensitivity to gravity were obtained, and their tolerance to microgravity was tested by parabolic flight using an airplane. The fish exhibited less looping and no differences in degree of looping between light and dark conditions. These are possible candidates for the first adult medaka (parent fish) to start a life cycle aboard ISS. Embryos were treated with a three-dimensional clinostat. Such simulated microgravity caused no differences in tissue architecture or in gene expression within the retina, nor in formation of cartilage (head skeleton). Otolith formation in embryos and fry was investigated for wild-type and mutant (*ha*) medaka. The *ha* embryos could not form utricular otoliths. They formed saccular otoliths but with a delay. Fry of the mutant fish lacking the utricular otoliths are highly light-dependent at the time of hatching, showing a perfect dorsal-light response (DLR). As they grow, they eventually shift from being light dependent to gravity dependent. Continuous treatment of the fry with altered light direction suppressed this shift to gravity dependence. Being less dependent on gravity, these fish can serve as model fish in studying the differences expected for the fish that have experienced a life cycle in microgravity.

*E-mail: ijiri@ric.u-tokyo.ac.jp

Introduction

The international space station (ISS) has now been in operation for some time, and it is now time for biologists to carry out long-term space experiments. For fish, the most exciting subject of all is a realization of fish life cycles in microgravity. Adult male and female medaka fish (*Oryzias latipes*) were the first vertebrate to successfully mate in space. Moreover, the eggs the fish laid in space developed normally and hatched as fry (baby fish) in space, being named "space-originated fry" (Ijiri, 1994, 1995a). Thus, at present, fish are the most likely candidates to be the first vertebrate to live their life cycle in space. This paper introduces recent research and considerations for realizing the life-cycle experiment of medaka fish.

Positive implications from previous space experiments

Normal number of germ cells were formed in embryos under microgravity

In the medaka experiment mentioned above, four out of the eight space-originated fry were fixed in formalin 6 h after the space shuttle landing. For each fry, the total number of germ cells was counted histologically. Figure 1 shows the germ cell number for each fry, together with the data for ground-kept and ground-born fry at the age of zero and 2 days after hatching (ground control). Two different ground control data sets are shown because we could not tell the exact date each space-originated fry hatched, though they should have hatched in space within 3 days before sacrifice. The number of germ cells of four space fry are in the range of those of the ground samples. Thus, embryos developed normally in external appearance and the embryos formed primordial germ cells normally in space. In one space fry, a germ cell in meiotic prophase was noted, similar to the phenomenon found in some of ground fry (Ijiri, 1998).

The two remaining space-originated fry became mature adults about half a year after returning to Earth. They were a male and a female, and they started laying eggs. These eggs were checked to determine if they were fertilized or not, and their development was traced until they hatched into fry. Percentages of fertilized eggs and hatched fry were obtained and were at the normal control levels. So far, no adverse effects have been detected in the offspring of the space-originated fish (Ijiri, 1995b). The offspring data proved that the germ cells formed in space were functionally normal, too. That is, those germ cells could produce fertile sperm and ova, having no lethal genetic damage impairing embryonic development.

Fry developed in space did not loop at all

Beside the results of fish mating and full development to hatching, there was another fish experiment carried out in the same 1994 mission. As reported

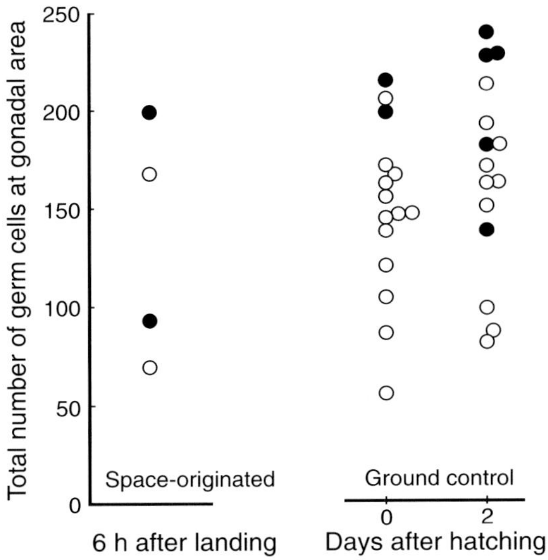

Fig. 1. Germ cell numbers of space-originated fry of medaka (killed six hours after landing), and of ground-kept fry (ground control, 0 and 2 days after hatching). One circle point corresponds to one fry. A fry having some germ cells in meiotic prophase is shown by a solid circle (Ijiri, 1998).

elsewhere (Ijiri, 1995c), it was an experiment to test whether fry that developed in space exhibit looping. Developing eggs of two different medaka strains (ccT and HO5) were collected on the ground and were loaded into the space shuttle. During the early part of the mission, the embryos hatched into fry. Together with the observation by astronauts (payload crew), we confirmed in video pictures that no hatched fry (neither ccT nor HO5 strain) looped in space. In space, all the fry positioned their backs to the light (a dorsal-light response, DLR).

During the microgravity generated by parabolic flight (using an airplane), fish are known to swim abnormally, such as looping and rolling (von Baumgarten *et al.*, 1972; de Jong *et al.*, 1996). Parabolic-flight experiments (20 s of microgravity) told us that in adult fish, there is a clear strain difference in the behavioral response of the medaka fish under microgravity. As seen in Figure 2, ccT and HNI-II strains exhibit very few or no loops in microgravity. In contrast, the HO5 strain loops many times (Ijiri, 2000).

Swimming behavior of the fry hatched in space indicated that, although adult fish were genetically determined to loop or not in microgravity, their fry developed in space did not loop under microgravity, regardless of their genetic traits.

Figure 3 shows that fry of two strains (HNI-II, *ha*) had few or no loops when exposed to a parabolic flight, while some fry of HO5 strain exhibited frequent loops. Such strain differences between the fry of HNI-II and HO5 were explained by the difference in their eyesight (Furukawa and Ijiri, 2002).

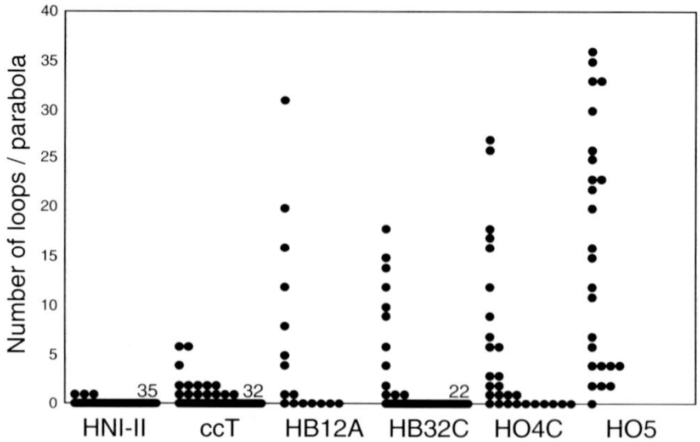

Fig. 2. Looping behavior of adult fish in microgravity for six different medaka strains. All the data plotted here were obtained in the first parabolic flight, i.e., the response when each fish experienced microgravity for the first time. Each dot represents an individual fish. The ordinate expresses the number of loops each fish made during a parabola (20 s of microgravity). For HNI-II, ccT and HB32C strains, fish that did not loop at all (0 loops) are indicated by a solid horizontal bar at 0 of the ordinate, with the total number of non-looping fish given on each bar. Illumination was applied laterally to the aquarium (Ijiri, 2000).

As described later in the text, the otolith formation takes place during embryonic development (first detected at stage 25, about 5 days before hatching at 25 °C). Even the fry just hatched have already developed much of the inner ear and eye structure. The visual acuity increases as fish grow (Rahmann *et al.*, 1979; Hairston *et al.*, 1982). In medaka, the visual acuity improves with growth during the first 20 days after hatching (Ohki and Aoki, 1985).

The visual acuity may play an important role in controlling fry posture. However, the space experiment on medaka fry told us that regardless of their visual acuity, no fish that developed in microgravity would loop. As will be explained later, the *ha* strain is a mutant strain that lacks the pair of otoliths responsible for gravisensing. Therefore, this strain may serve as a model for space-developed fry. Having ordinary eyesight, they exhibited very few loops in microgravity (parabolic flight), which resembled the behavior of the space-originated fry in space.

Hardware development for the long-term breeding of fish

Advancements in the hardware will also support the expectation that fish will be the first vertebrate to live a complete life cycle in space. There are a few outstanding facilities large enough to support the life of adult fish. For other aquatic facilities, see Slenzka (2002). Among them, VFEU, AAEU, and CEBAS-Minimodule have already demonstrated their capacity in real space-shuttle missions. The Vestibular Function Experiment Unit (VFEU) was the first aquatic facility developed by the Japanese Space Agency NASDA

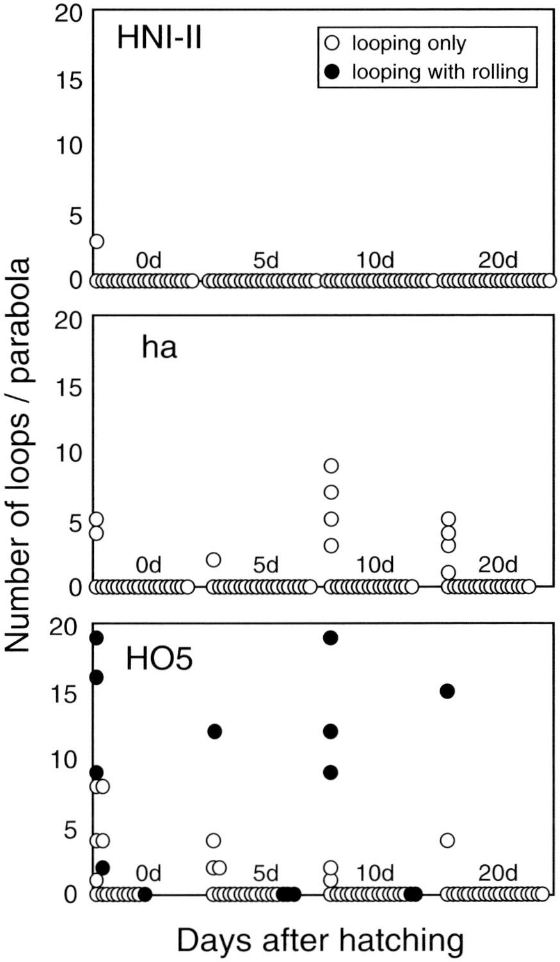

Fig. 3. Responses of medaka fry at four different ages to microgravity generated by a parabolic flight. The data for three strains (HNI-II, *ha*, HO5) are given. For each age (expressed in days after hatching) of each strain, 20 fry were tested. Illumination was applied laterally to the aquarium (redrawn from Furukawa and Ijiri, 2002).

(Sakimura *et al.*, 1999). It was used for sustaining Japanese carp (*Cyprinus carpio*, Japanese koi) and for obtaining their electroencephalogram (EEG) in microgravity during the SL-J mission (STS-47, 1992) (Mori, 1997; Mori *et al.*, 1996). Later, the VFEU was improved to accommodate marine fish (*Opsanus tau*, Oyster toadfish) (Nagaoka *et al.*, 1999; Uchida *et al.*, 1999) and to perform neurobiological experiments in Neurolab (STS-90, 1998) and STS-95 (1998) missions (Boyle *et al.*, 2001).

The Aquatic Animal Experiment Unit (AAEU) served for IML-2 (the Second International Microgravity Laboratory) mission (STS-65, 1994) and succeeded in maintaining the life of medaka and goldfish (*Carassius auratus*), as well as an amphibian (*Cynopus pyrrhogaster*, Japanese red-bellied newt). The goldfish were used for studying the adaptation of their behavior (Takabayashi *et al.*, 1997), and newts for observation of their embryonic development (Wiederhold *et al.*, 1997; Yamashita *et al.*, 2001) in microgravity.

The Closed Equilibrated Biological Aquatic System (CEBAS)-Minimodule can house various aquatic species (Bleum *et al.*, 1994; Slenzka, 2002). Together with plants, pond snails and baby swordtail fish (*Xiphophorus helleri*), adult swordtail fish were flown to space in STS-89 (1998) and in Neurolab (STS-90). In CEBAS-Minimodule, an ecosystem controlled water quality. In AAEU and the improved VFEU, a biological filter (using nitrifying bacteria, converting ammonia and nitrite to nitrate) has been employed.

The Aquatic Habitat (AQH) is a space habitat meant for long-term breeding of aquatic animals (Uchida *et al.*, 2002), and is especially suited for (or originally designed for) medaka fish (*Oryzias latipes*), zebrafish (*Danio rerio*), and *Xenopus* (*X. laevis, X. tropicalis*). Because it takes a long time for *Xenopus* to reach maturity, their development from fertilization of eggs to metamorphosis of tadpoles is presently planned for *Xenopus* life-cycle studies in microgravity. AQH has been designed to realize fish life cycles in space. The prototype AQH has already accomplished its objective of both medaka and zebrafish living one complete life cycle on the ground within 3 months (in the shortest case, within 2 months). The AQH is able to separate the eggs laid and developing larvae, and to place them in different compartments from adult fish.

In the life-cycle experiments of fish using a facility such as AQH, waste management, especially a denitrifying process to convert nitrate (NO_3^-) to nitrogen gas (N_2), is a most critical issue. Toxic concentration of nitrate when medaka fish were exposed to it chronically was studied. In the toxicity test using juveniles from just after hatching until 3 months old, gains in body weight were suppressed above nitrate concentrations of 50 mg/l (expressed in N of NO_3^- morphological abnormalities such as lordosis and scoliosis were observed in the 75–100 mg/l range. Mortality also increased above 100 mg/l. Furthermore, accumulated nitrate even at low concentrations of around 30 mg/l seems to affect the reproductive system (no effects on body growth). The concentration caused retardation of maturity and decreased the number of eggs laid each day. We therefore concluded that for a certain critical stage in their life cycle, the highest safe concentration of nitrate for assuring a complete life cycle of medaka fish is about 25 mg/l (Shimura *et al.*, 2002). Such a low concentration of nitrate was not expected from the acute toxicity test for medaka (Shimura *et al.*, 1999, 2000), and this should be kept in mind when conducting life-cycle experiments of medaka fish. Countermeasures are the replacement of the circulating water or addition of neutralizing agents to water when the nitrate level is elevated (i.e., when pH decreased).

Production of the first parent fish for life-cycle experiments in space

Adult fish might go to space first

There are two plans for realizing a life cycle of medaka fish in near-future space experiments. One is to send early embryos of any strain to space and let them develop under microgravity. The fry hatched there adapt well to microgravity (without looping), will continue to develop further to maturity, and will mate and lay eggs. Their embryonic development completes one life-cycle of the fish. The other plan is to send adult fish and let them lay eggs. When fry hatched have reached maturity and laid eggs, that realizes a complete life cycle.

For various reasons, the Japanese advisory committee for the NASDA aquatic habitat facility (chaired by K. Ijiri) has been inclined to sending adult medaka fish to space at the start of the life-cycle experiment aboard the ISS. Some reasons are that the step from fish mating to hatching of fry has already been proven in medaka under microgravity, and that in this plan many studies using abundant space-laid eggs are possible at the early phases of the life-cycle experiment. For the latter plan to work out well, adult fish should not loop in microgravity. Although, looping gradually decreases as observed in the fish *Fundulus* (von Baumgarten *et al.*, 1975; Hoffman e*t al.*, 1977), looping during the initial few days would exhaust the fish, and their mating behavior cannot be expected until their full recovery. It may take a long time, during which there is the risk of the adult fish dying.

Fish having good eyesight and less sensitivity to gravity

Studies on visual acuity of various strains of medaka and their behavior in microgravity (both in parabolic flights and in space) led us to conclude that heavily eye-dependent fish do not loop in microgravity, i.e., they are tolerant to microgravity. For their posture control on Earth (1G), they depend much on the visual organ and little on the gravisensing organ (otolith system). Because of this, they are not confused when exposed to microgravity, therefore exhibiting no looping.

A strain called HNI-II has been found to have good eyesight, and to have an ordinary sensitivity to gravity. Fish of a mutant strain *ha* (genotype *ha*/*ha*) have less sensitivity to gravity than normal fish, due to the absence of the otoliths responsible for gravisensing. The *ha* fish had no inferior eyesight, i.e., they have ordinary eyesight. Crosses (F1) between HNI-II and *ha* strains and between F1 (*ha*/-) were carried out to produce the fish F2 (*ha*/*ha*), whose phenotype was characterized by both good eyesight and less sensitivity to gravity (for details, see Ijiri, 2000).

Fish of HNI-II, *ha* and F2 (*ha*/*ha*) were exposed to microgravity (parabolic flights), and their looping behavior was observed. Figure 4 plots the number of loops of each fish during a 20-sec microgravity (parabolic flight). The data

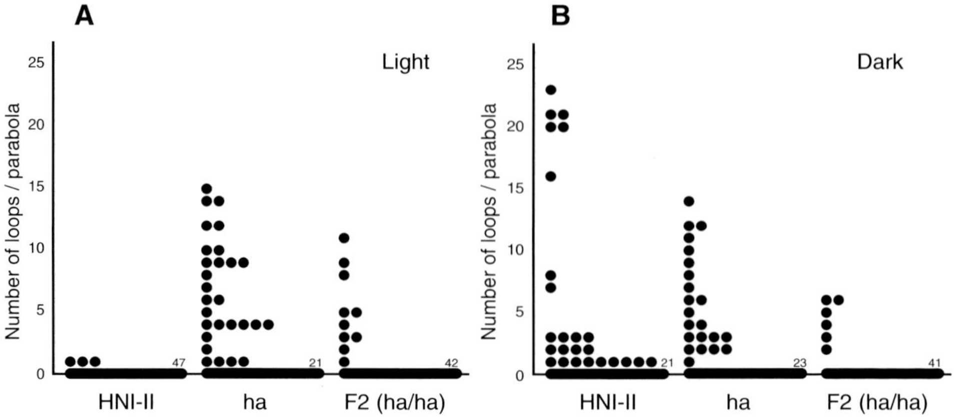

Fig. 4. Looping behavior of adult medaka fish in microgravity for three strains, HNI-II, *ha*, and F2 (*ha*/*ha*). A. Illumination applied laterally to the aquarium. B. Fish kept in complete darkness. All the data plotted here were obtained in the first parabolic flight. Each dot represents an individual fish. For other explanations, see Figure 2.

under light conditions (light applied laterally to the aquarium, Fig. 4A) and those in dark conditions (Fig. 4B) are shown. A comparison between the data of light and dark conditions tells us that the HNI-II strain remarkably increased loops when kept in darkness during microgravity. As expected, F2 (*ha*/*ha*) exhibited fewer loops than the *ha* strain. However, *ha* and F2 (*ha*/*ha*) exhibited no differences in degree of looping between the light and dark conditions. This implies that loss of eyesight (in darkness) is not a direct cause for looping behavior.

Later studies found that genotype (*ha*/*ha*) has different patterns of forming the utricular otoliths. Some fish form the utricular otoliths in later days. This may explain the individual differences observed for *ha* and F2 (*ha*/*ha*) data. The F2 (*ha*/*ha*) seem to be good candidates for the first parent medaka fish to realize their life cycle in space. When success in realizing the fish life cycle is of primary importance, the F2 (*ha*/*ha*) fish whose utricular otoliths are absent, should be the best parents. It is also interesting to see what changes occur when fish repeat their life cycle in microgravity. From this viewpoint, normal and microgravity-tolerant strains having full sets of otoliths, such as HNI-II and ccT (used in 1994), are also useful as the first parent fish.

Use of mutant fish for ground-based studies

Necessity for ground-based studies

Biological experiments in space have certain limitations in human resources as well as in spatial resources. Crew time, and facility and storage volume are less, and therefore the total flight samples obtained are far fewer when compared with the ground experiments as usually done in the laboratory. Such limitations still exist in the age of ISS. The ideal space experiment, therefore, should aim at gaining a sufficient number of samples for each data point, i.e., minimizing the number of data points (e.g., time points for sample collection). In order to plan such pin-point experiments in space, thorough preliminary studies in ground laboratories are necessary.

Several methods exist for realizing microgravity conditions on the ground. Free-falls in a drop tower, and parabolic flights of an airplane or a sounding rocket can generate a microgravity, but their durations are too short for studying the full development of fish and other animals. Simulated microgravity using an apparatus called a clinostat has been employed for fish development studies. Studies using a three-dimensional clinostat have been performed on medaka embryos. Both normal and clinostat-treated embryos developed on an almost identical time course. When the organogenesis of a complex tissue such as retina was studied, no differences were observed either in detailed tissue architecture (such as distribution of cells in photoreceptor layers) or in the expression of opsin genes (Nishiwaki *et al.*, 1999). Morphometry of the head skeleton could not reveal any differences between the cartilage formation of the control and clinostat-treated groups (Fig. 5). For zebrafish, simulated microgravity (using a bioreactor with a rotating wall) caused a delay in the otolith development (specifically for saccular otoliths) (Moorman *et al.*, 1999).

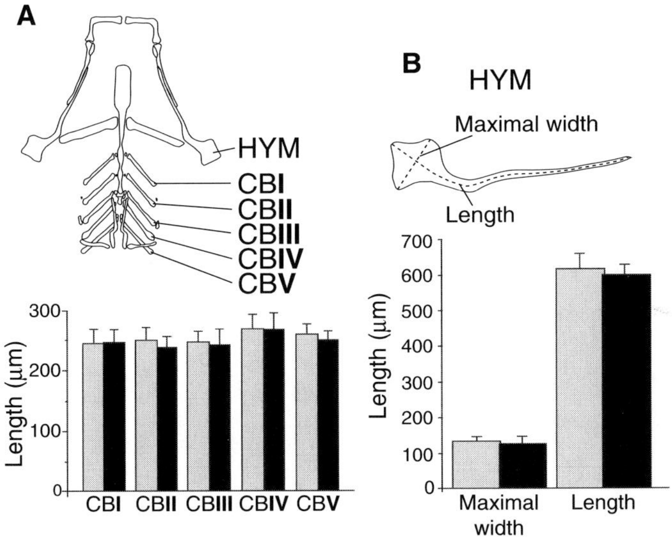

Fig. 5. Size of cartilage in the clinostat-treated and the control medaka embryos. Clinostat treatment started just after fertilization and stopped at stage 38 (just before hatching). A. Schematic illustration of ceratobranchial cartilage (CBI to CBV) and its length. B. Length and maximal width of hyomandibular cartilage (HYM). Grey bars indicate the data for clinostat-treated embryos, and black bars indicate the data for the control group. A three-dimensional clinostat was used (Xmax = 7 rpm, Ymax = 5 rpm).

However, for many studies on fish development, the clinostat treatment cannot be continued beyond the time of hatching because the rotation of the aquarium itself causes the fry to suffer motion sickness through visual and vestibular systems. Thus, on the ground there seem no measures for realizing microgravity conditions (including simulated-microgravity) long enough to cover a fish life cycle.

Beside the clinostat, the employment of specific biological samples, e.g., use of mutant fish, is another approach in ground-based studies. Different mutant fish may be used for studying what changes occur in different organs in microgravity. As an example, we introduce our recent studies using an otolith-deficient medaka mutant.

Otolith formation in normal medaka fish

Like most vertebrates, fish have three chambers of the inner ear (utriculus, sacculus, and lagena) on each side. In medaka, the utriculus is situated at the most anterior side (near the oral tip), the sacculus is next to it, and the lagena lies in the most posterior position. In medaka, usually one otolith exists in each chamber.

In embryonic development of normal medaka fish, both the utricular and the saccular otoliths are formed before fry hatch. The otoliths are observed in embryos for the first time at stage 25 (stage numbers given by Iwamatsu, 1994 are used throughout), which is the 18–19 somites stage, and the vitello-caudal vein and nasal sacs are also formed. Thus, in the ear vesicles of fry at hatching, two pairs of otoliths can be observed under a dissecting microscope through the semitransparent skin. Lagenar otoliths are formed later, i.e., about 30 days after hatching (see Fig. 6). The utricular otoliths are believed to be responsible for detecting the direction of gravity. One reason for this is that they are shaped like a small circular plate and lie perpendicular to the direction of gravity. Another reason comes from behavioral studies of normal and mutant medaka during parabolic flights. The saccular otoliths are assumed to detect the acceleration due to horizontal and vertical movements of the fish body. The lagenar otoliths are related to acoustic sensation.

Otolith formation in a mutant fish

Though head tilt (*het*) mutant mice completely lack otoliths (lacking both utricular and saccular otoliths) (Bergstrom *et al.*, 1998), fish that lack both pairs of otoliths have not been reported yet. The monolith mutant (mnl) of zebrafish is a dominant mutation that specifically inhibits formation of utricular otoliths (Riley and Moorman, 2000). The *ha* mutant of medaka was previously discovered by the late Dr. H. Tomita. He reported on their abnormal swimming behavior and responsibility of one recessive and autosomal gene, also implying a deficiency in their otolith formation. The strain has been

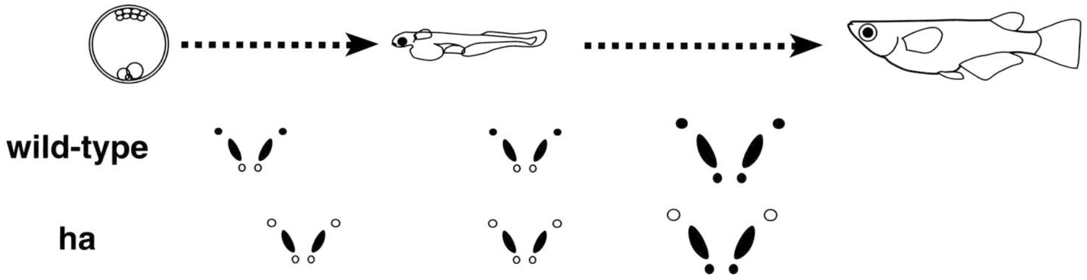

Fig. 6. Formation of otoliths in medaka. In wild-type (normal) fish, two pairs of otoliths (utricular and saccular otoliths) are formed before hatching. The lagenar otoliths appear after hatching. In mutant fish (*ha*), utricular otoliths are absent, and saccular otoliths are formed later than in the wild-type fish. The lagenar otoliths appear at the same time as those of the wild-type fish.

stocked in Nagoya University, without being used in any studies until we started a detailed investigation from the viewpoint of space biology, i.e., studies on their otolith formation together with their behavior in microgravity (Ijiri, 2000).

In *ha* mutant fish (genotype *ha/ha*), the utricular otoliths are not formed during embryonic stages, thus hatching fry have only one pair of the saccular otoliths. A delay of about a day is noted for the saccular otoliths to appear when compared with normal fish. The saccular otoliths first appear around stage 28 in the *ha* strain. The lagenar otoliths appear at the same time as in normal fish (Fig. 6). As the fish grow, it becomes difficult to see any otoliths through the skin, and X-ray photographs (using a microfocus X-ray apparatus) are useful for studying the shape and location of otoliths (Fig. 7).

What differences will be observed in the fish that have grown up or have experienced their life cycle in microgravity? For fry developed or hatched in space, differences in vestibular function and in enzyme levels of the brain and inner ear have been reported (Hoffman *et al.*, 1978; Horn and Sebastian, 2002; Krasnov, 1977; Nindl *et al.*, 1996; Rahmann *et al.*, 1994, 1996; Sebastian et al., 2001; Slenzka et al., 1995). The use of mutant fish (*ha*) with a deficiency in the vestibular system can help us to understand the differences we can expect. The detailed atlas of medaka (Anken and Bourrat, 1998; Ishikawa et al., 1999) may serve the purpose. This specific mutant fish can serve as a model fish for the life-cycle experiments in microgravity only when studies concern their vestibular system.

Additional treatment for establishing a perfect model fish

The medaka fry generally respond more to light than to gravity immediately after hatching. Such response to light is called the dorsal-light response (DLR), and it means that fish turn their dorsal side to the light. When the aquarium is illuminated laterally, fry of both normal and mutant fish show a perfect DLR, i.e., swimming upright with their back facing the light. However, a few days

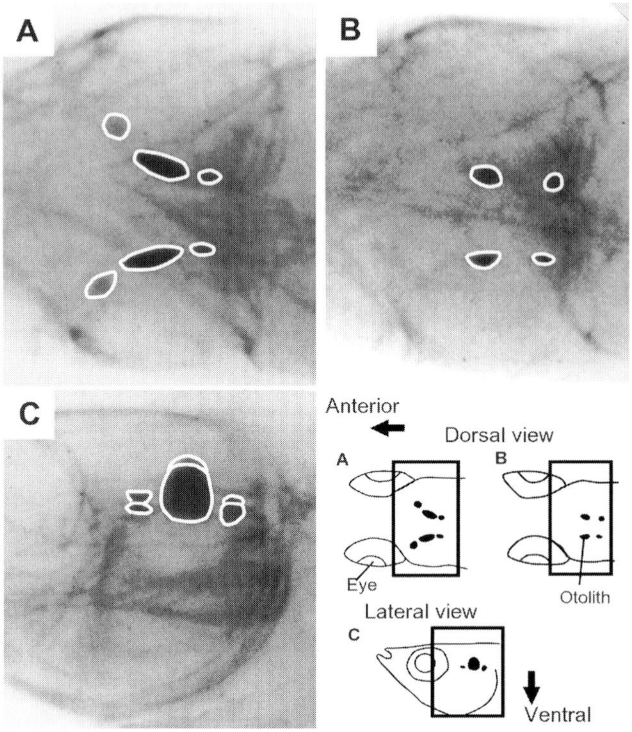

Fig. 7. X-ray photographs of the head portion of adult medaka. The wild-type and *ha* fish were examined. A. Dorsal view of wild-type fish. B. Dorsal view of *ha* mutant fish. Note the absence of utricular otoliths. C. Lateral view of wild-type fish.

after hatching, normal fish respond to the light for less time, and respond to gravity for more time. Even when illuminated laterally, fish now swim with their abdomen facing parallel to the bottom of the aquarium (i.e., the abdomen toward the direction of gravity).

In contrast, fry of the *ha* mutant continue to exhibit a complete DLR for a week or more. They always swim with their dorsal side oriented to the light. If the light is applied laterally, they swim upright as described above. If illuminated from the aquarium bottom, *ha* fry swim with their dorsal side oriented to the light, keeping their body parallel to the aquarium bottom. The degree of their light dependence can be obtained by illuminating the aquarium from the lateral side and measuring the total time they exhibit DLR in 10 min. When kept in an ordinary light regimen (light from the above; 16 h light and 8 h dark), *ha* fish eventually lose their light dependence as they grow, showing more gravity dependence instead. A shift from light dependence to gravity dependence takes place about 40 days after hatching (Fig. 8A). The reason for this shift is not known. The fish may begin to sense gravity using

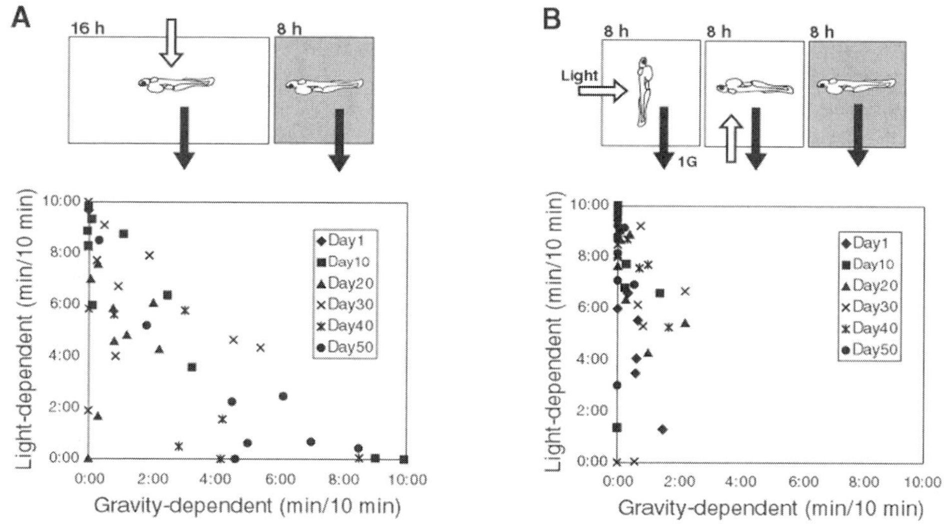

Fig. 8. Light dependence and gravity dependence of mutant medaka (*ha*). A. *ha* fish raised under an ordinary light regimen as illustrated. B. *ha* fish raised in altered light direction. Every 10 days starting at the day of hatching (Day 1), fish were tested for their light dependence and gravity dependence. Light was applied laterally to the aquarium, and fish behavior was observed for 10 min. The total time each fish spent with its back to the light (i.e., showing a perfect DLR) is given in the ordinate (light dependence). The total time each fish spent with its abdomen oriented to the aquarium bottom is given in the abscissa (gravity dependence).

proprioceptive sensation or they may have started to use the saccular otoliths for gravisensing.

In order to suppress this shift, mutant fish are continuously kept in the light regimen with the light direction altered from the time of hatching. We employed a daily light regimen such as 8 h of light from above, 8 h of lateral light, and 8 h of darkness. No shift to gravity-dependence occurred in the fish treated with altered light directions (Fig. 8B). If studies on the vestibular system (all the pathway from sensory organs to the brain) of these model fish detect some differences from the normal fish, similar differences may be observed for the fish that experienced a life cycle in microgravity.

References

Anken, R. and Bourrat, F. (1998) Brain atlas of the medaka fish, INRA Editions, Paris.

Bergstrom, R.A., You, Y., Erway, L.C., Lyon, M.F. and Schimenti, J.C. (1998) Deletion mapping of the head tilt (*het*) gene in mice: a vestibular mutation causing specific absence of otoliths. Genetics 150, 815–822.

Bleum, V., Stretzke, E. and Kreuzberg, K. (1994) C.E.B.A.S.-aquarack project: The mini-module as tool in artificial ecosystem research. Acta Astronautica 33, 167–177.

Boyle, R., Mensinger, A.F., Yoshida, K., Usui, S., Intravaia, A., Tricas, T. and Highstein, S.M. (2001) Neural readaptation to Earth's gravity following return from space. J. Neurophysiol. 86, 2118–2122.

de Jong, H.A., Sondag, E.N., Kuipers, A. and Oosterveld, W.J. (1996) Swimming behavior of fish during short periods of weightlessness. Aviat. Space Environ. Med. 67, 463–466.

Furukawa, R. and Ijiri, K. (2002) Swimming behavior of larval medaka fish under microgravity. Adv. Space Res. 30, 733–738.

Hairston, N.G., Jr, Li, K.T. and Easter, S.S., Jr. (1982) Fish vision and the detection of planktonic prey. Science 218, 1240–1242.

Hoffman, R.B., Salinas, G.A. and Baky, A.A. (1977) Behavioral analyses of killifish exposed to weightlessness in the Apollo-Soyuz test project. Aviat. Space Environ. Med. 48, 712–717.

Hoffman, R.B., Salinas, G.A., Boyd, J.F., von Baumgarten, R.J. and Baky, A.A. (1978) Effect of prehatching weightlessness on adult fish behavior in dynamic environments. Aviat. Space Environ. Med. 49, 576–581.

Horn, E. and Sebastian, C. (2002) Adaptation of the macular vestibuloocular reflex to altered gravitational conditions in a fish (*Oreochromis mossambicus*). Adv. Space Res. 30, 711–720.

Ijiri, K. (1994) A preliminary report on IML-2 medaka experiment: Mating behavior of the fish medaka and development of their eggs in space. Biol. Sci. Space 8, 231–233.

Ijiri, K. (1995a) Medaka fish had the honor to perform the first successful vertebrate mating in space. Fish Biol. J. MEDAKA 7, 1–10.

Ijiri, K. (1995b) The first vertebrate mating in space-a fish story, RICUT, Tokyo, Japan.

Ijiri, K. (1995c) Fish mating experiment in space-what it aimed at and how it was prepared. Biol. Sci. Space 9, 3–16.

Ijiri, K. (1998) Development of space-fertilized eggs and formation of primordial germ cells in the embryos of medaka fish. Adv. Space Res. 21, 1155–1158.

Ijiri, K. (2000) Vestibular and visual contribution to fish behavior under microgravity. Adv. Space Res. 25, 1997–2006.

Ishikawa, Y., Yoshimoto, M. and Ito, H. (1999) A brain atlas of a wild-type inbred strain of the medaka, *Oryzias latipes*. Fish Biol. J. MEDAKA 10, 1–26.

Iwamatsu, T. (1994) Stages of normal development in the medaka *Oryzias latipes*. Zool. Sci. 11, 825–839.

Krasnov, I.B. (1977) Quantitative histochemistry of the vestibular cerebellum of the fish Fundulus heteroclitus flown aboard the biosatellite Cosmos-782. Aviat. Space Environ. Med. 48, 808–811.

Moorman, S.J., Burress, C., Cordova, R. and Slater, J. (1999) Stimulus dependence of the development of the zebrafish (*Danio rerio*) vestibular system. J. Neurobiol. 38, 247–258.

Mori, S. (1997) Carp experiment in space microgravity-a visual-vestibular sensory conflict model. Biol. Sci. Space 11, 327–333.

Mori, S., Mitarai, G., Takabayashi, A., Usui, S., Sakakibara, M., Nagatomo, M. and von Baumgarten, R.J. (1996) Evidence of sensory conflict and recovery in carp exposed to prolonged weightlessness. Aviat. Space Environ. Med. 67, 256–261.

Nagaoka, S., Matsubara, S., Kato, M., Uchida, S., Uemura, M., Sakimura, T., Ogawa, N. and Nakamura, H.K. (1999) Water quality management for low temperature marine fishes in space. Biol. Sci. Space 13, 327–332.

Nindl, G., Koertje, K.H., Slenzka, K. and Rahmann, H. (1996) Comparative electronmicroscopical investigations on the influences of altered gravity on cytochrome oxidase in the inner ear of fish: a spaceflight study. J. Hirnforsch. 37, 291–300.

Nishiwaki, Y., Ijiri, K., Satoh, T., Tokunaga, F. and Morita, T. (1999) Retinal photoreceptor and related gene expression in normal and clinostat-treated fish embryos. Adv. Space Res. 23, 2045–2048.

Ohki, H. and Aoki, K. (1985) Development of visual acuity in the larval medaka, *Oryzias latipes*. Zool. Sci. 2, 123–126.

Rahmann, H., Jeserich, G. and Zentzius, I. (1979) Ontogeny of visual acuity of rainbow trout under normal conditions and light deprivation. Behaviour 68, 315–322.

Rahmann, H., Hilbig, R., Flemming, J. and Slenzka, K. (1996) Influence of long-term altered gravity on the swimming performance of developing cichlid fish: including results from the 2nd German Spacelab mission D-2. Adv. Space Res. 17, 121–124.

Rahmann, H., Slenzka, K., Hilbig, R., Flemming, J., Paulus, U., Kortje, K., Bauerle, A., Appel, R., Neubert, J., Briegleb, W., Schatz, A., Bromeis, B. (1994) Influence of hyper- and hypo-gravity on the early ontogenetic development of cichlid fish. Behavioral and ultrastructural investigations including first results of the 2nd German Spacelab mission D-2. Proc. Eur. Symp. Life Sci. Res. Space 5, 165–169.

Riley, B.B. and Moorman, S.J. (2000) Development of utricular otoliths, but not saccular otoliths, is necessary for vestibular function and survival in zebrafish. J. Neurobiol. 43, 329–337.

Sakimura, T., Suzuki, T., Matsubara, S., Uchida, S., Kato, M., Tanemura, R. and Honda, S. (1999) NASDA aquatic animal experiment facilities for space shuttle. Biol. Sci. Space 13, 314–320.

Sebastian, C., Esseling, K. and Horn, E. (2001) Altered gravitational forces affect the development of the static vestibuloocular reflex in fish (*Oreochromis mossambicus*). J. Neurobiol. 46, 59–72.

Shimura, R., Ijiri, K., Mizuno, R. and Nagaoka, S. (2002) Aquatic animal research in space station and its issues-focus on support technology on nitrate toxicity. Adv. Space Res. 30, 803–808.

Shimura, R., Kumagai, H., Kozu, H., Motoki, S., Ijiri, K. and Nagaoka, S. (1999) Application of nitrifying and denitrifying process to waste management of aquatic life support in space. Biol. Sci. Space 13, 351–360.

Shimura, R., Motoki, S., Mizuno, R., Ijiri, K. and Nagaoka, S. (2000) Long-term water environment study in a compact life support system for medaka-focusing on nitrate toxicity. Proc. 22nd Int. Symp.on Space Tech. and Sci., Vol. II, 2039–2043.

Slenzka, K. (2002) Life support for aquatic species-Past; present; future. Adv. Space Res. 30, 789–795.

Slenzka, K., Appel, R. and Rahmann, H. (1995) Development of altered gravity dependent changes in glucose-6-phosphate dehydrogenase activity in the brain of the cichlid fish *Oreochromis mossambicus*. Neurochem. Int. 26, 579–585.

Takabayashi, A., Ohara, K., Ohmura, T., Watanabe, S., Mori, S., Tanaka, M. and Sakuragi, S. (1997) Mechanism of vestibular adaptation of fish under microgravity. Biol. Sci. Space 11, 351–354.

Uchida, S., Masukawa, M. and Kamigaichi, S. (2002) NASDA aquatic animal experiment facilities for space shuttle and ISS. Adv. Space Res. 30, 797–802.

Uchida, S., Matsubara, S., Kato, M., Sakimura, T., Nakamura, H.K., Ogawa, N. and Nagaoka, S. (1999) VFEU water quality control in STS-95 mission. Biol. Sci. Space 13, 321–326.

von Baumgarten, R., Baldrighi, G. and Shillinger, G.L., Jr. (1972) Vestibular behavior of fish during diminished g-force and weightlessness, Aerospace Med., 43, 626–632.

von Baumgarten, R.J., Simmonds, R.C., Boyd, J.F. and Garriott, O.K. (1975) Effects of prolonged weightlessness on the swimming pattern of fish aboard Skylab 3. Aviat. Space Environ. Med. 46, 902–906.

Wiederhold, M.L., Gao, W.Y., Harrison, J.L. and Hejl, R. (1997) Development of gravity-sensing organs in altered gravity. Gravit. Space Biol. Bull. 10, 91–96.

Yamashita, M., Izumi-Kurotani, A., Imamizo, M., Koike, H., Okuno, M., Pfeiffer, C.J., Komazaki, S., Sasaki, F., Ohira, Y., Kashima, I., Kikuyama, S., Ohnishi, T., Mogami, Y. and Asashima, M. (2001) Japanese red-bellied newts in space-astronewt experiment on space shuttle IML-2 and space flyer unit, Biol. Sci. Space, 15, Suppl., s96-103.

Developmental Biology Research in Space
H.-J. Marthy (editor)

Mammalian Development in Space

April E. Ronca*

Life Sciences Division, NASA Ames Research Center, Moffett Field, CA 94035, USA

Abstract

Life on Earth, and thus the reproductive and ontogenetic processes of all extant species and their ancestors, evolved under the constant influence of the Earth's 1 *g* gravitational field. These considerations raise important questions about the ability of mammals to reproduce and develop in space. In this chapter, I review the current state of our knowledge of spaceflight effects on developing mammals. Recent studies are revealing the first insights into how the space environment affects critical phases of mammalian reproduction and development, viz., those events surrounding fertilization, embryogenesis, pregnancy, birth, postnatal maturation and parental care. This review emphasizes fetal and early postnatal life, the developmental epochs for which the greatest amounts of mammalian spaceflight data have been amassed. The *maternal–offspring system*, the coordinated aggregate of mother and young comprising mammalian development, is of primary importance during these early, formative developmental phases. The existing research supports the view that biologically meaningful interactions between mothers and offspring are changed in the weightlessness of space. These changes may, in turn, cloud interpretations of spaceflight effects on developing offspring. Whereas studies of mid-pregnant rats in space have been extraordinarily successful, studies of young rat litters launched at 9 days of postnatal age or earlier, have been encumbered with problems related to the design of in-flight caging and compromised maternal–offspring interactions. Possibilities for mammalian birth in space, an event that has not yet transpired, are considered. In the aggregate, the results indicate a strong need for new studies of mammalian reproduction and development in space. Habitat development and systematic ground-based testing are important prerequisites to future research with young postnatal rodents in space. Together, the findings support the view that the environment within which young mammals develop, comprised of its

* E-mail: aronca@mail.arc.nasa.gov

mother and siblings, is of paramount importance in interpreting spaceflight effects.

Introduction

Studying the life cycles of mammals in space promises to reveal astounding new insights into how gravity shaped life on Earth and will enable us to explore the adaptability of humans and animals to life beyond Earth. No mammalian (or even vertebrate) species have yet been raised from conception to sexual maturity in the absence of gravity (Miquel and Souza, 1991; Moody and Golden, 2000; Wassersug, 2001). However, experiments with fish, amphibians, birds and rodents flown at various phases during their reproductive and developmental cycles have shown that some aspects of vertebrate reproduction and development can proceed in space (e.g., Ijiri, 1995; Ronca and Alberts, 2000a). For example, ovulation was induced in *Xenopus laevis* females aboard the orbiting Space Shuttle and their eggs yielded larvae (Fejtek *et al.*, 1998). Ijiri (1995) provided evidence for mating, fertilization and hatching of *Medaka* fish in space. Pregnant rats flown in space were returned to Earth to give birth to healthy offspring (Ronca and Alberts, 2000a). These and other studies are beginning to answer fundamental questions as to whether vertebrates can reproduce and develop in space. The life cycles of different vertebrates are characterized by vastly differing adaptations that vary in number and complexity. In mammals, a myriad of intricate phases are required for successful reproduction and development to adulthood, viz., internal fertilization, pregnancy, placentation, embryogenesis, fetal development, birth, lactation, parental care and postnatal maturation. We are just beginning to learn about whether the sequence of events characterizing the mammalian life cycle can transpire under gravitational conditions that deviate from those experienced on Earth (1 *g*).

In this chapter, I review the current state of our knowledge of the effects of spaceflight on developing mammals. I will be primarily concerned with fetal and early postnatal life, the developmental epochs for which the greatest amounts of data have been amassed. The discussions will focus predominantly on the microgravity component of spaceflight. Although it is recognized that radiation, atmospheric pressure, launch and landing and other aspects of the space environment may affect the outcome of spaceflight studies, there is presently incomplete data available in these areas. Studies utilizing acute centrifugation have been useful for parceling out launch and landing effects, however, further systematic attention to the non-gravitational components of spaceflight awaits studies utilizing the International Space Station (ISS). In particular, the proposed onboard centrifuge module will enable simultaneous, direct comparisons of μ*g* and 1 *g* control specimens *within the space environment*. The first part of this chapter is concerned with the *maternal–offspring* *system*, the coordinated assemblage of mother and young that comprises mammalian development, and tracks the chronology of spaceflight experiments that have carried reproducing female mammals and their offspring. The second part of the chapter is devoted to review and integrative discussions of the published data, focuses primarily on hypogravity (spaceflight) data with relevant findings derived from ground-based models incorporated where appropriate.

The maternal–offspring system in mammals

Mammalian development progresses within the context of an intact, integrated system comprised of a mother and her young (Alberts and Ronca, 1999). This family system, as a unit, is characterized by coordinated changes within and between a mother and her young. As the offspring are developing within the womb and nest, the maternal–offspring system follows its own ontogenetic trajectory. Because mammalian development involves bi-directional linkages between mothers and infants, any factor that disrupts either the mother or her offspring is liable to disrupt the maternal–offspring system. If this occurs, the organizing influence of the system on its constituent parts is likely to change and the ontogenetic processes within the individuals can be disrupted.

The maternal–offspring system evolved under the constant force of the Earth's 1 *g* gravitational field. It is therefore reasonable to predict that the relationship between mothers and their young will change significantly in the weightlessness space environment. One consequence of these changes is that the expression of characteristic maternal behaviors may be impeded.

The importance of maternal behavior

Throughout the early formative phases of a young mammal's life, the mother is of paramount importance. This is particularly true for altricial species, such as the rat, that are born in a highly immature state with only relatively undeveloped sensory and motoric capabilities. The mammalian mother defines the major features of the infant's world, affording vital resources that lead to proper growth and development. Prenatally, the fetus grows and develops within the warm, protective enclave of its mothers' womb. In placental mammals, oxygen, nutrients and other resources are supplied automatically through the umbilical conduit. At birth, the mother protects her young and provides them with sustenance through the emergence of maternal behavior, a constituent set of behavioral activities including nursing, touching and carrying the young (Bridges, 1990). There is increasing recognition that maternal care-taking patterns play major roles in directing and regulating postnatal development. Perinatal stimulation, in particular, appears to play a predominant role in establishing neural, behavioral and physiological organization in the newborn. The neonate possesses few totally independent regulations, and relies on stimulation by the mother for induction and maintenance of normal behavior and physiology (Hofer, 1983).

In this regard, physical stimulation appears to have special biological relevance. Rat dams lick their pups incessantly, beginning at birth and through the preweanling period. Maternal licking has been implicated in: the onset of breathing at birth (Ronca *et al.*, 1996); adrenocortical activity (Smotherman, 1984); heart rate and blood pressure (Hofer, 1983; Hofer *et al.*, 1988); growth hormone (GH) and ornithine decarboxylase (ODC; a marker of growth and development) in brain, heart and liver (Kuhn and Schanberg, 1998); stimulation of the micturation reflex; numbers of motor neurons innervating anogenital area in male rats and adult male sexual behavior (Moore, 1992); maturation of the hypothalamic-pituitary-adrenal (HPA) axis (Francis *et al.*, 1999; Levine, 2001) and increased expression of brain-derived neurotrophic factor (BDNF) and other mRNA subunits involved in spatial learning and memory (Liu Diorio *et al.*, 2000).

Disruptions of the maternal–offspring system in space

Factors that disrupt physiological and behavioral interchanges between mothers and their offspring can produce gross changes in the trajectory of infant growth and development. Disturbances of the prenatal environment can result in intrauterine growth retardation (IUGR) and low birth weight (Langley-Evans, 2000), predictors of hypertension and other adult disease states (Godfrey and Barker, 2001). Postnatal maternal deprivation can have dire neural, physiological and behavioral consequences for the developing mammal, including starvation, dehydration, hypothermia, growth retardation, delayed neural development, impaired cognitive function and emotional disorders (Kuhn and Schanberg, 1998; Levine, 2001).

In spaceflight studies, disruptions of the maternal–offspring system can pose major interpretive problems by obscuring meaningful results or creating false impressions of how microgravity affects development. In rodents, the only developing mammals that have been flown in space so far, maternal behavior and physical relations between mother and pup may be altered in space (see Ronca *et al.*, 2001 for further discussion). For example, the mothers' maneuverability of pups and her proximity to them may be impeded during attempts to huddle and nurse. Pup licking may be diminished. Retrieving behavior, important for maintaining pups within a coherent nest, may not be possible as pups endlessly float out of the nest. These changes are likely to diminish pup milk intake, warmth and tactile stimulation normally provided by the mother. Altered development of brain, physiology and behavior may erroneously be attributed to effects of microgravity and spaceflight, rather than disruptions of the maternal–offspring system.

Sources of gravitational influence on developing mammals

A major challenge to accurate analysis of the effects of gravity on development is separating effects of gravity on mothers versus offspring. Because the maternal–fetal system is so highly intertwined in mammals, differentiation among these 'direct' and 'indirect' effects of gravity on the developing young is a critical, yet extremely difficult task. As discussed by Alberts and Ronca (1999), 'direct' effects of gravity are those that operate through a primary relation with the recipient organism, tissue or cell. In this case, applications of fractional g-loads would be expected to produce fractional differences from the 1 *g* phenotype. In contrast, 'indirect' effects of gravity are those expressed through avenues of the mammalian system. For example, if mothers cannot stabilize their bodies in the nursing posture or if pups cannot retain metabolic heat because their huddling behavior is disrupted, the altered growth effect would clearly be an indirect consequence of weightlessness.

In placental mammals, the potential for indirect influences of altered gravity on developing offspring may be even greater than the likelihood of direct gravity action at most developmental stages (Keefe, 1985). The phases spanning embryogenesis, perinatal life and the period of postnatal maturation are replete with maternal–offspring interactions. For these reasons, it is critical that gravitational studies of developing mammals incorporate careful, systematic analyses of changes in the maternal–offspring system in space. Only in this way can effects of gravity on development be accurately identified and studied.

A chronological overview of spaceflight experiments on mammalian development

Over 20 years has passed since the first reproducing mammals (rats) were launched into space by the former Soviet Union (Cosmos-1129; Serova and Denisova, 1982). Since this landmark event, we have begun to amass new data regarding the effects of spaceflight on reproducing and developing mammals (e.g., Burden *et al.*, 1997; Plaut, Maple, Vyas, Muniam, Darling, Casey, Alberts, 1998; Ronca and Alberts, 2000a; Serova *et al.*, 1984; Wong and DeSantis, 1998). Yet flight opportunities have been sparse. Table 1 shows the six flight missions that, to date, have carried either mating rats, pregnant mother rats or lactating mother rats and their offspring into space in relation to the duration and ages of the animals flown on each mission. These missions are: (1) Cosmos 1129; (2) Cosmos 1514; (3) NIH.Rodent (R)1; (4) NIH.R2; (5) NIH.R3; and (6) Neurolab. Pregnant mice have been flown only once thus far, on the Neurolab mission.

These six pioneering spaceflight missions, each representing important and difficult accomplishments, have revealed the first insights into whether mammalian reproduction and development can proceed in space. Collectively, these studies form the basis for the nascent field of mammalian developmental space biology. Although few in number, each of the flights involved multiple

Table 1

Chronological summary of spaceflight missions carrying reproducing and developing mammals

Mission	Year	Mission duration (days)	Age of offspring at launch[a]
Cosmos-1129	1979	18.5	Not applicable
Cosmos-1514	1983	4.5	Gestational days 13–18
NIH.Rodent 1 (STS-66)	1994	11	Gestational days 9–20
NIH.Rodent 2 (STS-70)	1995	9	Gestational days 11–20
NIH.Rodent 3 (STS-72)	1996	9	Postnatal days 5, 8 and 14
NEUROLAB (STS-90)[b]	1998	16	Postnatal days 8 and 15

[a]The day of conception is gestational day 1 =; birth occurs on G22/23 or postnatal day 0.
[b]Nursing rat litters and pregnant mice (*Mus musculus*) were flown this mission. The latter are not represented in this table.

analyses, thereby magnifying the value of each mission. To the credit of the teams of international researchers who have conducted space biology studies of developing mammals, much of the available data are detailed and systematic. In some cases, flight results have afforded less interpretable data but revealed important ‘lessons learned’. The maternal–offspring system is remarkably sensitive to environmental conditions, particularly during the early phases that encompass prenatal and neonatal life. For this reason, flight conditions and hardware, including caging, food delivery and waste removal, can exert extraordinary influences on the animals’ health and development. Because there have been so few attempts to maintain continuity across spaceflight experiments with regard to most variables, it is often difficult to make meaningful comparisons. In view of these observations, I first review the logistical details of each mission before discussing the data.

Cosmos 1129

Cosmos 1129 was an 18.5-day mating experiment in which male and female rats were launched while separated, then a partition was removed in flight to provide opportunities for copulation and conception (Serova and Denisova, 1982). The animals were flown in a single BIOS-Vivarium cage (Fig. 1, upper left-hand image) measuring 160 × 220 × 660 mm and equipped with automated feeders, watering devices, ventilation and waste management systems and a 16:8 light:dark cycle (Serova and Denisova, 1982). The animals were fed a paste diet.

Cosmos 1514

A fundamental goal of the Cosmos 1514 mission was to demonstrate the feasibility of pregnancy and prenatal development in microgravity (Serova, 1993). To accomplish this goal, the mission carried 10 pregnant rats into low Earth orbit on an unmanned biosatellite. The time-mated rat dams spent

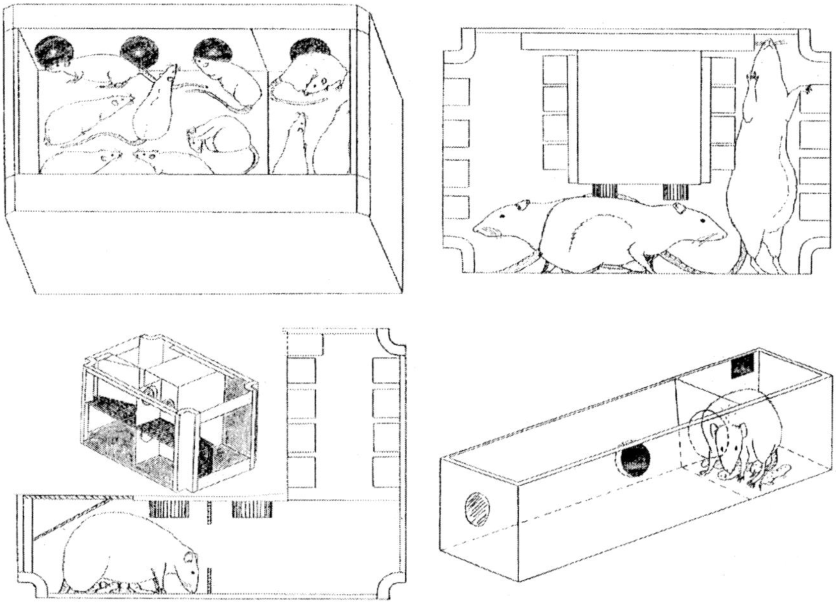

Fig. 1. Flight housing used for studies of reproducing and developing rats in space. For each image, rats are shown in the 1 *g* configuration. Upper Left: The Russian Bios-Vivarium cage was used for the rodent mating experiment flown on Cosmos 1129 and carried the first pregnant mammals (rats) into space on the 4.5 day Cosmos 1514 mission. For Cosmos-1129, a partition (demarcated by the solid interior lines), separated male and female rats prior to launch (Redrawn from Gazenko, 1993). Upper Right: The Animal Enclosure Module (AEM) carried mid-pregnant rats on the NASA Space Shuttle for 9 or 11 days, respectively, for the NIH.Rodent (R)1 and NIH Rodent (R)2 experiments. Lower Left: The Animal Enclosure Module-Nursing Facility (AEM-NF, a modified AEM) carried the first lactating dams and litters into space on either postnatal day 5, 8, or 15 for the NIH.Rodent (R)3 experiment. The AEM-NF was also used to house older litters (postnatal 16) on the Neurolab mission. Two dams and their litters were housed in each AEM-NF.The insert shows a perspective drawing of the AEM-NF with darker shading depicting the partition between litters and the lighter shading illustrating the habitable areas of a single compartment. Lower Right: The Research Animal Holding Facility (RAHF) was used to house rat litters launched on postnatal day 9 on the Neurolab mission. In all experiments, controls were housed in identical caging (synchronous controls) and in standard colony caging (vivarium controls). For some of the experiments, additional control groups were used that are beyond the scope of this discussion. Drawings by Kenneth E. Johnson.

4.5 days in space, spanning the 13–18 days of their 22-day pregnancies. Their offspring were the first mammalian specimens to undergo a portion of their gestation in the absence of Earth-normal gravity. All 10 dams were flown in the BIOS-Vivarium cage as described for Cosmos 1129.

NIH.Rodent1

Following the Cosmos mission, there was an 11-year hiatus until the next investigation of mammalian development in space. Then in 1994, NASA and the National Institutes of Health (NIH) jointly sponsored the first of a series of pioneering developmental biology experiments on the NASA Space Shuttle (the 'NIH.Rodent' experiments), the first of which was NIH.Rodent 1 (R1). The R1 mission was considerably longer in duration than Cosmos 1514 and provided an opportunity for far more detailed studies of the mammalian mother–offspring system during and after spaceflight. The mission carried 10 pregnant rats that were launched on the 9th day of gestation and landed 11 days later, just 48–72 h prior to the expected birth of the offspring.

Ten dams were group-housed five to a cage in an Animal Enclosure Module (AEM), shown in Figure 1 (upper right-hand image). The AEM has a habitable volume of 863 cm^2. A food bar diet lined the walls of the habitat and a central lixit system provided continuous access to water. Dams were maintained on a 12:12 h light:dark cycle. Environmental information was down-linked from the space shuttle to Kennedy Space Center where in-flight temperature, humidity and lighting were simulated in control AEMs.

The 10 dams flown in the R1 experiment underwent a surgical laparotomy 48 h prior to launch for the purpose of counting fertilization sites. Then, immediately after landing, dams underwent a 'unilateral hysterectomy' in which one of each dams' paired uterine horns was removed for immediate recovery of fetal specimens. The remaining uterine horn was left in place and the pregnancies of all 10 dams continued to term. This procedure yielded statistically valid numbers of litters from which both prenatal and postnatal offspring were derived (Alberts *et al.*, 1996).

NIH.Rodent (R)2

The NIH.R2 (R2) flight was in many respects similar to R1. Caging and pre-flight treatment of the dams was identical for the two flights. Launch occurred 2 days later than it had for R1, on the 11th gestational days with landing on gestational day 20 (Table 1). Unlike R1, unilateral hysterectomy was not performed on the dams upon recovery. Instead, four dams and fetuses were processed for study upon recovery; six dams underwent birth and their pups were studied postnatally. The two flights yielded complimentary data. The R1 labor and birth analyses were repeated in R2 dams that were not surgically manipulated following flight. This afforded a rare opportunity to validate the original findings (e.g., Ronca and Alberts, 2000a). Together, R1 and R2 yielded valuable new information on responses of the mother–offspring system to prenatal spaceflight.

NIH.Rodent (R)3

In 1996, the NIH.R3 (R3) experiment flown on STS-72 was the first flight of developing postnatal mammals. This experiment was a prerequisite to the Neurolab Mission (described in the following section), providing an opportunity to evaluate the health and survival of nursing litters during a 9-day spaceflight. A total of six lactating dams and their pups were flown. Litters (two per age group) were launched on either 5, 8 or 14 days of postnatal age. The litters were pair-housed in a single AEM, separated by a partition bisecting the apparatus. The AEM-NF is shown in Figure 1 (lower left-hand image). A small compartment at one end of the apparatus provided a circumscribed nursing area fitted with a 'wedge' to help support the dam during nursing.

Neurolab

A major priority for the Neurolab SpaceLab Mission, flown in 1998, was to advance understanding of the effects of space flight on the developing nervous system (Maese and Ostrach, 2002). The NIH.R3 experiment was intended to provide 'proof-of-concept' that litters of 8-day and older postnatal rats could be flown successfully in space using existing flight caging. For the Neurolab experiment, litters of rats were launched on the 16-day mission at either 8 or 15 days of age. In contrast with the NIH.R3 mission, 8-day-old litters were housed in the Research Animal Holding Facility (RAHF, Fig. 1, upper right-hand image), a rectangular design with a habitable volume of approximately 1100 cm^3. As in the NIH.R3 flight, the 15-day-old litters were housed in the AEM, with the nursing wedge removed. R3 and Neurolab led to a number of insights regarding the housing requirements of nursing rat litters in space.

Pregnant mice and embryos were also flown on the Neurolab mission and tissue was collected just a few days after launch. Although the animals fared well during flight, no detailed post-flight information is available on these dams or their offspring.

Gravity as a continuum

There is a growing recognition that, for many biological systems, gravity operates across a continuum (Phillips, 2002, in press). Many (but not all) biological systems respond to decreases versus increases in gravity with graded responses that are opposite in directionality. For example, antigravity muscles undergo hypotrophy in weightlessness, but hypertrophy in response to gravitational loading (Vasques *et al.*, 1998). Chronic acceleration (centrifugation) is a useful technique for establishing systematic relationships between gravity and particular biological responses, including those within the reproductive and developmental domain (Ronca *et al.*, 2001; Tou *et al.*, 2002). Centrifugation enables fractional increments in g-loads exceeding 1*g* to be applied to biological specimens, thereby providing the basis for a 'dose–response' curve for differing responses. The animal centrifuge planned for the ISS will enable researchers to make new comparisons between 0 and 1*g* in space and 1*g* on Earth. Comparisons across fractional increments in *g* spanning 0–1*g* will also be possible. Throughout this chapter, I discuss the results of ground-based centrifugation studies, where appropriate, to facilitate the interpretation of spaceflight studies.

Effects of spaceflight on conception, pregnancy and birth

Conception

The first major steps in mammalian reproduction are mating and conception. Serova and Denisova (1982) reported results of an 18.5-day spaceflight experiment (Cosmos 1129) in which five female rats and two male rats were sent into orbit in separate compartments, then allowed to intermingle in a common breeding chamber. No pregnancies resulted in flight dams, although post-flight laparotomies performed on the females revealed that ovulation had taken place. Two of the females were reported to have achieved pregnancy, but there was evidence for resorption of the embryos. Keefe (1985) noted that there were no pregnancies among ground controls for this mission, suggesting that factors other than spaceflight may have contributed to the experimental outcome. Yet, upon return to Earth, the male and female flight rats were mated with non-flight partners and all produced normal litters. While these experiments suggest that ovulation, copulation and fertilization can occur in space, some of the events apparently did not proceed normally. At this time, there is no equivocal evidence to indicate whether viable pregnancies either can or cannot result from rodent matings in space.

Studies conducted at the opposite end of the gravity continuum have been successful in demonstrating conception in altered *g*. Rats, mice and hamsters have been reported to reproduce and develop in hypergravity (Oyama and Platt, 1966; Oyama *et al.*, 1985; Sondag *et al.*, 1997). Oyama and colleagues (1966) adapted young rats and mice to 2.5, 3.6 or 4.7 *g* centrifugation for 2 months prior to mating. The rodents mated and conceived during continuous centrifugation at the lower (but not the highest) g-loads (Percent viable pregnancies: Rats, 86% at 2.5 *g*; 69% at 3.6 *g*, 0% at 4.7 *g*; Mice, 67% at 2.5 *g*). Together, findings derived from spaceflight and centrifugation suggest that initial phases of mammalian reproduction can occur in altered gravity, but definitive studies in space are needed to validate the generality of this conclusion. One caveat is that even if mammalian pregnancies can be established in space, it is not certain whether they can be sustained throughout critical phases of embryogenesis and implantation, or whether the events of gestation can proceed normally.

Pregnancy

As compared to studies of conception in space, far more research emphasis has focussed on the latter half of pregnancy during which many of the early, sensitive phases involving placentation, implantation and early organogenesis transpired on Earth. Thus for the three spaceflights that have carried pregnant rats into space (Table 1), the rats' pregnancies were relatively well-established prior to launch. Analyses of spaceflight effects on the early events of pregnancy comprise an important area for future research.

Upon recovery from space, the Cosmos 1514 flight dams, weighed approximately 18% less than synchronous control dams (Table 2; Serova *et al.*, 1984). However, food consumption by dams in the two groups was reported to be similar (Average food intake (g/day): flight, 54; synchronous, 57). In contrast, the R1 and R2 dams showed similar weight gains relative to synchronous and vivarium controls (Table 2; Difference from controls: NIH.R1, 0%; NIH.R2, −5%; Burden *et al.*, 1997; Ronca and Alberts, 2000a). For all three flights, fetal offspring that underwent a portion of their gestation in space were morphologically intact and appeared on a gross level to have grown and developed properly. For the R1 and R2 experiments, decidual swellings were counted and recorded during the pre-flight laparotomy. Post-flight inspections revealed that spaceflight was not associated with fetal wastage (Burden *et al.*, 1999).

The Cosmos 1514 flight fetuses weighed less than control fetuses at recovery (Table 2; Serova *et al.*, 1993) whereas for R1 and R2, flight fetuses did not differ from controls (Burden *et al.*, 1999 Dumars *et al.*, 1996). Crown-rump measurements were smaller in flight and synchronous fetuses relative to vivarium controls, suggesting a caging effect. For Cosmos 1514, R1 and R2, there were no differences in birth weights of the flight and synchronous control neonates born several days post-flight, however flight pups weighed significantly less than vivarium pups (Table 2; Serova *et al.*, 1984, 1993c; Ronca and Alberts, 2000a).

Together, these observations suggest that housing, diet or other condition(s) specific to the Cosmos 1514 flight may have led to the reduced body weights of the dams. One striking finding of the Cosmos experiment is that fetal body mass was relatively unaffected in relation to the small (5 *g*) total weight gain of dams during the 4.5 day flight. This observation suggests that mothers directed available resources to their developing fetuses. Taken together, these data provide clear evidence that rat dams flown in space during the latter half of their pregnancies are able to sustain normal body weight gains in space and support the growth and development of their gestating offspring.

During the R1 and R2 missions, we had the unique opportunity to obtain in-flight video footage of the pregnant rats that revealed their condition within the AEMs during flight (Alberts and Ronca, 1997). The dams were observed to eat, drink, groom and interact with one another. Species-typical behaviors and movements were observed in both flight and synchronous control dams. It was clear that under flight conditions, the dams ambulated about the full volume of the AEM, using their legs against the walls and surfaces to propel themselves throughout the habitat. The in-flight videos provided evidence that the pregnant dams adapted well to the weightless space environment, within just a few days of launch.

Table 2

Summary of effects of prenatal spaceflight on rat pregnancy and birth[a]

	Cosmos-1514[b]			NIH.R1[c]			NIH.R2[c]		
	Flight	Synch	Viv	Flight	Synch	Viv	Flight	Synch	Viv
Pregnancy outcome									
No. fetuses at recovery	14.8 ± 0.7	14.4 ± 0.7	13.0 ± 0.8	13.1 ± 0.3	13.1 ± 0.2	–	11.0 ± 0.82	11.7 ± 1.2	11.0 ± 0.0
No. neonates born	13.6 ± 0.8	13.4 ± 1.0	12.4 ± 0.9	6.2 ± 0.5	5.8 ± 0.4	5.8 ± 0.6	12.5 ± 0.6	13.2 ± 0.4	12.4 ± 0.6
Weights at recovery									
Dam body weights	293 ± 7[b,d]	355 ± 6	349 ± 11	317 ± 17[c,d]	318 ± 18	–	293 ± 16[c,d]	313 ± 12	317 ± 36
Fetal body weights	0.84 ± 0.03[b,d]	0.94 ± 0.02	0.92 ± 0.03	2.54 ± 0.09[c,d]	2.54 ± 0.05	2.39 ± 0.06	2.25 ± 0.08[c,d]	2.37 ± 0.08	2.60 ± 0.04
Neonatal birth weights	5.92	5.6	6.48	5.65 ± 0.54[d]	5.95 ± 0.56	6.13 ± 0.72	5.6 ± 0.1[d]	6.2 ± 0.1	6.4 ± 0.1
Birth									
No. labor contractions	–	–	–	93.1 ± 20.4	42.6 ± 21.7	–	104.5 ± 22	54.7 ± 11.3	–
Myometrial Connexin-43	–	–	–	–	–	–	2.3 ± 0.8[e]	4.0 ± 0.7	5.1 ± 0.2
Pregnancy duration (h)	–	–	–	528.8 ± 2.5	533.1 ± 3.1	532.6 ± 3.0	531.1 ± 4.0	530.8 ± 3.4	536.3 ± 3.6
Birth duration (min)	–	–	–	57.8 ± 11.9[f]	42.1 ± 2.8	–	97.5 ± 7.9	87.3 ± 17.3	–

[a]From Ronca and Alberts (2000a) and Burden *et al.* (1999).
[b]Dam and fetal body weights at recovery on gestational day 18.
[c]Dam and fetal body weights at recovery on gestational day 20.
[d]g ± s.d.
[e]HSCORE analysis (Burden *et al.*, 1999).
[f]Reduced numbers of neonates and shorter birth durations for NIH.R1 due to removal of one paired uterine horn at recovery.

Soon after recovery, the dams were housed individually in Plexiglas observation cages and videotaped continuously throughout the remainder of their pregnancies (Ronca and Alberts, 2000a). This enabled us to quantify the dams' post-flight behavior. Pregnant flight dams ambulated far less than synchronous control dams during the dark phase of the circadian cycle. Still, the amount of time that dams spent eating and drinking was similar across groups and the typical circadian rhythm of feeding behavior was observed. In summary, there were relatively few noticeable effects of spaceflight on the behavior of pregnant dams either during or following flight.

Birth

Although several invertebrate and vertebrate species have reproduced in space (e.g., Fejtek *et al.*, 1998; Ijiri, 1995), no mammal has yet undergone birth in space. Because of the complexity of mammalian birth, the first extra-terrestrial birth of a mammal will be a truly remarkable event. Several potential problem areas were recognized at the outset of the rodent studies. One concern focussed on the pregnant dams' abdominal musculature. It was thought that expulsion of the fetus from the uterus might be compromised by deconditioning of the transverse abdominus, an antigravity muscle in tetrapods (Fejtek and Wassersug, 1999). Studies of female rats flown in space during the latter half of their pregnancies and returned to Earth to undergo parturition have yielded a host of interesting and promising findings. It is encouraging that many of our initial concerns were not realized.

Following landing, four out of five of the Cosmos 1514 flight dams had successful vaginal births. However, two of the four births were reported to be long and difficult and the fifth litter was stillborn after a 2-day labor, possibly due to a blockage within the birth canal. The litters were similar in size relative to controls (Table 2). Far more detailed analyses of parturition were made of the R1 and R2 flight dams, the major finding for both experiments being that flight dams had successful, uncomplicated vaginal deliveries (Ronca and Alberts, 2000a). Spaceflight had no effect on the duration of pregnancy, and litter sizes were similar in flight and control dams. A detailed analysis of labor and birth in the R1 and R2 dams revealed no effects on the overall duration of the birth process or the temporal patterning of individual births (Table 2). One surprising finding was that the flight dams had twice as many labor contractions as to ground controls, as measured by behavioral expressions of contractions (Table 2; Burden *et al.*, 1999; Ronca and Alberts, 2000a). No effects on contraction duration were observed. It is known that uterine contractions play vital roles in facilitating the transition from prenatal to postnatal life (Ronca *et al.*, 1996; Lagercrantz and Slotkin, 1986). Studies of pregnant women have shown that although the likelihood of preterm delivery increases with increased uterine contraction frequency, contraction frequency is not clinically useful for predicting preterm delivery (Iams *et al.*, 2002).

Although the effects of increased contractions on the fetus are unknown, changes in connexin proteins that form gap junctions in myometrium at term may underlie the changes in contractile activity (Table 2; Burden *et al.*, 1999). Spaceflight did not affect uterine connexin 26, localized primarily in epithelial cells of the endometrium, but was associated with a decrease in connexin 43, the major gap junction protein in the myometrium. Decreased connexin 43 may have altered the synchronization and coordination of labor contractions, resulting in a requirement for more than the usual number of contractions to complete the delivery process.

A strong correspondence was observed between the R1 and R2 birth data despite differences in experimental design, including flight duration. This is particularly striking because, unlike the R1 dams, the R2 dams were not surgically manipulated post-flight prior to collection of the birth data thus their data are not confounded by this post-flight treatment. Together, these findings show that pregnant rat dams flown in space during the latter half of gestation for periods up to 11 days are capable of giving birth following return to 1 *g* to well-formed, healthy offspring.

The flight dams were also competent mothers. Systematic analyses of pup-oriented behaviors shown by the R1 and R2 dams indicated that maternal licking and handling of neonates during parturition, the consumption of birth products (e.g., placentophagia), and the onset of nursing were indistinguishable in flight and synchronous control dams (Ronca and Alberts, 2000a). Although the pups were placed with foster mothers soon after birth, studies of the mammary metabolism suggest that the dams were capable of lactating following spaceflight during pregnancy (Plaut *et al.*, 1999). Additional evidence is derived from observations of nursing and of milk in the gut during the first few postpartum hours.

The R1 and R2 missions offered some preliminary clues as to changes that may occur in space. The data are promising but do not shed light on whether mammalian birth can occur successfully in the weightless space environment. There may be problems with expelling the fetus from the womb. Given the increased numbers of labor contractions exhibited by the R1 and R2 flight dams, it is reasonable to assume that labor contractions may, in fact, be less forceful or less organized following a period of spaceflight exposure during pregnancy. If this is true, there may be a need for surgical cesarean births in space. Because labor contractions play vital roles in helping the neonate adapt to the extrauterine environment (Ronca *et al.*, 1996; Lagercrantz and Slotkin, 1986), such considerations should not be taken lightly.

In conclusion, the existing studies have established feasibility of exposing late prenatal rats to spaceflight conditions and pave the way for further spaceflight studies of fetal development. Many questions remain unanswered regarding the possibility of human birth and births of other mammals in space. At an approximate 3-week duration, the rats' pregnancy is relatively brief, compared to species higher on the phylogenetic scale. There is little basis at this time for predicting how reproductive capacities of different species may differ under spaceflight conditions. Still, as we embark on a future that includes a sustained human presence in space, it is promising that, based on the rat studies, certain fundamental phases of mammalian reproduction and development are preserved. It may be possible for humans to undergo pregnancy and prenatal development in space.

Effects of prenatal spaceflight on developing mammals

Postnatal growth and development

Numerous analyses were conducted on postnatal offspring following 4.5–11 days of spaceflight during late gestation, then born on Earth. For Cosmos 1514, newborn pups were derived from four of the 10 flight rats, for R1, newborn rats were studied from 10 dams that gave birth following unilateral hysterectomy, and for R2, newborn rats were available from six of 10 flight dams. In all three cases, litter sizes were similar across flight and control conditions (Table 2). The external appearance of the pups was appropriate to their calendar age. Postnatal body weights of flight pups did not differ from synchronous controls at birth (Ronca and Alberts, 2000a), but for R2, were significantly less than synchronous controls until the 14th postnatal day. No differences in body weight were observed thereafter through postnatal day 90 for either R1 or R2 offspring (Fig. 2; Hoban-Higgins *et al.*, 1999; Wong and DeSantis, 1997). There were also no observable changes in the development or behavior of the prenatally-flown offspring.

Physiology

A fundamental objective of the Cosmos 1514 experiment was to determine whether homeostasis would be maintained in the maternal–fetal system in microgravity. This led to a large number of physiological analyses of many different fetal systems (Gazenko, 1993; Serova and Denisova, 1982), rather than specific hypothesis-driven studies. In many systems, no differences were observed. There were a few differences in major organ systems. Liver weights of the flight pups were reduced compared to controls. There were no differences in concentrations of sodium, potassium or calcium in the liver nor were differences in liver concentration of DNA, RNA, or protein. Myocardial weight was increased in flight pups relative to controls, but there were no water or electrolyte changes. Elevated kidney weights were also reported with no water or electrolyte differences except for increased sodium concentrations among the flight pups. Importantly, it is not known whether the mothers' small in-flight weight gains affected fetal physiological profiles or how meaningful the findings are to the pups' development. Often opposite results were reported in fetuses at landing and neonates born several days post-landing. For example, on

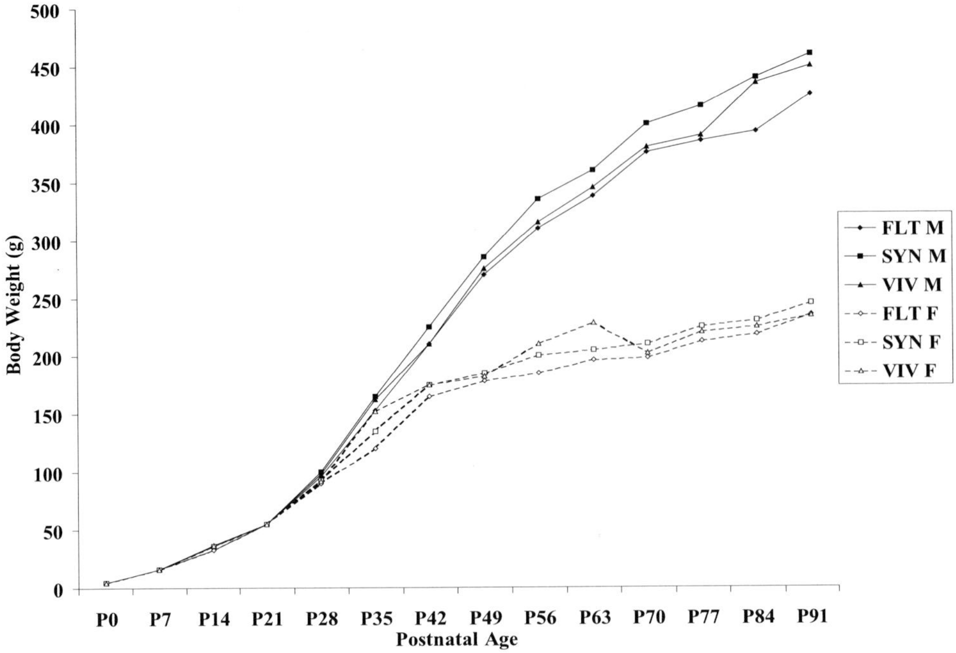

Fig. 2. Postnatal body weight gains of rat offspring exposed to 9-day spaceflight from gestational day 11–20 (NIH R2 mission) of the rats' 21-day gestation period. Adapted from Hoban-Higgins *et al.* (1999) Note: Due to assignments to experimental protocols at different postnatal ages, subject numbers decrease across time.

gestational day 18, numbers of ossification sites in flight fetuses lagged behind controls whereas ossification sites were actually increased in newborn flight neonates, possibly due to post-flight compensation.

Multiple analyses were carried out on the R1 and R2 pups as well. These were largely unrelated studies since NASA selected 16 different research proposals from a solicited announcement. Atrial natriuretic peptide (ANP) is a hormone important in the cardiovascular system via its regulatory roles in sodium and water excretion, and in vasodilatation. ANP is elevated in astronauts, probably in response to in-flight fluid shifts (Leach *et al.*, 1988). Rat fetuses from the R1 mission that were dissected on gestational day 20 displayed an altered maturation of cardiac ANP activity as evidenced by a significant, sixfold increase in mRNA biosynthesis (Davet *et al.*, 1999).

It has been reported that spaceflight impairs immunological responses of adult mammals. Sonnenfeld and colleagues studied immune responses of the R1 pregnant dams and their fetuses (Sonnenfeld *et al.*, 1998). Changes observed in the dams were similar to those reported previously for adult, non-pregnant animals, including inhibited blastogenesis of spleen cells in response to mitogen, a trend toward reduced interferon-production by spleen cells and a trend toward inhibited response of bone marrow cells to colony-stimulating factor. It is noteworthy that none of the examined immune parameters that were altered in rat dams after spaceflight were changed in their offspring. The second half of the rat's gestation is the time when major development and maturation of the immune system occurs (Silverstein, 1995). The emergence and establishment of immune responses were not studied and therefore could have been dramatically altered during the observed time interval (Sonnenfeld *et al.*, 1998).

Exposure to zero gravity has been shown to cause a decrease in bone formation in adults, and therefore, fetal bone was studied in gestational day 20 fetuses on the R2 mission (Davis *et al.*, 1998). Components of the fetuses' skeletal systems, namely osteoblasts, that produce neutral proteinases, including collagenase and tissue plasminogen activator (tPA), are important in bone development and remodeling. After flight, calvaria of G20 fetuses and postnatal days 1, 8, 14, 21, 28 and 35 pups were analyzed using immunohistochemistry for the presence and locations of the two proteinases. The developmental expression of the two proteinases was completely unaffected by spaceflight.

Together, these findings, derived from three different spaceflight missions, provide little evidence for structural-anatomical abnormalities among offspring flown prenatally in space, with relatively few changes in the offsprings' growth and rate of development. Few differences were observed in organ system physiology. Some differences identified at landing (e.g., skeletal retardation) were not observed in the neonates examined several days after landing, suggesting an astounding capacity for either re-adaptation or compensation by the offspring.

Brain and behavior

Neural and behavioral development begin *in utero*. A portion of the offspring derived from Cosmos 1514, R1 and R2 were studied prenatally (upon recovery from spaceflight), and the remainder were studied as pups following several days of re-adaptation and birth at 1 *g*. A range of sensory, motor and neural analyses were made.

Sensory function

General sensory function was studied in the Cosmos 1514 pups beginning on postnatal day 1 and extending to day 20 (Alberts *et al.*, 1985). The pups were evaluated for tactile sensitivity to von Frey hairs on different body regions, vestibular function, olfactory sensitivity, auditory responses, and visual acuity. No gross deficits were detected in any sensory system. The only differences that emerged from these studies were delayed development of responsivity to 40 kHz tones in the flight pups and a trend suggesting increased vestibular sensitivity to rotation in the flight pups. The significance of these observations is not clear.

Because gravity plays special roles in the development and maintenance of the vestibular system, spaceflight during gestation presents a unique opportunity to deprive the developing organism of vestibular input. In the rat, the cellular components of the vestibular system begin to form and make interconnections from around the 9th to the 19th gestational days, with further maturation throughout the first few postnatal weeks (Ronca and Alberts, 2000b) Even during the last few days of rat gestation, the immature neural substrates are capable of transducing vestibular input. For the R1 and R2 spaceflight mission, neurovestibular connectivity and onset of vestibular function were established within the microgravity of space. Vestibular-mediated responses were analyzed (Ronca and Alberts, 2000b) from the 1st to the 5th postnatal days. The contact righting reflex, composed of stereotyped movements that rotate the body from supine to prone on a solid surface, did not differ in flight and control pups, revealing that the flight pups were capable of performing the movements required for righting. However, the presence of tactile and proprioceptive cues of the surface against the dorsum may have compensated for impaired vestibular processing. Water immersion righting tests, in which tactile and proprioceptive cues were removed, revealed less frequent righting in flight neonates compared with controls on postnatal days 1 and 3. This deficit was profound, but short-lived as evidenced by complete response recovery on P5. Collectively, these findings provide evidence for a selective disruption of vestibular-mediated responses after prenatal exposure to spaceflight. Longer periods of spaceflight spanning the period of vestibular system maturation may be associated with permanent changes in behavioral responses to vestibular cues. A complementary developmental study mapping vestibular connections from the gravity receptors of the inner ear to the brain in the R1 fetuses (Bruce and Frizsch, 1997) revealed neuroanatomical changes consistent with the behavioral profiles (see *Neuroanatomical Studies,* below).

Data derived from centrifugation reported permanent behavioral deficits in hamsters following prolonged hypergravity exposure during development (conception until weaning on P20) whereas similar exposures in adult animals are transient (Sondag *et al.*, 1997). Together, these data suggest that a critical period may exist during the early phases of vestibular system formation and maturation. Many more studies are needed to identify the presence of a critical period and determine how it is correlated with changes in anatomy and behavior.

Motor function

There are no reported effects of prenatal spaceflight on somatic motor development. The emergence and development of locomotion was studied in the R1 offspring for over 81 days after birth (Wong and DeSantis, 1997). Walking was achieved by postnatal day 13 in flight and control pups and preceded eye-opening, which was complete in all pups by postnatal day 17.

Characteristic changes in hindlimb step length and gait width were observed as the pups grew which occurred at the same time in all groups. Prenatal spaceflight did not interfere with the establishment of normal patterns for hindlimb placement during walking. Exposure to microgravity from days 9 to 20 of gestation had no effects somatic motor skills.

Neuroanatomical studies

Macroscopic evaluation of the fixed brains of the flight fetuses revealed no gross differences between flight and control specimens (Olenev *et al.*, 1989). In the Cosmos flight, neuroanatomical evaluation of the day 18 fetuses revealed that mitotic figures were present in the cortical layers of the brains of flight fetuses, but not in synchronous or vivarium controls. This was interpreted to indicate either an aberrant or decelerated schedule of neuronal births (Keefe, 1985). Ocular morphology was also disrupted, and vestibular nuclei showed anomalies. The blood vessels and capillaries of the brains were reported to be smaller and thinner as measured by alkaline phosphatase (Olenev *et al.*, 1989). Brain tissue derived from the R1 pups submitted to histological analysis exhibited foci of neuronal degeneration characterized by the absence or decreased number of neurons in 70% of the specimens plus fewer glial cells and individual capillaries as compared to controls (Proshchina *et al.*, 2000). These foci were detected throughout the neuraxis: in hippocampus, neocortex, the dorsolateral nucleus of the thalamus, caudate nucleus, hypothalamus and spinal cord. Together, these findings suggest that prenatal spaceflight may exert significant effects on brain morphology during development.

Astronauts experience circadian rhythm disturbances, sleep loss, decrements in neurobehavioral performance, and post-flight changes in REM sleep (Dijk *et al.*, 2001). With extended human visits to space, understanding the gravitational biology of the circadian timing system and its development acquires even greater importance. In rats, the circadian timing system becomes established prenatally (Weaver and Reppert, 1989). Studies conducted on the R2 offspring of the developing superchiasmatic nucleus (SCN), the reception area for photic information that entrains circadian rhythms to the light/dark cycle, provide additional evidence for developmental delay following gestation in space (Murakami *et al.*, 1997).

Choroid plexuses, organs that secrete 70–90% of cerebrospinal fluid (CSF), are involved in brain homeostasis. Studies of adult rats flown in space suggest modified CSF production in rats during 2 weeks exposures to microgravity. Analyses of the choroid plexus of the R1 fetuses (Mani-Ponset *et al.*, 1997) provide evidence for developmental delay in a different system. Choroidal epithelial cells, derived from the lateral, third and fourth ventricles, and apical microvilli were studied in the R1 flight and control fetuses at recovery (gestational day 20). The main differences involved immunocytochemical changes in the distribution of ezrin, a cytoskeletal protein involved in apical cell differentiation, and carbonic anhydrase II, involved in cerebrospinal fluid production. Ezrin was strongly distributed within the choroidal cytoplasm, but not in the membrane in apical microvilli. Together, the changes observed in flight fetuses are characteristic of delayed maturation of the choroid plexus.

Bruce and Frizsch (1997) conducted a neuroanatomical study of the R1 fetuses that complemented analyses of vestibular-mediated responses of the R1 neonates (Ronca and Alberts, 2000b). Upon recovery, the fixed ears of the fetuses were infiltrated with DiI, a lipophilic dye that diffuses along the lipid bilayer of membranes. Gravistatic vestibular receptors (saccule) and one of the non-gravistatic vestibular receptors (posterior vertical canal) were targeted for analysis. Whereas peripheral branching was similar in flight and control fetuses, central branching from gravistatic receptors into medial vestibular nucleus showed that in flight fetuses, most axons ended in small growth cones and rarely had side branches. When the sections with labeled axons were photo-converted and examined with light microscopy, no structures were observed in the flight animals that could unequivocally be identified as a synapse. However, in control animals, small but definitive synaptic boutons were observed. If the cues that guide the development of the visual system also apply to the development of the gravity sensing system, there will be a 'critical period' during which the ambient stimulus of gravity will govern synapse formation and behavior mediated by the vestibular system. The scope of the R1 study did not permit identification of this critical period, but together with the behavioral studies suggests an important direction for future research.

In the aggregate, the data that has emerged from spaceflight studies of prenatal rats consistently report brain changes indicative of maturational delay. This is true despite varying mission lengths (4.5–11 days), differences in maternal in-flight weight gains and similarities in the body weights of flight and synchronous control fetuses. General maturational delay may have occurred in R2, as evidenced by differences in postnatal body weight (up to postnatal day 14). Specific changes in vestibular neuroanatomy and behavior were observed. The relative absence of physiological and behavioral effects is striking in light of the corpus of data suggesting developmental retardation reported in offspring exposed prenatally to spaceflight. The most noteworthy changes relate to cardiac ANP and specific changes in vestibular perception. Although studies of many other physiological processes remain to be done, the seemingly minor behavioral and physiological changes reported attests to the profound adaptive capacity and neural plasticity inherent to the developing mammal.

Effects of postnatal spaceflight on developing mammals

The R3 and Neurolab flights that carried nursing litters into space were encumbered by a series of offspring survival, and in the case of Neurolab, health problems. Still, some meaningful data was collected from these flight experiments.

Growth and development

For the R3 flight, maternal body weights were lower at launch relative to controls. Of the three postnatal ages (5, 8 and 14) flown on the 9-day R3 mission, only 10% of the pups launched at 5 days of age survived. Whereas 90% of the pups launched at 8 days of age survived, their body weights averaged approximately 25% less than those of synchronous controls (Fig. 3, left-hand panel). All of the litters launched at 14 days of postnatal age survived and gained comparable amounts of weight relative to controls. For the 16-day Neurolab Mission, the pups launched at 15 days of postnatal age survived the mission in apparent good health. However, more than half of the young rats launched at 8 days of age died or were moribund and therefore were euthanized during the flight (Maese and Ostrach, 2002). The survivors had dramatically reduced body weights at landing, approximately 50% of the controls (Fig. 3, right-hand panel, see also Maese and Ostrach, 2002). Due to what appears to have been extensive disruption of the maternal–offspring system, many of the younger Neurolab pups suffered from malnutrition, dehydration and hypothermia during the flight. Such effects were not observed in the older pups. Given the fundamental importance of the mother in processes of mammalian

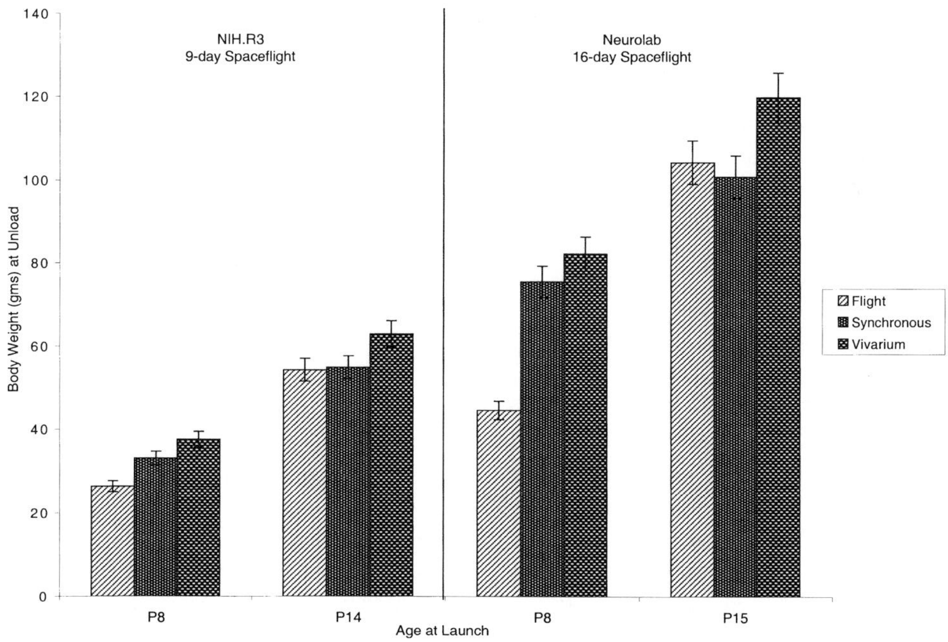

Fig. 3. Percent body weight gain of postnatal day (P) 8 and P14 rat pups flown on the 9-day NIH.R3 mission and P8 and P15 rat pups flown in the 16-day Neurolab mission.

development, it is astounding that no correlative studies were conducted on the lactating dams flown in these experiments. Analyses of physiological status including stress hormones, reproductive hormones involved in milk production (prolactin and oxytocin) and maternal behavior would have yielded important data, particularly in regard to interpreting effects on offspring. A truly important opportunity was missed. Still, these flights have led to important 'lessons learned' and have significantly raised our collective awareness regarding the importance of flight caging. The Neurolab dams, housed in the RAHF (Fig. 1 lower right-hand panel) had an especially difficult time relative to the R3 dams in the AEM-NF (Fig. 1, lower left-hand panel) retaining the pups in the habitat nesting area. This led to many of the Neurolab pups being abandoned or destroyed by the dams (Adams *et al.*, 2000a). During the flight, the crew members provided subcutaneous liquids and antibiotics to dehydrated and sick pups, re-distributed pups to the most proficient mothers, and euthanized moribund pups. From the experimental perspective, it is unfortunate that none of these measures were mimicked in the control groups. Post-flight, the mothers and pups of the 15-day group were reported to be "soaking wet and hypothermic". Post-flight, NASA veterinarians further eliminated unhealthy pups from the experimental pool. The RAHF housing is completely unsuitable for flying litters of neonatal rats and the AEM-NF appears to be of limited usefulness. There is clearly a serious need for habitat development and careful ground-based testing before new studies of nursing rodents can be conducted in space.

A number of interesting and important empirical analyses were planned for the R3 and Neurolab Mammalian Development payloads. Some reports have been published. It is important, however, to interpret with caution data collected from the younger age group flown on the Neurolab mission, given the potential confounds related to possibilities for malnourishment, extensive disruptions in maternal–offspring relations, litter re-distributions and crew handling and medical treatment of flight pups but not controls. Only healthy survivors of flight were tested in the Neurolab studies—animals that represent a select sample from the 96 pups flown.

The conditions experienced by the Neurolab pups launched on postnatal day 8 might have potential associations with a number of postnatal impairments or delays. For example, undernourishment in early life causes significant delays in physical growth of the young and long-lasting morphological changes, particularly in brain areas undergoing cell proliferation (Altman and McCrady, 1972; Fish and Winick, 1979). Reduced neocortical dendrogenesis, number of dendritic spines, and reduced synapse-to-neuron ratio have been reported following neonatal malnourishment (Escobar and Salas, 1995; Anokin, 1964). These changes are frequently correlated with reflexive and long-term behavioral abnormalities, including aberrant locomotion characterized as "prolonged persistent of infantile patterns" (Altman *et al.*, 1971).

Physiology

Cardiovascular

Exposure to microgravity causes significant changes in the cardiovascular system (Nicogossian *et al.*, 1989). One important problem to be solved in the coming space era is to understand the development of the regulation mechanism of blood circulation in animals developing in space. Shimizu (1999) conducted a histological examination of the aortic nerve, the afferent limb of the aortic baroreflex, in the Neurolab pups. The number of unmyelinated fibers in flight pups was significantly less than those in the control groups, although no differences were observed between flight and control pups in either axon diameter or myelin thickness. It is not clear whether the pups' nutritional status, rather than microgravity, may have affected myelination of the aortic nerve.

Muscle

Muscle is a primary target for the effects of microgravity (Nicogossian *et al.*, 1989). The first month of postnatal life in the rat encompasses a period of rapid growth and development during which skeletal muscle mass increases exponentially and the adult pattern of muscle contractile protein expression becomes fully established (Adams *et al.*, 2000a). Coincident increases in thyroid hormone (TH) help coordinate growth and thyroid deficiency during the neonatal period is associated with heart and skeletal muscles retardation, possibly mediated by suppression of insulin-like growth factor I (IGF-I). Based on these findings, Baldwin and colleagues (Adams *et al.*, 2000a,b; Inobe *et al.*, 2002) reported an interesting series of studies on muscle derived from the pups that underwent early postnatal development on the Neurolab mission. The researchers hypothesized that thyroid deficiency and exposure to spaceflight would interact to reduce skeletal muscle growth via a reduction in IGF-I levels. Neurolab pups (thyroid-normal) launched on day 8 showed dramatic reductions body weight (Fig. 3) and muscle weight (fast muscles, ~60% less than controls). Thyroid deficient pups flown on the same mission did not show additional reductions in body weight relative to flight-only pups. However, in the soleus muscle, either spaceflight or thyroid deficiency alone resulted in growth reductions that were augmented by interactions between the two. Strong correlations were observed between muscle mass and muscle IGF-I levels and between circulating IGF-I and body weight, suggesting that exposure to spaceflight limited the somatic and muscle-specific growth of neonatal rats through an IGF-I mechanism. The postnatal day 15 pups did not show the same deficiency in somatic growth or growth of the tibialis anterior or medial gastrocnemius, suggesting the possibility of a critical period for this response. Nonetheless, it is possible that the inadequate housing and care of the younger pups as compared to the older ones contributed to these age differences.

The myosin heavy chain (MHC) molecule is the principal structure and regulatory protein that serves as the molecular motor to control the intrinsic contractile properties of muscle (Adams *et al.*, 2000b). The separate and combined effects of spaceflight and thyroid deficiency on myosin heavy chain (MHC) gene expression (protein and mRNA) were examined in muscles of the postnatal day 8 and 15 Neurolab pups. It was found that spaceflight markedly reduced expression of the slow, type I MHC gene by approximately 55%, whereas expression of the fast IIx and IIb MHCs in antigravity skeletal muscles was enhanced. In fast muscles, spaceflight caused subtle increases in the fast IIb MHC relative to the other adult MHCs. Although the body growth of these younger pups was greatly reduced relative to controls, α-MHC was not appreciably diminished in this group. The similarity in response profiles of the younger and older pups, with the latter showing small differences in body and muscle weight relative to controls, suggest that the pups studied in these experiments may not have been nutritionally deprived.

These researchers also reported that the expression of certain myogenic factors, particularly the MyoD-family transcription factors was reduced in the younger Neurolab flight pups relative to controls (Inobe *et al.*, 2002). These results suggest that MyoD and myogen, but not MRF4, may be sensitive to gravity-related stimuli in some skeletal muscles during postnatal development.

Together, these studies provide initial evidence for changes in rat muscle following early postnatal exposure to spaceflight. Questions remain regarding the critical periods for these responses. In addition, disruptions in maternal–offspring relations may have altered the developmental trajectories of the pups flown on the Neurolab mission. These noteworthy experiments deserve to be repeated and the findings validated once suitable habitats are developed for supporting nursing rat litters.

Endocrine

The hypothalamo–neurohypophyseal system and the autonomic nervous system are involved in homeostatic responses associated with changes in head position and the orthostatic reflex. There is evidence for increased activity of the hypothalamo–neurohypophyseal system after spaceflight. Because environmental modifications in young animals can cause permanent alterations in neuroendocrine function (Francis *et al.*, 2002), it was of considerable interest to analyze spaceflight effects on the key hypothalamic neurons of the developing rats flown on the Neurolab mission (Garcia-Ovejero *et al.*, 2001). Oxytocinergic and vasopressinergic magnocellular hypothalamic neurons of prepuberal rats of the 15-day-old rat pups flown on the Neurolab mission were analyzed either at landing or 18 weeks post-landing. Enhanced transcriptional and biosynthetic activity was observed in magnocellular supraoptic neurons of flight animals on the day of landing as compared to control rats, including increased c-Fos expression, enlarged nucleoli and cytoplasm, and increased volume in the neuronal perikaryon of mitochondriae, endoplasmic reticulum, Golgi apparatus, lysosomes and cytoplasmic inclusions (nematosomes). In contrast, the volume occupied by neurosecretory vesicles in the supraoptic neuronal perikarya was significantly decreased in flight rats and associated with decreased oxytocin and vasopressin immunoreactive levels, suggesting increased hormonal release. Vasopressin levels, cytoplasmic volume and c-Fos expression returned to control levels by 18 weeks after landing. Whereas the reversible effects may have been associated with shifts in extracellular fluid and plasma volume during flight and landing, the flight rats' oxytocin levels remained suppressed at 18 weeks post-flight. The results suggest that spaceflight during the preweanling period may induce irreversible modifications in the regulation of oxytocinergic neurons. Importantly, central pathways for oxytocin and vasopressin have been implicated in the neurobiology of anxiety and social behaviors. Variations in maternal care and gender are associated with differential expression of oxytocin and vasopressin receptors in developing brain (Francis *et al.*, 2002). In addition, changes in maternal–offspring interactions are likely to occur under altered gravity conditions (Ronca *et al.*, 2001). For these reasons, it would be informative in future studies to analyze maternal care and gender in relation to changes in the hypothalamo–neurohypophyseal system.

Brain and behavior

Sensory systems

How the developing nervous system responds and adapts to environmental influences and the mechanisms underlying this plasticity is an important topic in gravitational biology. In an effort to identify critical periods in peripheral vestibular system development, plastic changes in the organization of the vestibular efferent network of the rat utricle were studied in the postnatal rats launched on day 9 of the Neurolab flight (Dememes *et al.*, 2001). Immunofluorescence experiments were performed with a specific biochemical marker of the efferent system, the calcitonin gene-related peptide (CGRP). The utricles were analyzed by confocal microscopy. Maturation of the vestibular efferent system was similar in spaceflight and control rats. The absence of changes in the development of vestibular efferents is encouraging in view of the dramatically reduced body weights of the pups studied.

Motor system

Work conducted on the developing motor system in relation to spaceflight has focussed predominantly on identification of a critical period for motor development (Walton, 1998; Walton *et al.*, 1997), similar to the critical period that has been established for the developing visual system (Hubel and Wiesel, 1998).

These studies have incorporated either spaceflight (R3) or the hindlimb suspension (HLS) models (Morey-Holton and Globus, 2002) and have applied contemporary techniques for measuring movement and gait (Walton *et al.*, 1997; Walton, 1998).

Ground studies were first conducted using the hindlimb suspension model of weightlessness in rats ranging in age from postnatal day 2 to postnatal day 13 at the start of the study for varying durations spanning 5–18 days. Comparisons were made between HLS and control pups, with two control conditions: (1) a 'companion' littermate housed with each suspended pup; or (2) a vivarium control housed with the dams. HLS and 'companion' pups were returned to the dams every 4 h for 1 h to suckle. Analysis of daily videotapes indicated that hindlimb unloading led to significant changes in motor performance. Relative to controls, HLS pups spent less time swimming, had more difficulty maintaining their balance, and exhibited increased ankle extension which was associated with an increase in stroke duration. The effects were most pronounced in pups suspended from P8 to P13.

The motor function of pups launched on the 9-day R3 flight study was consistent with that seen in the HLS studies, being marked by exaggerated limb joint extension. However, there were additional, more profound differences in righting, swimming, and locomotion believed to be associated with unloading of gravistatic receptors in flight, the longer period of unloading in flight, and the inability to evaluate motor function during the flight.

One concern regarding these behavioral studies relates to disruptions in maternal–offspring interactions as described in pervious sections. Both flight and HLS pups were subject to significant reductions in pup body weight and growth (Fig. 3; also see Huckstorf *et al.*, 2000). Nursing, maternal licking and pup huddling may have been atypical in flight. Using the HLS model, young pups are clearly deprived of maternal licking and warming, and the species-typical rhythm of milk ingestion is significantly changed. The use of 'companion' controls facilitated interpretation of the HLS findings, but, in the absence of more appropriate flight housing to support nursing litters, the incorporation of pair-fed controls in spaceflight studies of postnatal rats is an important next step in validating many of the Neurolab findings.

Neuroanatomical studies

A study of the effects of rearing environment on size and complexity of dendritic architecture of spinal motor neurons, particularly those that are likely to participate in postural control, was conducted in postnatal day 8 rats flown on the Neurolab mission (Inglis *et al.*, 2000). There was a reduction in dendritic branching in motor neurons of flight relative to control animals. This pattern was observed only in medial neurons, those that are more likely to receive vestibular input, but not in lateral neurons. The specificity of the neuronal populations affected suggests that general factors related to nutrition or stress may have played little role in the pattern of results, but again, future studies will be important in validating the results.

A study of the postnatal day 15 pups flown on the Neurolab mission suggests that development in microgravity leads to a constellation of interesting changes in the number and morphology of cortical synapses and does so in a laminar-specific manner (DeFelipe *et al.*, 2002). In the layers II/III and Va, the synaptic cross-sectional lengths were significantly larger in flight animals than in ground control animals. Flight animals also showed significantly lower synaptic densities in layers II/III, IV and Va. The greatest difference was found in layer II/III, where there was a difference of 344 million synapses per mm^3 (15.6% decrease). After a 4-month period of re-adaptation, some changes disappeared (i.e. the alterations were transient), while conversely, some new differences also appeared. For example, significant differences in synaptic density in layers II/III and Va after re-adaptation were no longer observed, whereas in layer IV the density of synapses increased notably in flight animals (a difference of 185 million synapses per mm^3 or 13.4%). In addition, all the changes were in asymmetrical synapses, known to be excitatory. These results suggest that gravity may be an important environmental parameter for normal cortical synaptogenesis.

Learning and memory

This study evaluated the cognitive mapping abilities of the postnatal day 8 or 15 Neurolab pups. Upon recovery, pups were tested in three different tasks that measure spatial learning ability: the Morris water maze, a radial arm maze, and an open field apparatus to measure general activity and exploratory activity. Performance and search strategies evaluated for each task revealed remarkably few differences between the flight groups and their Earth-bound controls in these tasks. Flight pups learned the water maze and radial arm maze as rapidly as did controls, although subtle, but transient, differences in patterns of exploration and in the strategies used to solve the tasks were initially observed. Together, these data suggest that development in an environment without gravity has minimal long-term impact on spatial learning and memory abilities. In view of the high levels of in-flight mortality and reduced body weights of Neurolab pups, the relative absence of effects on established spatial learning tasks is surprising. However this outcome is encouraging for interpreting data derived from the younger animals flown on the Neurolab Mission.

Interpreting studies of mammalian development in space

Development proceeds as a constant interaction between genetic and environmental factors. Gene–environment interactions establish developmental trajectories and determine both short and long-term outcomes. The early environment influences development, in part, through a process of 'programming'. There is a growing appreciation for the role(s) of early environmental factors in sensory ontogenesis (Hubel and Wiesel, 1998), development and regulation of the neuroendocrine system (Francis *et al.*, 1999) and in the emergence of adult disease states (Godfrey and Barker, 2001).

The mammalian mother forms the environmental context within which the fetus and newborn grow and develop. Changes in maternal–offspring interactions can exert important formative influences on the young. The quality of infant care regulates early growth and adaptation during development (Hofer, 1983). Periods of maternal deprivation result in a precipitous decline in growth, heightened stress reactivity, alterations in markers of growth and development. Given the considerations reviewed in this chapter, it is clear that studies of developing mammals in space must be meticulously planned to preserve as closely as possible characteristic maternal behaviors and species-typical interchanges between mothers and pups of vital resources, thereby retaining the integrity of the maternal–offspring system. The Cosmos, NIH.R1 and R2 experiments provided a balanced aggregate of maternal and offspring studies. However, critical studies of the lactating dams were overlooked in the NIH.R3 and Neurolab missions, thereby limiting interpretations of the data and our understanding of postnatal maternal changes in space. Only though comprehensive studies of offspring *in relation to their mothers* can gravitational biologists make proper interpretations of the developmental data and retain confidence that the developmental effects they are studying are truly due to *direct* influences of the space environment rather than *indirect* effects arising from disrupted maternal–offspring relations.

Conclusions and summary

Life on Earth, and thus the reproductive and ontogenetic processes of all extant species and their ancestors, evolved under the constant influence of the Earth's 1 g gravitational field. This raises the question of whether mammalian reproduction and offspring development can successfully take place in the absence of gravity. Successful reproduction is the hallmark of a species' adaptation to its environment and must be realized to sustain extra-terrestrial populations. Before taking this immense step, we need to understand the effects of gravity on critical phases of mammalian reproduction, viz. those events surrounding fertilization, embryogenesis, pregnancy, birth, postnatal maturation and parental care.

Developmental analyses of mammals exposed to spaceflight have focussed almost exclusively on rats, with only one study that has utilized mice (see Table 1). Cross-species comparisons are crucial for establishing mammalian developmental biology as a meaningful field of research. Data on mammalian development have been collected only on very brief epochs of reproductive and developmental time. We are a long way from achieving a complete mammalian life cycle in space. Because so few spaceflights have been dedicated to developing mammals, our knowledge at the present time is limited. For example, it is not known whether spaceflight might preclude early phases of development encompassing fertilization and placentation. The most common effects of spaceflight, viz. headward fluid shifts, alterations in bone and calcium metabolism and muscular deconditioning (Edgerton and Roy, 1994; Johnson, 1998; Nicogossian *et al.*, 1989) may provide formidable obstacles to sustaining the gravid state and impede the ability of mothers to give birth. Long-duration spaceflights could affect important physiological or behavioral responses of mothers and developing fetuses during pregnancy. The bones and muscles of developing fetus may not form properly, and the brain may become wired in completely different ways. A mammal developed in space may not be able to return to Earth due to an inability to survive the Earth's gravitational field.

Still, a number of conclusions can be drawn from the available data. Later phases of the rats' relatively brief gestation period can proceed in the microgravity of space. There is no evidence for morphological abnormalities in the offspring. Very few and seemingly insignificant physiological and behavioral changes have been reported for rats exposed prenatally to spaceflight. The R1 and R2 studies of prenatal development in space have been extremely successful, that it is now appropriate to proceed with more detailed and longer spaceflight missions of pregnant rodents and their offspring. These studies support the view that the maternal–fetal system is highly adaptable, as evidenced by its adjustment to these unique environmental conditions during ontogenesis.

Delayed brain maturation is suggested by the prenatal presence of mitotic figures (Gazenko, 1993; Keefe, 1985; Olenev *et al.*, 1989), neuronal degeneration (Proshchina *et al.*, 2000), immature patterns of c-fos activation in the suprachiasmatic nucleus (Murakami *et al.*, 1997) and delayed development of the choroid plexus (Mani-Poinset *et al.*, 1997). Postnatally, changes in dendritic arborization (Inglis *et al.*, 2000) and cortical structure (DeFelipe *et al.*, 2002) were reported. Together, these changes strongly suggest a need for systematic analyses of the brain changes that occur following exposure to spaceflight during development.

As the International Space Station (ISS) unfolds, our ability to grow healthy, viable colonies of breeding rodents in space is critically dependent upon our understanding of how pregnancy, birth and early life within the nest are affected by deviations from earth-normal gravity. Plans for the ISS animal laboratories include an onboard centrifuge, capable of delivering variable g-loads ranging from 0 to $2g$, thus spanning g-loads across the hypogravity to hypergravity continuum. The ISS centrifuge capabilities will provide unique scientific opportunities to answer sophisticated questions of gravitational influence on mammalian development. However, if we are to support young mammals in space and during centrifugation, and accurately interpret neural, physiological and behavioral data derived from them, much work remains to be done to further our understanding effects of altered gravity environments on the maternal–offspring system.

Acknowledgements

I would like to thank Jeff Alberts for introducing me to space biology, and Charles Wade and Lisa Baer for teaching me about gravity's effects on biological systems. I would also like to thank Ken Souza for encouraging me in this endeavor. Charles Wade, Lisa Baer and Nicole Mills provided important comments on an earlier draft of this manuscript. I would like to acknowledge Kenneth E. Johnson for creating the drawings of the rodent flight housing.

References

Adams, G.R., McCue, S.A., Bodell, P.W., Zeng, M. and Baldwin, K.M. (2000a) Effects of spaceflight and thyroid deficiency on hindlimb development. I. Muscle mass and IGF-I expression. J. Appl. Physiol. 88, 894–903.

Adams, G.R., Haddad, F., McCue, S.A., Bodell, P.W., Zeng, M., Qin, L., Qin, A.X. and Baldwin, K.M. (2000b) Effects of spaceflight and thyroid deficiency on rat hindlimb development. II. Expression of MHC isoforms. J. Appl. Physiol. 88, 904–916.

Alberts, J.R., Burden, H.W., Hawes, N. and Ronca, A.E. (1996) Sampling pre- and postnatal offspring from individual rat dams enhances animal use without compromising development. Contemp. Top. Lab. Anim. Sci. 35, 61–65.

Alberts, J.R. and Ronca, A.E. (1997) Rat pregnancy and parturition survive microgravity challenge: new considerations of developmental consequences. J. Gravit. Physiol. 4, 55–58.

Alberts, J.R. and Ronca, A.E. (1999) Effects of gravity on development: the importance of behavior. J. Gravit. Physiol. 6, 5–8.

Alberts, J.R., Serova, L.V., Keefe, J.R. and Apanasenko, Z. (1985) Early postnatal development of rats gestated during flight of Cosmos 1514. Physiol. 28, 81–82.

Altman, J. and McCrady, B. (1972) The influence of nutrition on neural and behavioral development. IV. Effects of infantile undernutrition on the growth of the cerebellum. Dev. Psychobiol. 5, 111–122.

Altman, J., Sudarshan, K., Das, G.D., McCormick, N. and Barnes, D. (1971) The influence of nutrition on neural and behavioral development. 3. Development of some motor, particularly locomotor patterns during infancy. Dev. Psychobiol. 4, 97–114.

Anohkin, P.K. (1964) Synaptogenesis as a general regulator of brain development. Prog. Br. Res. 9, 54–86.

Bridges, R.S. (1990) Endocrine regulation of parental behavior in rodents. *In:* Krasnegor, N.A. and Bridges, R.S. (eds.). *Mammalian Parenting: Biochemical, Neurobiological, and Behavioral Determinants.* Oxford University Press, New York.

Bruce, L.L. and Frizsch, B. (1997) The development of vestibular connections in rat embryos in microgravity. J. Gravit. Physiol. 4, 59–61.

Burden, H.W., Zary, J.T. and Alberts, J.R. (1999) Effects of spaceflight on the immunohistochemical demonstration of connexin 26 and connexin 43 in the postpartum uterus of rats. J. Reprod. Fertil. 116, 98–102.

Burden, H.W., Zary, J.T., Lawrence, I.E., Jonnalagadda, P., Davis, M. and Hodson, C.A. (1997) Effects of spaceflight on ovarian-hypophyseal function in postpartum rats. J. Reprod. Fertil. 109, 193–197.

Davet, J., Fagette, S., Mani-Ponset, L., Bayard, B., Dumars, P., Reiss-Bubenheim, D., Guell, A., Gharib, C. and Gabrion, J. (1999) Cardiac atrial natriuretic peptide (ANP) in rat dams and fetuses developed in space (NIH-R1 and NIH-R2 experiments). Life Sci. 64, 1533–1541.

Davis, B.A., Sipe, B., Gershan, L.A., Fiacco, G.J., Lorenz, T.C., Jeffrey, J.J. and Partridge, N.C. (1998) Collagenase and tissue plasminogen activator production in developing rat calvariae: normal progression despite fetal exposure to microgravity. Calcif. Tissue Int. 63, 416–422.

DeFelipe, J., Arellano, J.I., Merchan-Perez, A., Gonzalez-Albo, M.C., Walton, K. and Llinas, R. (2002) Spaceflight induces changes in the synaptic circuitry of the postnatal developing neocortex. Cereb Cortex. 12, 83–91.

Dememes, D., Dechesne, C.J., Venteo, S., Gaven, F. and Raymond, J. (2001) Development of the rat efferent vestibular system on the ground and in microgravity. Dev. Brain Res. 128, 35–44.

Dijk, D.J., Neri, D.F., Wyatt, J.K., Ronda, J.M., Riel, E., Ritz-De Cecco, A., Hughes, R.J., Elliott, A.R., Prisk, G.K., West, J.B., Czeisler, C.A. (2001) Sleep, performance, circadian rhythms, and light-dark cycles during two space shuttle flights. Am. J. Physiol. Regul. Integr. Comp. Physiol. 281, 1647–1664.

Dumars, P., Elland, C., Hinds, K., Katen, C., Schnepp, T. and Reiss-Bubenheim, D. (1996) National Institutes of Health. Rodent 2 (NIH R2) study of the effect of microgravity on the development of rats: A collaborative series with NASA and NIH. ASGSB Bull. 10, 47.

Edgerton, V.R. and Roy, R.R. (1994) Neuromuscular adaptation to actual and simulated weightlessness. Adv. Space Biol. Med. 4, 33–67.

Escobar, C. and Salas, M. (1995) Dendritic branching of claustral neurons in neonatally undernourished rats. Biol. Neonate 68, 47–54.

Fejtek, M., Souza, K., Neff, A. and Wassersug, R. (1998) Swimming kinematics and respiratory behaviour of *Xenopus laevis* larvae raised in altered gravity. J. Exp. Biol. 201, 1917–1926.

Fejtek, M. and Wassersug, R. (1999) Effects of laparotomy, cage type, gestation period and spaceflight on abdominal muscles of pregnant rodents. J. Exp. Zool. 284, 252–264.

Fish, I. and Winick, M. (1969) Effect of malnutrition on regional growth of the developing rat brain. Exp. Neurol. 25, 534–540.

Francis, D.D., Champagne, F.A., Liu, D. and Meaney, M.J. (1999) Maternal care, gene expression, and the development of individual differences in stress reactivity. Ann. N.Y. Acad. Sci. 896, 66–84.

Francis, D.D., Young, L.J., Meaney, M.J. and Insel, T.R. (2002) Naturally occurring differences in maternal care are associated with the expression of oxytocin and vasopressin (V1a) receptors: gender differences. J. Neuroendocrinol. 14, 349–353.

Gabrion, J., Herbute, S., Oliver, J., Maurel, D., Davet, J., Clavel, B., Gharib, C., Fareh, J., Fagette, S., Nguyen, B. (1995) Choroidal responses in microgravity. (SLS-1, SLS-2 and hindlimb-suspension experiments). Acta Astronaut. 36, 439–448.

Garcia-Ovejero, D., Trejo, J.L., Ciriza, I., Walton, K.D. and Garcia-Segura, L.M. (2001) Space flight affects magnocellular supraoptic neurons of young prepuberal rats: transient and permanent effects. Dev. Brain Res. 130, 91–205.

Gazenko, O.G. (1993) Ontogenesis of mammals in microgravity, NASA TM-103978.

Godfrey, K.M. and Barker, D.J. (2001) Fetal programming and adult health. Public Health Nutr. 4, 611–624.

Hoban-Higgins, T.M., Murakami, D.M., Tang, I.H., Fuller, P.M. and Fuller, C.A. (1999) Development of circadian rhythms in rat pups exposed to microgravity during gestation. J. Gravit. Physiol. 6, 71–79.

Hofer, M.A. (1983) The mother–infant interaction as a regulator of infant physiology and behavior. pp. 61–75. *In* Rosenblum, A. and Moltz, H. (eds.) *Symbiosis in Parent–Offspring Interaction*. Plenum Press, New York.

Hofer, M.A., Leaf, J.P. and Shair, H.N. (1988) Baroreceptor sensitivity in two-week-old rat pups: effects of nutrient deprivation and sino-aortic denervation. Physiol. Behav. 43, 823–830.

Hubel, D.H. and Wiesel, T.N. (1998) Early exploration of the visual cortex. Neuron 20, 401–412.

Huckstorf, B.L., Slocum, G.R., Bain, J.L., Reiser, P.M., Sedlak, F.R., Wong-Riley, M.T. and Riley, D.A. (2000) Effects of hindlimb unloading on neuromuscular development of neonatal rats. Dev. Brain Res. 119, 169–178.

Iams, J.D., Newman, R.B., Thom, E.A., Goldenberg, R.L., Mueller-Heubach, E., Moawad, A., Sibai, B.M., Caritis, S.N., Miodovnik, M., Paul, R.H., Dombrowski, M.P., Thurnau, G., McNellis, D. (2002) Frequency of uterine contractions and the risk of spontaneous preterm delivery. N. Engl. J. Med. 346, 250–255.

Ijiri, K. (1995) Fish mating experiment in space—what it aimed at and how it was prepared. Biol. Sci. Space 9, 3–16.

Inobe, M., Inobe, I., Adams, G.R., Baldwin, K.M. and Takeda, S. (2002) Effects of microgravity on myogenic factor expressions during postnatal development of rat skeletal muscle. J. Appl. Physiol. 92, 1936–1942.

Inglis, F.M., Zuckerman, K.E. and Kalb, R.G. (2000) Experience-dependent development of spinal motor neurons. Neuron 26, 299–305.

Johnson, R.B. (1998) The bearable lightness of being: Bones, muscles and spaceflight. Anat Rec. 253, 24–27.

Keefe, J.R. (1985) Final Report of the NASA Mammalian Developmental Biology Working group. In: K.A. Souza and T.W. Halstead, (eds.) *NASA Developmental Biology Workshop*. pp. 46–63, NASA TM-86756, Washington, DC.

Kuhn, C.M. and Schanberg, S.M. (1998) Responses to maternal separation: mechanisms and mediators. Int. J. Dev. Neurosci. 16, 3–4.

Lagercrantz, H. and Slotkin, T.A. (1986) The "stress" of being born. Sci. Am. 254, 100–107.

Langley-Evans, S.C. (2001) Fetal programming of cardiovascular function through exposure to maternal undernutrition. Proc. Nutr. Soc. 60, 505–513.

Leach, C.S., Johnson, P.C. and Cintron, N.M. (1988) The endocrine system in space flight. Acta Astronaut. 17, 161–166.

Levine, S. (2001) Primary social relationships influence the development of the hypothalamic–pituitary–adrenal axis in the rat. Physiol. Behav. 73, 255–260.

Liu, D., Diorio, J., Day, J.C., Francis, D.D. and Meaney, M.J. (2000) Maternal care, hippocampal synaptogenesis and cognitive development in rats. Nat. Neurosci. 3, 799–806.

Maese, A.C. and Ostrach, L.H. (2002) Neurolab: Final report for the Ames Research Center Payload. NASA TM-211841.

Mani-Ponset, L., Masseguin, C., Davet, J., Herbute, S., Maurel, D., Ghandour, M.S., Reiss-Bubenheim, D., Guell, A. and Gabrion, J. (1997) Effects of an 11-day spaceflight on the choroid plexus of developing rats. Dev. Brain Res. 99, 187–200.

Miquel, J. and Souza, K.A. (1991) Gravity effects on reproduction, development, and aging. Adv. Space. Biol. Med. 1, 71–97.

Moody, S.A. and Golden, C. (2000) Developmental biology research in space: issues and directions in the era of the international space station. Dev. Biol. 228, 1–5.

Moore, C.L. (1992) The role of maternal stimulation in the development of sexual behavior and its neural basis. Ann. N.Y. Acad. Sci. 662, 160–177.

Morey-Holton, E.R. and Globus, R.K. (2002) Hindlimb unloading rodent model: technical aspects. J. Appl. Physiol. 9, 1367–1377.

Murakami, D.M., Hoban-Higgins, T.M., Tang, I.-H. and Fuller, C.A. (1997) The effect of spaceflight on retino-hypothalamic tract development. J. Gravit. Physiol. 4, 67–70.

Nicogossian, A.E., Huntoon, C.L. and Pool, S.L. (1989) *Space physiology and medicine*, 2nd edn., Lea & Febiger, Philadelphia.

Olenev, S.N., Danilov, A.R., Kryuchkova, T.A., Sorokina, L.M. and Krasnov, I.B. (1989) Effect of weightlessness on some indices of brain development (Results of residence of pregnant rats aboard the Cosmos-1514 biosatellite and investigation of subsequent development of their offspring on Earth). Neurosci. Behav. Physiol. 19, 191–195.

Oyama, J. and Platt, W.T. (1967) Reproduction and growth of mice and rats under conditions of simulated increased gravity. Am. J. Physiol. 212, 164–166.

Oyama, J., Solgaard, L., Corrales, J. and Monson, C.B. (1985) Growth and development of mice and rats conceived and reared at different g-intensities during chronic centrifugation. The Physiol. 28, 83–84.

Phillips, R. (2002) Gravity: It's the law. J. Gravit. Physiol. 2002, in press.

Plaut, K., Maple, R.L., Vyas, C., Munaim, S., Darling, A.J., Casey, T.M. and Alberts, J.R. (1999) The effects of spaceflight on mammary metabolism in pregnant rats. Proc. Soc. Expt. Biol. Med. 222, 85–89.

Proshchina, A.E., Besova, N.V., Voronov, K.A., Gulimova, V.I., Serova, L.V. and Savel'ev, S.V. (2000) Morphogenesis of asymmetry of rat brain nuclei under normal conditions and during exposure to microgravitation. Bull. Exp. Biol. Med. 9, 908–911.

Ronca, A.E., Abel, R.A. and Alberts, J.R. (1996) Perinatal stimulation and adaptation of the neonate. Acta Pediat. 415, 8–15.

Ronca, A.E. and Alberts, J.R. (2000a) Physiology of a microgravity environment selected contribution: Effects of spaceflight during pregnancy on labor and birth at 1 G. J. Appl. Physiol. 89, 849–854.

Ronca, A.E. and Alberts, J.R. (2000b) Effects of spaceflight on postnatal vestibular responses in rats. J. Appl. Physiol. 89, 2318–2324.

Ronca, A.E., Baer, L.A., Daunton, N.G. and Wade, C.M. (2001) Maternal reproductive experience enhances birth outcome and early postnatal survival following gestation and birth of rats in hypergravity. Biol. Reprod. 65, 805–813.

Serova, L.V. (1993) The objectives of the experiment. A comprehensive program of post-flight examinations of animals. *In*: O.G. Gazenko, Ontogenesis of mammals in microgravity. NASA TM-103978, pp. 15, Washington, DC.

Serova, L.V. and Denisova, L.A. (1982) The effect of weightlessness on the reproductive function of mammals. Physiol. (Suppl.) 25, S9–S12.

Serova, L.V., Denisova, L.A., Lavrova, E.A., Makeyeva,,V.F., Natochin, Yu. V., Pustynnikova, A.M., Shakhmatova, E.I. (1993) Parameters of the reproductive function of the animals: Fetal and placental characteristics. In O.G. Gazenko, Ontogenesis of mammals in microgravity. NASA TM-103978, pp. 35–36, Washington, DC.

Serova, L.V., Denisova, L.A., Makeeva, V.F., Chel'naya, N.A. and Pustynnikova, A.M. (1984) The effect of microgravity on the prenatal development of mammals. Physiol. (Suppl.) 27, S107–S110.

Serova, L.V., Vacek, A., Denisova, L.A., Lavrova, E.A., Makeyeva,,V.F., Natochin, Yu.V., Chel'naya, N.A., Shakhmatova, E.I. (1993c) State of the neonates. In O.G. Gazenko, Ontogenesis of mammals in microgravity. NASA TM-103978, pp. 37–38, Washington, DC.

Shimizu, T. (1999) Development of the aortic baroreflex system under conditions of microgravity. J. Gravit. Physiol. 6, 555–558.

Silverstein, A.M. (1995) From the forehead of Zeus: the ontogeny of the immune response. Eye 9, 147–151.

Smotherman, W.P. (1983) Mother–infant interaction and the modulation of pituitary-adrenal activity in rat pups after early stimulation. Dev. Psychobiol. 16, 169–176.

Sondag, H.N., de Jong, H.A. and Oosterveld, W.J. (1997) Altered behaviour in hamsters conceived and born in hypergravity. Brain Res. Bull. 43, 289–294.

Sonnenfeld, G., Foster, M., Morton, D., Bailliard, F., Fowler, N.A., Hakenewerth, A.M., Bates, R. and Miller, E.S., Jr. (1998) Spaceflight and development of immune responses. J. Appl. Physiol. 85, 1429–1433.

Temple, M.D., Kosik, K.S. and Steward, O. (2002) Spatial learning and memory is preserved in rats after early development in a microgravity environment. Neurobiol. Learn. Mem. 78, 199–216.

Tou, J., Grindeland, R.E., Ronca, A.E. and Wade, C.E. (2002) Review: Models to study gravitational biology of mammalian reproduction. Biol. Reprod. 67, 1681–1687.

Vasques, M., Lang, C., Grindeland, R.E., Roy, R.R., Daunton, N., Bigbee, A.J. and Wade, C.E. (1998) Comparison of hyper- and microgravity on rat muscle, organ weights and selected plasma constituents. Aviat Space Environ. Med. 69, 2–8.

Walton, K. (1998) Postnatal development under conditions of simulated weightlessness and space flight. Br. Res. Rev. 28, 25–34.

Walton, K., Heffernan, C., Sulica, D. and Benavides, L. (1997) Changes in gravity influence rat postnatal motor system development: from simulation to space flight. Grav. Space Biol. Bull. 10, 111–118.

Wassersug, R.J. (2001) Vertebrate biology in microgravity. Am. Sci. 89, 46–52.

Weaver, D.R. and Reppert, S.M. (1989) Direct in utero perception of light by the mammalian fetus. Dev. Brain Res. 47, 151–155.

Wolgemuth, D.J. and Murashov, A.K. (1995) Models and molecular approaches to assessing the effects of the microgravity environment on vertebrate development. ASGSB Bull. 8, 63–72.

Wong, A.M. and DeSantis, M. (1997) Rat gestation during space flight: outcomes for dams and their offspring born after return to Earth. Integr. Physiol. Behav. Sci. 32, 322–342.

Developmental Biology Research in Space
H.-J. Marthy (editor)

New Facilities and Instruments for Developmental Biology Research in Space

Enno Brinckmann*

European Space Agency, ESTEC/MSM-GF, Keplerlaan 1, NL-2201 AZ, Noordwijk, The Netherlands

Abstract

Three new ESA facilities will be available for biological experiments in Space, Biopack on the Space Shuttle and two instruments on the International Space Station (ISS): BIOLAB in the European "Columbus" Laboratory and the European Modular Cultivation System (EMCS) in the US Lab "Destiny". The experiments are housed in standard Experiment Containers, allowing either research in microgravity or acceleration studies with variable *g*-levels, if mounted on the centrifuges. While Biopack provides only thermal control, BIOLAB and EMCS supply each container with a dedicated atmosphere (controlled CO_2, O_2 concentration and relative humidity, trace gas removal); EMCS contains also fresh and wastewater reservoirs on its rotors. Power and data lines are available in all the described facilities. Highly automated systems, like BIOLAB's Handling Mechanism and Analysis Instruments, support the telescience concept and help reducing crew time in orbit. A BioGlovebox with its support instruments allows unique research possibilities in Space. The feasibility of experiment hardware inside the containers has been studied by ESA for several kinds of Experiment Support Equipment with potential use for research in Developmental Biology: design concepts for experiments with small eggs, cells and tissues, with small aquatic animals, with insects and with plants are described in this article.

Introduction

Research in Developmental Biology in Space is not only a scientific enterprise, but also a challenge for the hardware developer who has to provide facilities for a large variety of organisms, many of them already established as research objects in Space (Table 1).

* E-mail: Enno.brinckmann@esa.int

Table 1

Typical organisms for space research in developmental btiology

Plants	Algae
	Protoplasts
	Higher plant seedlings
Invertebrates	Jellyfish
	Sea urchin larvae
	Nematodes
	Rotifers
	Molluscs
	Insect larvae and adults
Vertebrates	Fish
	Amphibian embryos and larvae
	Bird eggs
	Mice
	Rats

This spectrum of animal and plant objects requires in several cases specialised facilities like the Animal Enclosure Module (AEM; Ronca, this volume) for rats, the Avian Development Facility (ADF; Ronca, this volume) for bird eggs, FERTILE for amphibians (Durnon, this volume), the Insect habitat (IH) for insects, the CEBAS module for aquatic animals, or the plant Growth Unit and its precursors for plant research (PGU; Musgrave, this volume). A different approach is the development of multi-user facilities with standard interfaces allowing investigations with experiment specific hardware, usually enclosed in standard experiment containers. This approach is followed by the European Space Agency (ESA) with two new facilities for the International Space Station (ISS): BIOLAB as part of ESA's Columbus Laboratory and EMCS (European Modular Cultivation System) as part of the Early ISS Utilisation Programme with accommodation in NASA's Destiny Laboratory. The Japanese Space Agency NASDA realized also the concept of a multi-user facility in its Cell Biology Experiment Facility (CBEF; Izumi, this volume).

Important for flight hardware design is the kind of information or result the experiment shall produce: can the scientific output only be obtained on the live sample after return to ground, can it be chemically fixed, refrigerated or frozen? This leads to logistics questions like usage of transport containers including late access time before launch and early retrieval after landing. If the experiment cannot be transported live into Space, dormant stages of the organisms must be used, e.g., plant seeds, eggs, dry or freeze-dried organisms or deep frozen samples. After a revival process, the samples are cultivated in their appropriate environment until experiment termination, but an on-board analysis may be required in addition to the final disposition of the specimen: observation (video, photography), microscopy, spectro-photometry and biometrics measurements.

The recording and transmission of facility housekeeping data to ground is also mandatory to have a complete picture of the experimental conditions on board, which are usually repeated on ground in a reference experiment. A further, in some cases indispensable in-flight control is performed with a 1-*g* reference centrifuge, which should help to discriminate between microgravity effects and those evoked by other factors of the Space environment, like Space radiation.

Especially with long lasting investigations, a life support system is required to keep cultivation conditions close to ground standards, of course with the usual Space flight constraints such as low mass, little volume, little power and high safety standards when crew operations are involved. However, one of the first experiments in Developmental Space Biology was done during a few minutes long sounding rocket flight: the moment of fertilization of *Xenopus* eggs was captured in microgravity and the fixed material was analysed on ground (Ubbels *et al.*, 1989; Willemsen, 1994). The early development of these embryos until the blastula stage was documented later during Spacelab experiments in the Biorack facility (De Maziére *et al.*, 1996; Ubbels, 1997). For that kind of experiments, only thermal control, power supply and data transfer were required in pressurised payloads, followed by refrigerated storage for the longer flight in Spacelab. This experimental environment and the presence of a 1-*g* reference centrifuge led to 82 biological investigations in Biorack, ESA's first multi-user facility dedicated to Space Biology research (Longdon and David, 1987; Mattok, 1995; Cogoli, 1996; Snyder, 1997; Perry, 1999).

The Biorack tradition is continued now by a more compact facility, Biopack (Van Loon and Riesselmann, 1998), designed mainly for Space Shuttle flights and for integration in the Shuttle's Middeck or in Spacehab (Table 2). It contains three thermally controlled volumes: an incubator, a refrigerator and a freezer (also usable as refrigerator). The incubator houses static racks and three small centrifuges for standard Type I and Type II containers (Fig. 1); since all positions are supplied with power and data lines, commanding of automated experimental steps can be done from ground. This facility will have its maiden flight on STS-107 in January 2003 with the typical Space Shuttle flight scenario: handover of the experiments to ESA 17–24 h before launch (Middeck Lockers), activation of the first experiment 2:30 h after launch, individual experiment performance between 4 h and 9 days in Space (followed by a ground control with a 2 h delay), and return of the experiments back to the investigators several hours after landing.

Experiments on the International Space Station have a different logistics approach, since the transport of the samples to their dedicated facility lasts already two days and the combination with other factors may lead to even longer periods in the transport container without 1-*g* control—this makes a scenario more likely where samples should be transported in a dormant state, in which they are more or less insensitive to the (microgravity) environment. Frozen cells, for example, seem to be good candidates for this scenario and they will be re-activated during a controlled thawing process in BIOLAB.

Table 2

Technical data of the Biopack facility

Incubator	22–37 °C ± 1.5 °C (spatial)
	Static positions: 10 Type I/E and 2 Type II/E containers
	Power to each position: ± 12 V dc/1 A, + 5 V dc/1.2 A; total power < 30 W
	Data transfer via 4 analogue lines or via RS-422 (3060–19200 bits/s)
	Commanding possibility from ground
Centrifuge	3; 2 with 4 Type I/E positions, 1 with 2 Type I/E and 2 Type II/E positions
	0.1—2-*g*; 1-*g* at 110 rpm, 2-*g* at 210 rpm
Refrigerator	+ 4 ± 1 °C
	2 rapid cool down Type I positions, 8 storage Type I positions (+ 4 °C); cool down from + 37 to + 4 °C within 2 h
Freezer	15 ± 1 °C (or used as refrigerator at + 4 °C)
	2 rapid cool down Type I positions, 8 storage Type I positions (−15 °C); cool down from + 37 to −15 °C within 4 h
Experiment container	Type I internal volume: 20.4 × 40.4 × 83 mm
	Type II internal volume: 63.2 × 63.2 × 87 mm
	Type I/0 and II/0 without, Type I/E and II/E with electrical connectors

Fig. 1. Biopack incubator tray with static racks and three centrifuges, one of them spinning. Refrigerator and freezer are located above the incubator. A small portable glovebox (background) can be used in-flight for experimental operations in a safe environment (e.g., for chemical fixation).

Another example is ESA's EMCS with its water reservoirs and illumination system, which allows seed germination and subsequent plant growth besides other features for research in Developmental Space Biology on the ISS. This potential draw back by the critical transport phase is compensated by the fact, that the ISS provides a unique environment for biological research in Space: the experiment duration of several weeks up to 3 months or more is a prerequisite for performing long lasting studies in Developmental Biology (Rahmann and Anken, 2001). This article describes the two facilities in more detail and gives some examples, how experiments could be performed in the field of Developmental Biology.

Biolab in the Columbus laboratory

BIOLAB is a facility dedicated for biological research in Space, mainly with small organisms, tissues and cell cultures. It is composed of several subsystems for automatic experiment performance (Table 3a) and for manual interface with the crew (Table 3b, Fig. 2). The experiment itself is housed in the standard Experiment Container (EC) or in the larger Advanced Experiment Container (AEC), both connected to the Life Support System and to power and data lines. The biological material could be accommodated either in a kind of standard hardware, like ESA's technology demonstrators for the Experiment Support Equipment (ESE), or in experiment specific hardware developed individually by the investigators.

Incubator

The incubator provides the thermal environment (18–40 °C ± 0.5 °C, selectable in 0.5 °C steps) to the Experiment Containers; its thermal control system allows a warming-up time of one EC from +4 to +37 °C within 30 min. Six EC's each are locked by the crew on two centrifuge rotors (60 cm diameter), four of each exposed to illumination and video cameras (NTSC format, resolution 0.18 mm). The illumination is parallel to the *g*-vector along the radius of the rotor; video observation is directed perpendicular to it towards the rotor platter. Sixty white LED's and six red LED's (650 nm) are used for the "white" illumination, adjustable from 1.7 to 11.1 W/m^2; the red LED's have been added to the light panel to improve the spectrum of the white LED's allowing e.g., better germination of plant seeds (Fig. 3). Alternatively, infrared illumination at 935 nm with eight LED's per panel is possible for all light panels together on the same rotor. The individual observation of the four EC's on each rotor is done by two colour video cameras, pointing with mirrors to the selected EC with a field of view of 40 × 40 mm. Focus can be adjusted remotely to three planes of the EC interior. One video signal per rotor will be routed at a time either to the video recorder (Hi8, NTSC, 2 h tape) or to ground or to a laptop for examination by the crew.

The centrifuge can be used for microgravity experiments when not rotating (gravity levels are expected to be less than 2.1×10^{-4}-*g* at 1.42 Hz and 5×10^{-4}-*g* at 1000 Hz) or for acceleration experiments (0.001-*g* up to 2.0-*g*, equal to 1.9 and 83.8 rpm, respectively) or for 1-*g* controls (at 59 rpm).

Table 3a

BIOLAB subsystems and their features in the automatic section

Incubator	18–40 °C ± 0.5 °C (spatial and temporal)
	Accommodates centrifuges, LSS and Smoke Detector
Centrifuge	2, each with 6 EC or 4 EC + 2 AEC positions
	4 positions each with illumination/observation system (not for AEC)
	Radial acceleration 0.001—2.0-*g* (in center of EC)
LSS	Connected to all (2 × 6) EC positions through filters (0.2 μm pore size)
	Oxygen: 15–22% ± 1% (selectable in 1% steps)
	Carbon dioxide: low range 0–0.2% ± < 0.004% (0.01% steps)
	Carbon dioxide: high range 0.2–5.5% ± 0.2% (0.1% steps)
	Nitrogen: up to 100%; air flow into EC 80 ± 20 ml/min
	Relative humidity: 60–85% ± 5%, 85–90% ± < 4%
	Trace gas (ethylene) removal through $KMnO_4$ and active charcoal
Observation/Illumination	Observation with rotor based colour NTSC cameras, 1 for 2 EC's
	Field of view on top of EC: 40 × 40 mm, 0.18 mm resolution
	Illumination with white + red LED's or with infrared LED's
Experiment container	Internal volume 60 × 60 × 100 mm (*g*-vector: 60 mm)
	Sterilisation on ground: ozone/ethanol, cover: autoclave > 125 °C
	Power: ± 12V dc/5 W and + 5V dc/10 W
	Data: 5 analogue (2 for P and T sensors), 3 digital I/O, RS-485 line
	Gas supply from LSS through internal filters (0.2 μm pore size)
	Interface plate to HM: access for syringes and tools
	Built-in pressure (± 7 hPa of cabin pressure) and temperature sensor
Advanced exp. container	External volume 125 × 147 × 175 mm (*g*-vector: 175 mm)
	Power, gas and data as for EC, but additional analogue NTSC video line
	Optional interface plate to HM
Automatic ambient stowage	Temperature not controlled, expected range + 16– + 30 °C
	Insert with 101 syringes (1 ml), 1 double containment syringe, 1 tool, 1 wiper, 2 bags for waste and cleaning fluid; or custom made contents
Automatic temperature controlled stowage	2, each selectable from −20– + 10 °C ± 1 °C (1 °C steps); each contains 1 insert with 89 vials, 2 ml volume each
	Can be stowed before/after use in TCU at the same temperature
Microscope	Bright field, dark field, phase contrast; B&W NTSC video camera
	Objectives: Olympus UPFL 4 × , 10 × PH, 20 × PH, 40 × PH
	Large field of view: 1.015 mm diameter, resolution 1.23 μm. Small field of view: 0.25 mm diameter, resolution 0.3 μm
	X–Y scanning, autofocus with manual override (Z scan)
	Flow-through cuvettes with 100/500 μm thickness, volume 28/33 μl
Spectrophotometer	Range 220–900 nm, resolution 10 nm
	Photometric range 0.2–1.5 ± 0.005 Absorbance Units
	Flow through cuvette with 10 mm optical path, volume 17.7 μl
Handling Mechanism	7 axes robotic mechanism, using syringes or a push/pull/turn tool
	Syringe volume 1 ml, accuracy 2% of ejected volume (50 μl to 1 ml), piston speed selectable from 2.2 to 220 μl/s, needle inner diameter 0.5 mm, positioning accuracy ± 1 mm
	Double containment syringe, volume 1 ml
	Wiper to sterilise septum surface
	Rotating tool: 4–120 rpm with 5 rpm intervals, torque 0.1 Nm
	Push/pull tool: 1–20 mm/s, 1 mm/s intervals, 20 mm depth, 20 N

Table 3b

BIOLAB subsystems and their features in the manual section

Temperature controlled units	2, each selectable from −19 to +10 °C ± 1 °C (1 °C steps) Usable volume ~20 l: 12 EC, 2 AEC or 10 ATCS inserts each Cooling capacity: 1 EC from +37 to −19°C in <60 min
BioGlovebox	Working volume 32 l; main door, glove doors and window removable Maximum port size: 220 × 130 (glove doors), 220 × 160 mm (main door), 420 × 115 or 180 × 205 mm (top window), 105 mm diameter (air lock) Illumination adjustable 0–300 Lux; spot light and back-light panel Temperature control +21–+38 °C ± 1 °C Filtered air 99.97% particle free, flow 0.3–0.5 m/s, adjustable Internal pressure 0–1.3 hPa below ambient, adjustable Ozone generator: 100 ppm, sterilisation in 4–10 h Interface plate for 2 EC's with power, data and video lines 2 external video cameras for monitoring BGB activities Accommodates *Experiment Preparation Unit* (EPU) for automatic preparation of fresh cell cultures from frozen stock
Crew command station	Standard ISS Laptop Computer, connected to BIOLAB Standard Payload Computer (SPLC) BIOLAB facility housekeeping data: sampling rate 1 Hz, 50 Kbps transfer Data storage capacity: 45 min of all BIOLAB data sampled at 1 Hz Simultaneous data and video display on laptop TEAC video recorder, Hi8 NTSC format, tape length 120 min

The radial acceleration can be selected between 10^{-3}- and 2-*g* by a total of 40 steps, 10 steps per order of magnitude. The tangential acceleration during spinning up and down of the centrifuge is selectable in three steps (0.96/1.3/3.8 × 10^{-3}-*g*) resulting in a de- and acceleration ramp of 4/3/1 min, respectively.

Experiment container (EC) and life support system (LSS)

The standard Experiment Container is vacuum tight and has a transparent or black cover and two interface plates: one is in contact with the rotor platter and connects power, data and gas supply lines, the other faces the Handling Mechanism (HM), which can insert syringes or a tool into specific ports of the EC (Fig. 4). The syringe interfaces with one or more septa (diameter 6–8 mm), the tool with a rotating or a push–pull mechanism, which can be used for automatic actions on the experiment specific hardware without opening the EC. This operation is of course only possible when the centrifuge has stopped: experiments requiring automatic actions during centrifugation have to provide their own robotic systems inside the EC. To allow an easier accommodation of these automatic systems (e.g., automatic medium exchange or fixation), each rotor has two positions for Advanced Experiment Containers (AEC's). The AEC has a larger volume than the EC (~3.0 l vs. 0.36 l), but no rotor-based

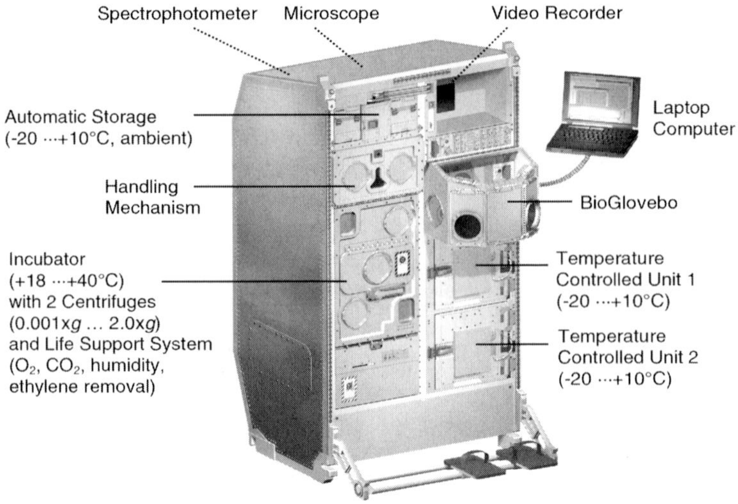

Fig. 2. BIOLAB facility with its subsystems: the automatic section is on the left side with the HM as central part for robotic operations, the manual section dedicated for crew operations is on the right side. Experiment Containers can be placed in the Incubator through a small EC exchange door in the center of the Incubator. The circular structures on the doors of HM and Incubator are ports for ozone flushing from the BGB during the disinfection process. The BGB can be pushed back into the rack for stowage. The Analysis Instruments (microscope and spectrophotometer) behind the Automatic Storage are only accessible, when the HM drawer is extracted from the rack.

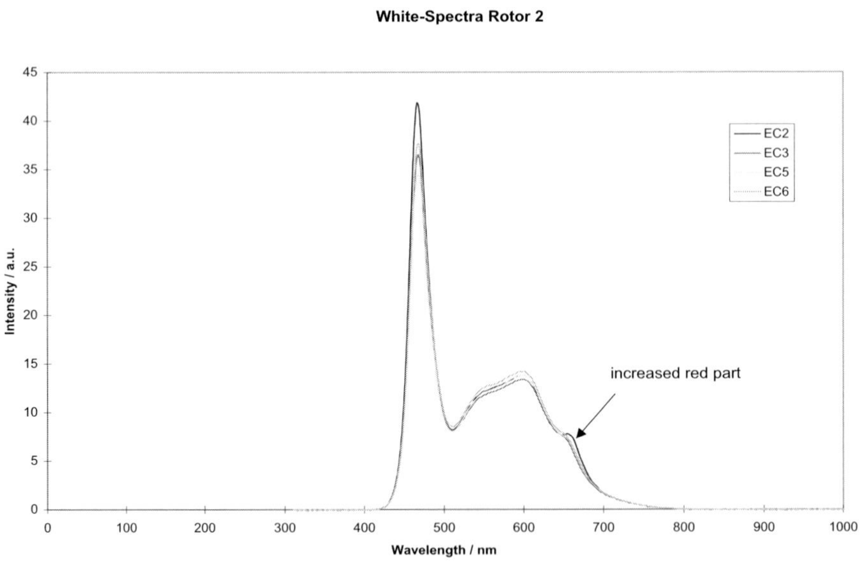

Fig. 3. Spectrum of the illumination panels on the BIOLAB centrifuge rotor 2, supplying four different Experiment Container (EC) positions with "white" light. The panels are composed of 60 white Light Emitting Diodes (LED's) and six red LED's.

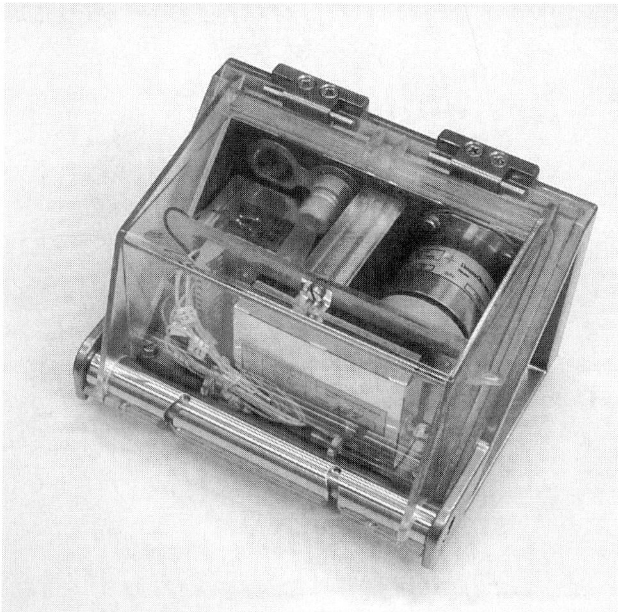

Fig. 4. BIOLAB Reference Experiment Container (REC) with visible built-in test units: septum interfaces and vial, aluminium block for heat capacity measurements, push/pull interface to the HM and electronic boards for testing the RS-485 data line.

illumination/observation system; instead, an individual NTSC video line to the AEC can be connected to a dedicated custom-made observation system inside the AEC. The HM can also interact with the AEC, e.g., for replacement of spent medium as a preparation for the next automatic run on the centrifuge.

Each container is connected through microbial filters (0.2 μm pore size) to the Life Support System (LSS) which provides nitrogen, oxygen, carbon dioxide, controlled air humidity and trace gas removal.

The airflow at the container inlet is 80 ± 20 ml/min, resulting in a pressure of ± 7 hPa above/below the cabin pressure. The containers can be isolated individually from the LSS by electromagnetic valves, if required. In case of failure or depleted gas supply bottles, cabin air can also be pumped through the LSS to ensure a basic gas exchange for the enclosed samples. Compressed carbon dioxide (40 l) and a gas mixture of 30% oxygen with 70% nitrogen (85 l) are transported in 0.5 l pressure bottles to the ISS. For cleaning and disinfection runs, the incubator volume and the LSS can be flushed with pure nitrogen or with 100 ppm ozone, the latter generated in the BioGlovebox.

The Life Support System can be set to low CO_2 concentrations (0–0.2%) e.g., for experiments with plants in an Earth like environment, or to a higher concentration (up to 5.5%) for cell culture experiments requiring a CO_2 incubator. The LSS supplies all EC's on both rotors with the same gas

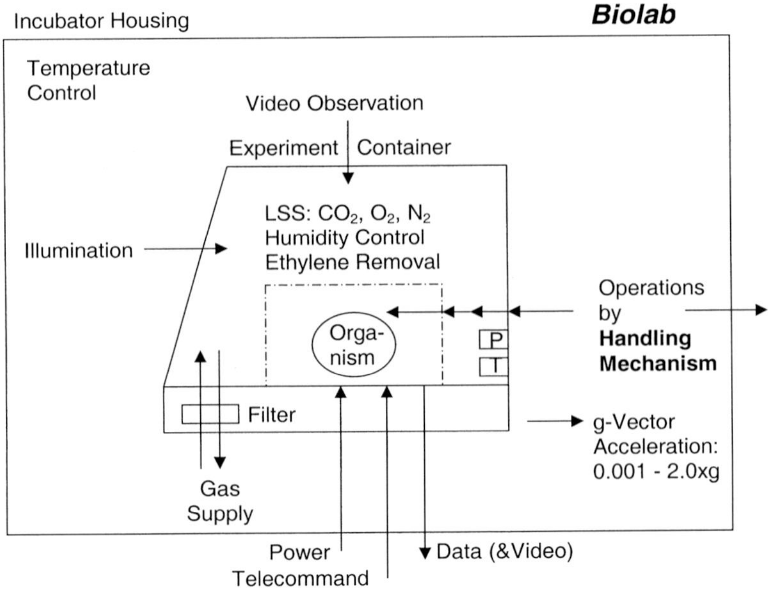

Fig. 5. Schematics of environmental control, services and automatic operations of the Experiment Container in the BIOLAB Incubator. Video output line is only available for the Advanced Experiment Container (AEC).

composition; there is no individual gas control foreseen in a single EC, except the total isolation of each EC from the LSS. Air humidity control is possible between 60 and 90% with an accuracy of 5% in the lower range; 94% shall not be exceeded to avoid condensation in the EC's or in the system itself. The excess of carbon dioxide will be absorbed by a CaOH cartridge (13 ml CO_2/h). Trace gases in the LSS, especially ethylene, which is critical for the development of plant seedlings, can be removed with 99.95% efficiency by a filter composed of $KMnO_4$ and activated charcoal. Figure 5 summarises the environmental conditions and the services provided to the EC's in BIOLAB.

Handling mechanism (HM) and automatic stowage

The drawer located above the incubator contains the HM, which is a robotic manipulator with seven controlled axes (Fig. 6). Its main automatic activities will be liquid transport in syringes, mechanical manipulation of experiments with a tool and cleaning of septum surfaces with a wiper.

The operations of the HM are programmed on ground in a specific experiment schedule, which can be adapted to the real experiment performance during flight. A typical sequence would be the extraction of a sample from the Experiment Container: the gripper moves upwards to a dedicated syringe position in the Automatic Ambient Stowage (AAS), pulls the syringe out of the

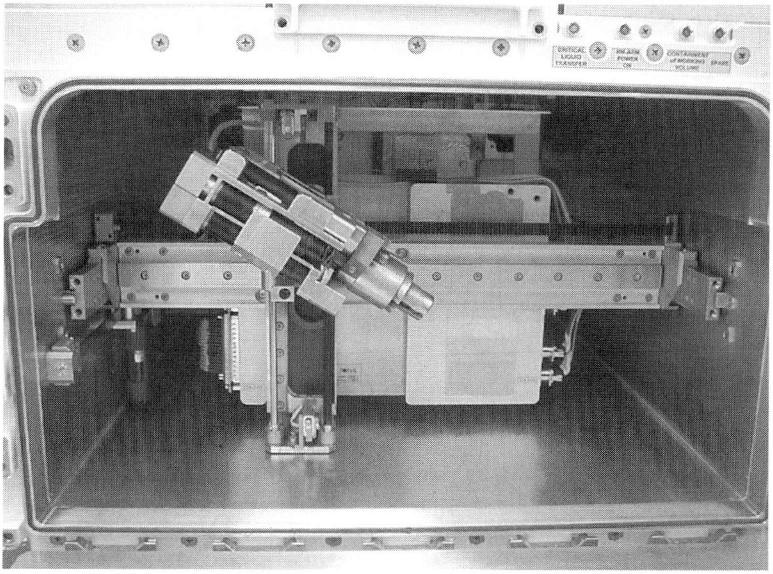

Fig. 6. View into the open HM drawer, showing the tilted gripper and the motor-controlled arms of the *X*-, *Y*- and *Z*-axis. The gripper can fetch syringes or a tool to access the Experiment Containers through two windows in the bottom plate or to transfer samples to the Automatic Stowage and to the Analysis Instruments above.

AAS insert, moves twice to an infrared light bridge for checking the accurate position of the needle tip, rotates the syringe and inserts it down into a specific septum of the EC through a window in the incubator housing; this window has also a bellow released to the EC to minimise the thermal impact on both, the incubator and the HM. The gripper pulls then the piston of the syringe up until the desired volume is reached and injects it either for storage into the Automatic Temperature Controlled Stowage (ATCS, set between -20 °C and $+10$ °C) or into one or both of the two Analysis Instruments for immediate analysis. To avoid cross-contamination, each syringe shall be used only once; it contains in addition a sheath over the needle to ensure sterility before use and to minimise any spillage. After this operation, the empty syringe is returned into the AAS insert.

Mechanical operations inside the EC can be executed through specific interface mechanisms by using the push/pull/rotate tool of the HM. The tool is stowed in the AAS insert, next to the wiper and the vial with the wiping fluid. By means of the wiper, the septum surface (interface to the EC or to the Analysis Instrument) can be disinfected automatically, e.g., with ethanol. All these operations will help reducing the crew time considerably and leave only the basic steps to the crew, i.e., the exchange of EC's in the incubator and the exchange of inserts in the Automatic Stowage.

Analysis instruments (AI)

Two Analysis Instruments (AI's) are available for immediate in-flight analyses in an automatic manner: a microscope (MS) and a spectrophotometer (SP). The samples are injected with a pressure up to 70 hPa through a septum into the AI: about 0.25 ml are required to fill the liquid handling system (Distribution & Cleaning cassette, D&C) until the sample is visible in the analysis spot, the flow-through cuvette. Before that, the sample passes a de-bubbling device to remove air bubbles. The microscope has two cuvettes with different heights: one with 100 μm (volume: 28 μl), the other with 500 μm (volume: 33 μl), the SP only one cuvette with 10 mm light path and 17.7 μl volume. It is obvious, that only suspended samples or clear media can be analysed in the MS or SP, respectively. After the observation, the samples are flushed with distilled water into the D&C waste reservoir. The system is cleaned after the analyses with a disinfection fluid and rinsed with water to be ready for the next analysis cycle. The D&C cassette contains 81 ml water and 35 ml cleaning solution. To replace the D&C cassette, the crew has to slide the entire HM drawer out of the rack, since the AI's are located above the HM and behind the Automatic Stowage compartments.

The microscope can be used in three modes: bright field, dark field and phase contrast. Four magnifications can be chosen remotely: 4 ×, 10 ×, 20 × and 40 × with a maximum resolution of 0.3 μm. Illumination is provided by blue LED arrays, but the transmitted picture is a black and white video image (NTSC format). The sample can be scanned automatically in *X*- and *Y*-direction over an area of 4 × 4 mm, including an autofocus function. All optical controls can also be performed from ground, i.e., change of objectives, condenser lenses and LED arrays as well as scanning, focussing and a stepwise sample transport through the observation area. The only task for the crew is finally the exchange of the D&C cassette or in special cases the monitoring of a sample on the laptop computer screen. Since the Analysis Instruments are closed systems with very limited volume, no internal temperature control could be fitted in the housing, resulting in a sample temperature of up to 37 °C in the worst case.

Spectroscopic and photometric analyses are possible between 220 and 900 nm in the single-beam spectrophotometer. A deuterium and a halogen lamp provide the required spectral range, the light of which is routed with fibre optics to the flow-through cell and from there into the Zeiss MMS 1 detectors. The D&C cassette of the SP has similar performance as the one of the MS. After a lamp stabilisation period of 2 h, the dark current is measured, followed by a measurement series of the blank D&C water sample and of the experiment sample injected previously by the HM (repeated 16 times to reduce the statistical noise of the signals). This measurement cycle can also be repeated to monitor reaction kinetics with a maximum frequency of 1 scan/min.

Bioglovebox (BGB) and experiment preparation unit (EPU)

The BGB has been designed for biological research in Space, i.e., it has an important feature, which is not available in other kinds of closed compartments: the working volume can be disinfected with ozone to guarantee the highest level of cleanliness e.g., for conducting cell culture experiments in BIOLAB. One hundred parts per million ozone gas are produced in an ozone generator, which is a subsystem of the BGB; this ozone flow can also be routed with hoses into the HM and into the Incubator, allowing a disinfection of these BIOLAB compartments to prevent microbial growth on its internal surfaces and inside the Life Support System.

Besides this new feature and the thermal control, the BGB provides all services required for working in a clean environment: adjustable airflow, filtered up to 99.97% particle free, adjustable underpressure in the working volume, adjustable illumination with three different sources (light tubes, spot light, back light panel), power, data and video lines for two Experiment Containers, scavenger and exchangeable filters (through an airlock). The glove doors, the front door and the top window can be opened and removed, thus allowing insertion of larger items for treatment in the enclosed working volume. As an option, the glovebox activities can be followed on ground by video cameras mounted above the BGB window.

Using the clean area in the BGB, the Experiment Preparation Unit (EPU) will provide an automatic platform to prepare new cell cultures from deep-frozen stock samples. This is done after manual loading of vials with the frozen specimen onto a rotating platform, which provides a thermally controlled 1-*g* environment to the thawing sample and replaces the cryo-preservative by fresh medium through automatic washing cycles. Finally, the crew has to transfer the new cell suspension to the experiment hardware (done inside the BGB by means of a specific, syringe-type tool), which may be incubated first in 1-*g* conditions for adaptation to the new culture environment before the actual microgravity experiment starts. This procedure shall minimise the uncontrolled sample exposure to microgravity, which can have an effect on fast intracellular processes.

Crew command station (CCS) and ground control

Besides the experiment control from ground (telescience) and the automatic experiment schedule performed by the HM or by the custom-made hardware, the crew may interfere with the experiment protocol. This is done by means of a dedicated laptop computer, which displays to the crew the same information as transmitted to ground. The crew can check for example the video pictures from the observation cameras on the rotors or from the microscope and discuss these observations with the scientist on the ground. Also technical data of the

BIOLAB facility and of the experiment can be displayed, e.g., for maintenance activities.

Close to the laptop computer is a video recorder with Hi8 NTSC format. The crew has to change the tapes with 2 h recording capacity. This recorder serves also as a buffer, when video data cannot be sent to ground. Video data can then be played back from the recorder (manually or by telecommand) or the tapes return to ground as part of the scientific experiment analysis.

As mentioned before, all functions of the HM, the Analysis Instruments and all parameters of BIOLAB (Life Support System, rotor speed, etc.) can be controlled from ground. This holds also for any kind of automated hardware inside the Experiment Containers. Data retrieval and command sending will be co-ordinated by the BIOLAB User Support and Operations Centre (USOC), located at DLR in Cologne (Germany); there is also the possibility that the scientists can monitor and control their experiment from their own lab in the User Home Base (UHB).

Science reference model (SRM)

The ground control experiment will be performed in the Science Reference Model (SRM), which is a copy of all Flight Model subsystems performances with some modifications for ground use. The 1-*g* gravity vector of the centrifuges is represented by a static shelf in an incubator, i.e., no acceleration experiments are foreseen in the presence of the overlaying 1-*g* vector on ground. The Analysis Instruments are represented by commercial instruments of a similar quality: Olympus CX 40 microscope and Varian CARY 50 spectrophotometer. Also, the thermal environment of the microscope can be simulated by a heated stage. The BGB is simulated by a safety cabinet with controlled airflow; it contains all interfaces for the EPU and the Experiment Containers, but has no internal ozone sterilisation capability. The TCU's are replaced on ground by commercial temperature chambers with the same thermal profile as the flight units. The observation system on the flight rotor is realised in the SRM without mirrors, i.e., there are only two video cameras on the shelf (one for four EC's) with a direct view into each EC. The Life Support System is identical in the SRM as compared to the Flight Model, which is assumed to be the most critical part of the reference experiment. The other interfaces to the EC, i.e., illumination, power, data and—for AEC's only—video lines, are identical with the Flight Model.

The HM is represented on ground by a powered tool system, i.e., the syringe will be loaded by hand on a gripper with subsequent automatic insertion into the EC interface plate, followed by automatic liquid transfer. The content of the syringe will then be injected manually either into the flow-through cells of the Analysis Instruments or into the storage vials simulating the ATCU inserts;

these inserts are stored in the TCU simulators, i.e., in temperature chambers. The implementation of this semiautomatic procedure is considered a compromise to avoid the construction of a very complex replica of the HM with its specific features for the ground environment.

The use of the ozonizer in flight is meant to take care of potential contaminations and microbial growth during the long stay of BIOLAB in orbit with all its implications concerning crew time, safety and operational aspects, e.g., no open use of ethanol for sterilisation on the ISS, which shall be not applicable for the ground model. Standard cleaning and sterilisation processes are foreseen for the SRM on ground. Specific tests for the compatibility of the flight procedures with the experiment protocol are possible in the BIOLAB Engineering Model (EM), which has a flight-identical configuration with only minor differences to the Flight Model in some components. In this model, critical liquid transfer steps or automatic actions of the HM push/pull/turn tool can be tested before flight. It is also available for fit checks and specific crew training, which is in general performed on a dedicated Training Model.

The SRM and the BIOLAB Engineering Model (EM) will be located at the BIOLAB Facility Responsible Centre (FRC) at DLR in Cologne (Germany). The SRM will be used for the ground reference experiment and for verification of the experiment interface to the BIOLAB facility in an Experiment Sequence Test, simulating the full timeline of the experiment. However, to facilitate the development of the experiment specific hardware, a Facility Support Centre (FSC) has been established at "Biotesc" at the ETH in Zürich (Switzerland), providing knowledge of Space experimentation and interface models to the BIOLAB EC's, which allows the investigators to test their new hardware during all steps of the development.

European modular cultivation system (EMCS)

To provide European investigators with access to the ISS for research in the early assembly phase, ESA and NASA had agreed on an Early Utilisation Programme, which led to the development of the European Modular Cultivation System (EMCS) for accommodation in NASA's Destiny laboratory. EMCS is mainly designed for experiments with plants, especially multi-generation (seed-to-seed) experiments and studies on gravity effects on early plant development and growth, on signal perception and transduction in plant tropisms. Experiments with insects or amphibians as well as studies with cell and tissue cultures are also foreseen in EMCS. EMCS has many features similar to BIOLAB, but other than BIOLAB it consists only of an incubator with its life support and control systems while the other resources required for Space experiments are located elsewhere in the US lab. Table 4 summarises the technical data of EMCS.

Table 4

EMCS subsystems and their features

Incubator	18–40 °C ± 0.5 °C (spatial and temporal) Accommodates centrifuges, LSS and Smoke Detector
Centrifuge	2, each with 4 EC positions Each position with illumination/observation system Radial acceleration 0.001—2.0-*g* (at outer rotor radius) Tangential acceleration is < 10% of radial acceleration (< 25% for radial accelerations below 0.01-*g*)
LSS (static)	Controls gas composition for entire incubator volume Oxygen: 15–22% ± 1% (selectable in 1% steps) Carbon dioxide: low range 0–0.2% ± < 0.008% (0.01% steps) Carbon dioxide: high range 0.2–5.5% ± 0.2% (0.1% steps) Nitrogen: up to 100%; air flow into EC: 390 ml/min Trace gas (ethylene) removal through $KMnO_4$ and active charcoal
LSS (rotor)	Relative air humidity: 50–85% ± 5%, 85–90% ± 2.5% Relative air humidity in drying mode: 30% at 20 °C Water supply to EC from reservoirs (250 ml per rotor) Condensed water from LSS can be re-routed to the EC Waste water reservoir (250 ml per rotor)
Observation/Illumination	Observation with rotor based colour NTSC cameras, 1 for 2 EC's Field of view on top of EC (max. zoom): 30 × 22 mm, < 0.1 mm resolution; wide angle: 80–100% of EC visible, ~ 0.3 mm resolution Illumination with white + red LED's or with infrared LED's
Experiment container	Internal volume 60 × 60 × 160 mm (*g*-vector: 160 mm) Sterilisation on ground: ozone/ethanol, cover: autoclave > 125 °C Power: + 12 V dc/5 W short term only; − 12 V dc/1 W and + 5 V dc/10 W continuously (thermal constraints based on 1 W per EC); maximum simultaneous power consumption per rotor: + 12 V dc/10 W, − 12 V dc/4 W, + 5 V dc/10 W Data: 5 analogue (3 for P, T and RH sensors), 3 digital I/O, RS-485 line; NTSC video line (video lines for 2 rotor positions only) Gas supply from LSS through internal filters (0.2 μm pore size) Rotor-based water supply through internal filters (0.2 μm pore size) Built-in pressure sensor (< 60 hPa above cabin) Built-in temperature and relative humidity sensors
Crew command station	EXPRESS Laptop Computer, connected to EMCS Standard Payload Computer (SPLC) EMCS facility housekeeping data: sampling rate 1 Hz, 50 Kbps transfer Data storage capacity: 45 min of all EMCS data sampled at 1 Hz Simultaneous data and video display on laptop TEAC video recorder, Hi8 NTSC format, tape length 120 min Frame grabber of individual video scenes

Incubator and life support system (LSS)

The EMCS incubator provides the thermal environment (18–40 °C ± 0.5 °C, selectable in 0.5 °C steps) to the Experiment Containers. Four EC's each are placed by the crew on the two centrifuge rotors (60 cm diameter). All EC's are exposed to illumination and video observation systems (NTSC colour camera, resolution 0.1 mm). The illumination is parallel to the *g*-vector along the radius of the rotor; video observation is perpendicular to it towards the rotor platter. One hundred and thirty-two white LED's and six red LED's (650 nm) are used for the "white" illumination, selectable between either 2/3 or full intensity ($= 75\,W/m^2$); the red LED's have been added to the light panel to improve the spectrum of the white LED's supporting e.g., better germination of plant seeds (Fig. 3). Alternatively, infrared illumination at 935 nm with five LED's per panel is possible for all light panels together on the same rotor. The individual observation of the four EC's on each rotor is done by two video cameras, pointing with a mirror to the selected EC. Focus and zoom can be adjusted remotely to observe either the entire EC or details in the EC interior. One video signal per rotor will be routed at a time either to the video recorder (Hi8, analogue NTSC format, 2 h tape) or to ground or to the laptop for examination by the crew. A frame grabber allows the generation of digitised still images and their download to the ground, when only still photos are required for experiment evaluation.

The centrifuge can be used for microgravity experiments when not rotating or for acceleration experiments (0.001-*g* up to 2.0-*g*, equal to 1.73 and 77.2 rpm, respectively; *g*-levels are given for the outer centrifuge radius) or for 1-*g* controls (at 54.6 rpm). The corresponding rotation for a 1-*g* level in the center of the EC is reached at 66 rpm. The time needed for spinning up from 0- to 1-*g* can be chosen between 30 s and more, whilst breaking from 1- to 0-*g* can be done in ≤ 10 s. Tangential acceleration during spin up or down is less than 10% of the radial acceleration ($<25\%$ below 0.01-*g*), with a minimum of 0.006-*g*.

When the incubator door is closed and locked, the internal air volume is replaced by pure nitrogen for flushing and a new, customised atmosphere is composed by the Life Support System (LSS). CO_2 and O_2 levels can be adjusted through telecommand or automatically by the experiment scheduler; trace gases, like ethylene, are removed by a cartridge containing $KMnO_4$ and activated charcoal. The air humidity in the incubator is controlled around 50% relative humidity, but will be adjusted to the desired level for each individual Experiment Container on the rotors.

Experiment container (EC) and rotor-based life support system (RB-LSS)

The standard EMCS Experiment Container is vacuum tight and has a transparent, fully removable cover and a baseplate, where power, data, video, gas and water supply lines are connected to the rotor platter. The larger volume of $60 \times 60 \times 160$ mm shall allow installation of automated experiment specific hardware, which can handle experiment procedures without opening the EC (Fig. 7). The experiment should run therefore automatically after installation of the EC's on the rotors by the crew. Each rotor accommodates also a water reservoir and a RB-LSS. The water reservoir contains 250 ml sterile, distilled

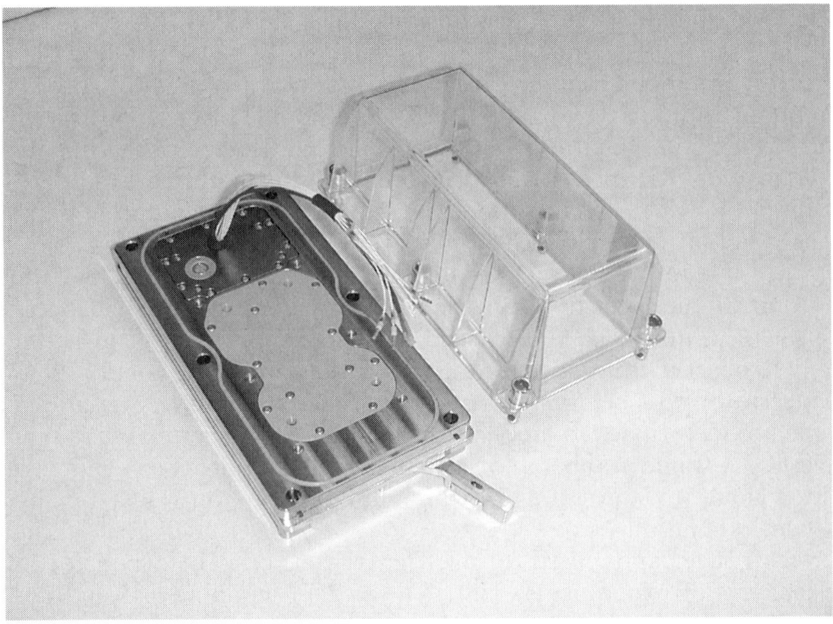

Fig. 7. EMCS Experiment Container (EC) with cover open and removed. The lever of the cover locking mechanism extends at the lower right from the EC baseplate. The baseplate contains the cables for power, data and video lines, fixation points for experiment specific hardware and the in-/outlets of gas and water supply lines.

water and a wastewater reservoir of the same volume. The water is routed to each EC through microbial filters (0.2 μm pore size) and shall be mixed inside the EC e.g., with nutrient stock solutions. The atmosphere of the incubator interior volume is sucked into the rotor-based LSS and conditioned for relative humidity at the individual EC level. This allows adjustment of the air humidity according to the experimental needs, e.g., to reduce the potential condensation at the interior surface of an EC cover when a plant is transpiring at a high level. Also a drying mode with 30% relative humidity (at 20 °C) is available, which can be used to remove water vapour from the experiment, e.g., before harvesting seeds. Figure 8 summarises the environment and services provided to the EMCS container.

The described features differ in several aspects between EMCS and BIOLAB (Table 5): it is obvious from those data, especially with respect to EC size, water supply, illumination level and humidity control, that EMCS is more favourable for the growth of small plants than BIOLAB. The lack of a HM is compensated somehow by the larger EC dimensions in EMCS, allowing the installation of highly automated experiment hardware performing the required experimental steps with control from ground. Ancillary equipment, however, like cooler, freezer, glovebox or analysis instruments, has to be used somewhere else

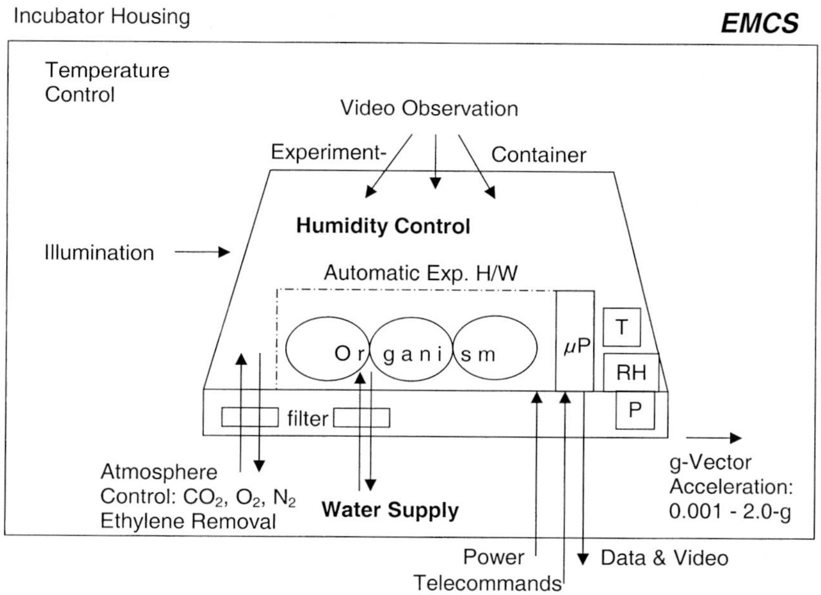

Fig. 8. Schematics of environmental control and services supplied to the Experiment Container in the EMCS Incubator. The temperature and atmospheric composition is identical for the entire incubator volume, but humidity control is done individually on container level. T, RH and P are sensors for temperature, relative humidity and pressure, respectively, μP represents the microprocessor of the automatic experiment hardware.

Table 5

Main differences between BIOLAB and EMCS on incubator level

Features	BIOLAB	EMCS
Experiment container volume	60 × 60 × 100 mm = 360 ml	60 × 60 × 160 mm = 576 ml
EC height in direction of *g*-vector	60 mm	160 mm
EC electrical interface	Power, data (video only for AEC)	Power, data, video
Illumination	10 W/m^2	75 W/m^2
Water supply to EC	None (but through syringes via HM)	Connection to rotor-based fresh and waste water reservoirs
Gaseous life support system	Common atmosphere for all EC's	EC-individual control of relative air humidity
Video format	Analogue NTSC format	Analogue NTSC format and video frame grabber
Video resolution	< 0.18 mm, fixed focal length	< 0.1 mm, zoom capability

in the ISS. However, the overall design of EMCS and its LSS provides not only research capabilities for plant experiments, but also for investigations on cell and tissue cultures and on organisms used for studies in developmental biology, like insects, amphibians and fish.

Experiment reference model (ERM)

Similar to BIOLAB, the EMCS ground reference experiments and preparations will be performed in a facility without centrifuges, i.e., only the static 1-*g* environment is available for ground controls. To have the same direction of the *g*-vector in the EC on ground as on the flight centrifuges, the Engineering Model (EM) of EMCS with its two centrifuges cannot be used and a different design for the ERM was necessary (Fig. 9). All eight EC's are mounted on a static shelf with two video cameras between them, observing four EC's each without the use of mirrors. Thermal control, water supply and life support including humidity control are the same as in the Flight Model, but to ease the air composition on ground, a closed tank with the same volume as the air

Fig. 9. EMCS Experiment Reference Model (ERM) designed for experiment development and ground reference experiments. The incubator doors are opened to allow a view on the two movable video cameras located on the upper shelf together with 2×4 experiment containers, illuminated by Light Emitting Diodes (LED's). The space below the experiment shelf accommodates electronic devices, pumps and water reservoirs of the LSS and a stainless steel tank representing the gas volume of the incubator atmosphere in the Flight Model.

volume of the flight incubator is mounted in the ERM. In this tank, the mixing process of the air can be followed with the same time profile as in flight, preventing any potential leakage problems of the commercial incubator on ground. There will be two ERM's available for testing and reference experiments: one will be located at the FMCS Facility Responsible Center (FRC) in Trondheim (Norway), the other at NASA Ames Research Center in Moffett Field, CA (USA).

Experiment support equipment (ESE) and mission scenario

The concept of multi-user facilities for both, BIOLAB and EMCS, has to focus finally on the experiment specific hardware inside the standard Experiment Containers. With this concept, the responsibility for the development of the experiment hardware is in general with the Principle Investigator and its sponsoring agency. However, due to the high complexity of the interface between the facility and the EC as well as between the EC and the experiment specific hardware, the potential design of this hardware has been studied by ESA for several experiment scenarios, resulting in 10 breadboards of the Experiment Support Equipment (ESE), which demonstrate the technical feasibility for:

- Multi-generation cultivation system for insects
- Multi-generation cultivation system for small aquatic species
- Microtiter plates
- Active fluid processing system
- Passive fluid processing system
- Experiment controller
- Basic miniature microscope
- Advanced miniature microscope
- Agitated cell cultivation
- Mini aquarium system

It shall be recalled that all these systems are inside the Experiment Container, i.e., the cultivation and observation can be done automatically at the same time in microgravity and on the running centrifuge during acceleration experiments or in the 1-*g* controls.

Experiments with small eggs, cells and tissues

With respect to research in Developmental Biology, several of these breadboards could be used for early development studies on small eggs, cells and tissues: the two microscopes provide objectives with a magnification up to 40 × and observation in bright field or phase contrast. The specimen is housed in a flow-through cell with connections to fresh and waste medium reservoirs, which can be accessed through septa in the EC interface plates from outside.

Fig. 10. Breadboard model of the "Agitated Cell Cultivation System", showing one of the two tube-shaped culture chambers and the roller, which is driven by a screw rod to mix gently the culture medium in the tube. This design fits into the BIOLAB Experiment Container.

The HM of BIOLAB could exchange automatically these liquids or transfer specimens from one container to another, while EMCS needs the crew to do this service during an experiment. The microtiter plates, the fluid processing systems and the agitated cell cultivation system are designed mainly for cell and tissue cultures. Figure 10 shows the gentle agitation system, which keeps the culture medium in slow movement around the suspended or attached cells and organisms with neglectable influence on the microgravity environment. This kind of culture system shall compensate the depletion of the cultivated specimen from nutrients, which may in general be one reason for changed behaviour in Space due to the lack of convection.

Experiments with aquatic animals

For studies on small aquatic species like sea urchin larvae, jellyfish ephyrae, tadpoles or fish youngsters, a "Mini-aquarium system" has been developed and tested on ground with sea urchin larvae: swimming behaviour and development was normal during 14 days even without any water exchange, proving the biocompatibility of this system. A connection module is the interface between the EC baseplate and the aquarium (Fig. 11): fresh and wastewater can be supplied to/from the EMCS reservoirs on the rotor; in BIOLAB, the HM could inject or withdraw liquids through a septum port. The connection module can

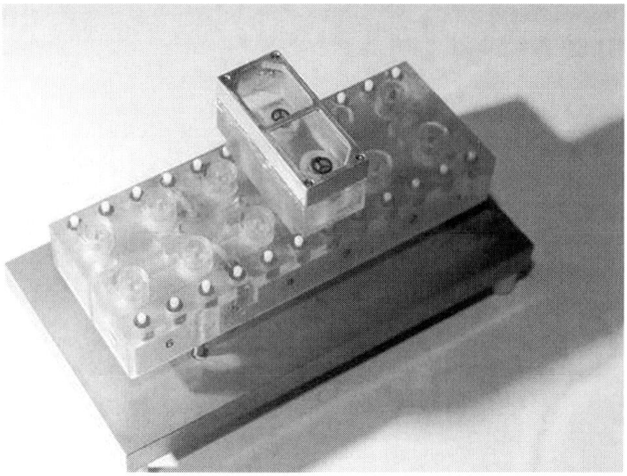

Fig. 11. Breadboard model of the "Aquatic Animals Cultivation System" as designed for EMCS. One of the six mini-aquaria is sitting on top of the connection module, which provides contact through variable tube routings to the water supply system of EMCS. The white dots on each side of the connection module are lateral pressure pads allowing a quick and safe attachment of the mini-aquaria or other modules like pumps, filters or nutrient supply systems.

accommodate 6/4/3/2/1 mini-aquaria with water flowing through a volume of 30/45/60/90/180 ml each. By different lay-outs of the interconnecting tubes, the aquaria can be supplied e.g., in a parallel or serial mode; also the use of an EC specific pump module or filter/nutrient module is possible by replacing one or more aquaria on the connection module. Each mini-aquarium is covered by a biofoil membrane, supporting gas exchange of the medium to the atmosphere in the EC, which is conditioned by the LSS. This transparent foil allows also the visual control of the small animals by the rotor-based video observation system.

A similar kind of small aquarium (volume $\sim$10 ml) has been chosen for the "Active fluid processing system" (Fig. 12). The specimen can be observed through the biofoil membrane, which allows also gas exchange with the atmosphere in the Experiment Container. The fluid in the aquarium is pumped through tubes from a medium container to a waste reservoir; both of them can be accessed from outside through septa, allowing automatic medium replacement by the HM in BIOLAB or manually by the crew in EMCS. In addition to this medium flow, samples can be taken by telecommand: five vacuum filled syringes can be inserted automatically through the septum bottom of the aquarium to remove up to five 1.3 ml samples. A sixth syringe has an overpressure and can release any kind of chemical (activator, fixative) into the aquarium.

A third type of culture system for aquatic animals had been designed for rotifers and nematodes, but a more detailed description of this hardware is given by Ricci (in this volume).

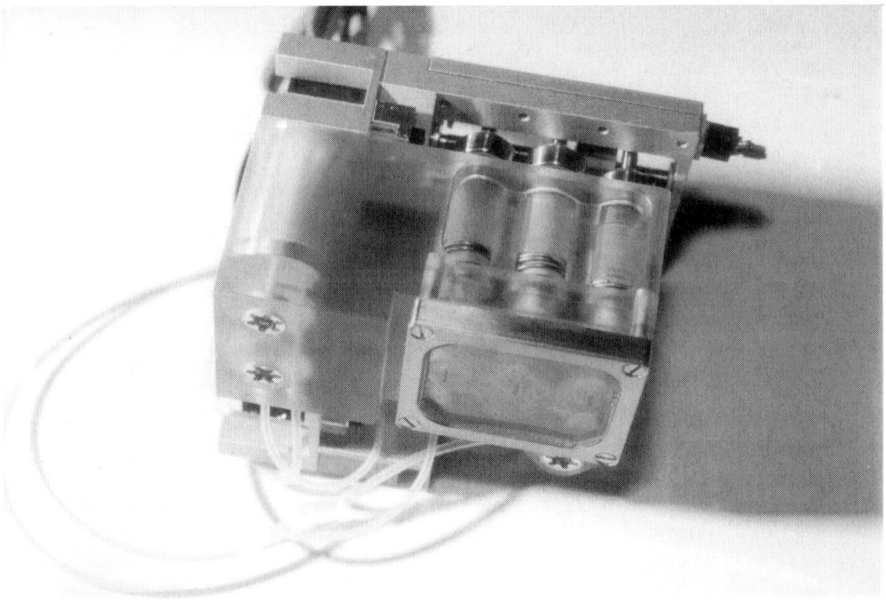

Fig. 12. Breadboard model of the "Active Fluid Processing System" in the BIOLAB configuration. The culture chamber (or aquarium) is visible on the bottom right, adjacent to the sampling syringes. A transparent biofoil on the culture chamber allows observation with BIOLAB's rotor-based video system. Reservoirs for fresh and spent medium are visible on the left side, connected with tubes to a pump and the culture chamber.

Experiments with insects

The prototype of an automated culture system for multi-generation experiments on insects has been tested on ground with *Drosophila melanogaster* (Marco *et al.*, 1999; also this volume). This module has two transparent chambers containing the adults, which can feed on the food trays and can lay eggs there on the food surface (Fig. 13). When the food tray is overlaid with embryos, the drum is rotated and the embryos may either develop in the other adult chamber for a second generation, or they are preserved by heat shock (50 °C) for later analysis on ground. A drum with six new food trays can easily be replaced by the crew in the glovebox, where the experiment container can be opened in a safe environment. A photoelectric sensor in the adult chambers is used for monitoring the activity of the flies in addition to the video observation provided by BIOLAB or EMCS on the centrifuge rotors. One of the adult chambers contains also a receptacle compartment, which can be opened to fix a population chemically with ethanol. All functions of this insect cultivation system are handled by a microcontroller, which fits together with it into the BIOLAB or the EMCS container. The microcontroller can be commanded

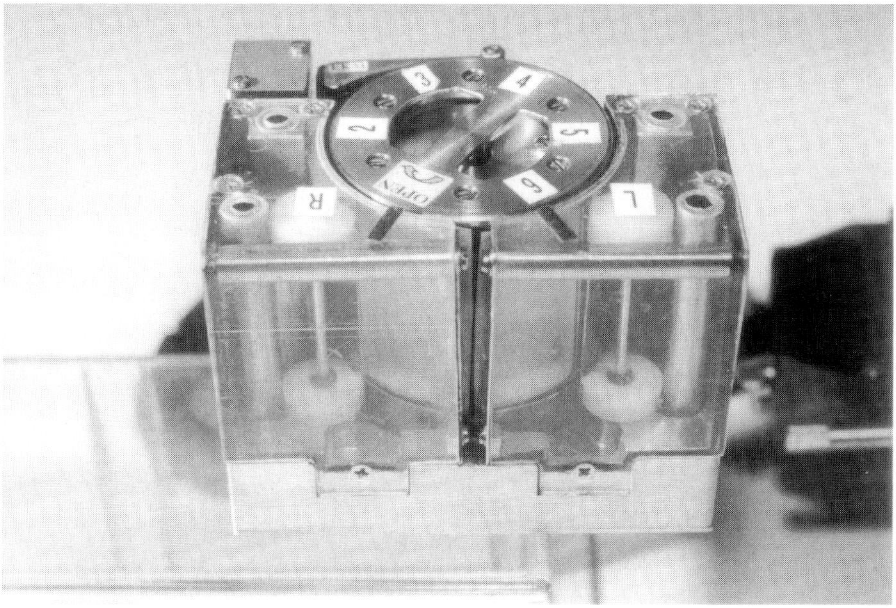

Fig. 13. "Automatic Culture System for Insects" with two compartments for adults, one drum with six food trays and a heat fixation device. The compartments for the adults are covered by transparent windows to allow visual checks with the BIOLAB video observation system.

from ground and transmits science and status data to the experimenter on ground, thus allowing telescience operations.

Experiments with plants

One of the main goals of EMCS, to allow seed-to-seed experiments on small plants, needs a careful design of both, the facility and the experiment specific hardware in the EC. The EMCS facility development was closely followed by a team of scientists at the University of Trondheim, led by T.H. Iversen (NTNU, Department of Botany, Plantebiosenteret): a test bench was used to prove biocompatibility and performance of several EMCS subsystems and is still available to test the environmental parameters applied to a plant in the Experiment Container. The plant growth chamber, however, is another challenge due to the problems with water and nutrient supply, with assimilation and transpiration and finally with observation of the plant in its container under microgravity conditions. The same team studied under ESA contract a Functional Model to verify the scientific concept and biological performance of EMCS by using the available test bench and additional adapters to the EMCS EC (Iversen *et al.*, 2002). This model is shown in Figure 14 with cress seedlings. The seed-to-seed growth of *Arabidopsis* was successfully demonstrated earlier on ground. The plant seeds germinate in a water absorbent layer on top of the

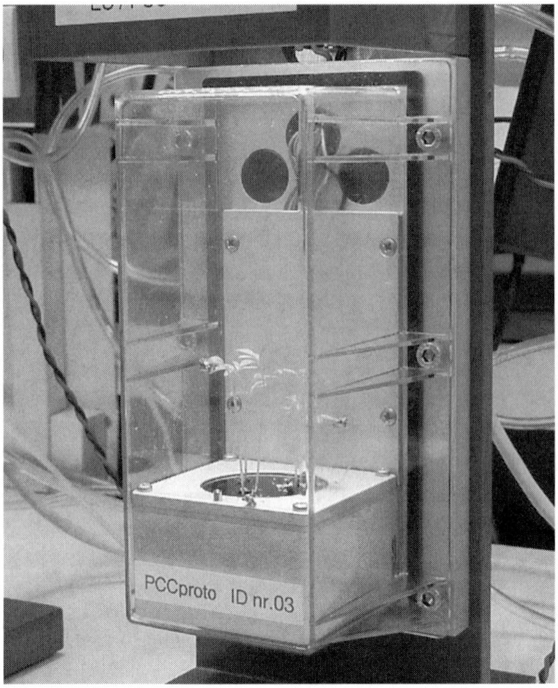

Fig. 14. Plant Cultivation Chamber (PCC) in an EMCS-like container on its test stand with cress seedlings (courtesy of T.H. Iversen, University of Trondheim).

Plant Cultivation Container (PCC) by means of water supplied by the distribution system on the EMCS rotor. After germination, the roots protrude into the growth medium, which has its own humidity control system to avoid water clogging. After harvesting by the crew, the PCC can easily be exchanged by a new one to start an additional growth period, may be with the second generation of plants in Space.

Mission scenario

After selection of the experiment in the frame of an international Life Sciences Research Announcement (LSRA), the scientist coordinates with his/her sponsoring agency and ESA the development of the experiment specific hardware inside the Experiment Container (EC); the EC will be provided by the European Space Agency. The hardware and experiment protocol development for BIOLAB shall be supported by the Facility Support Center. After development and testing of the experiment specific hardware (e.g., biocompatibility, vibrations, off-gassing, electromagnetic forces), a final interface test will be done in the SRM/ERM to assure flight readiness of the experiment.

Sample preparation prior to flight can be done in the home lab of the investigator, at the Facility Responsible Center (FRC) or at the launch site: it should be emphasised that the samples have to stay in a safe transport condition for several days before they can be loaded into BIOLAB or EMCS due to the critical transport time to the Space Station. During flight, a ground control shall be performed in the SRM/ERM together with potential 1-*g* controls on board. Experiment control (telescience) will be possible from the FRC. When the experiment has been terminated, the evaluation had either been done on board by means of in-flight analyses or will be done on ground. In this case, specimens have to be returned to ground in a transport-safe state i.e., chemically fixed, frozen or in a dry state. When the Minus Eighty degrees Laboratory Freezer for ISS (MELFI) is used, deep frozen (−80 or −26 °C) or refrigerated (+4 °C) samples can be launched, stored during flight and returned to ground with less than 10 °C temperature increase due to the large thermal capacity of the well isolated storage containers. Life samples, however, have to travel back from Space with the Space Shuttle, which may last at minimum one day between removal from the incubator and retrieval on ground. After landing, the experiment is handed back to the investigator for further evaluation.

Conclusions and summary

The two new facilities on the International Space Station, BIOLAB and EMCS, will contribute considerably to the possibilities for research in Developmental Space Biology. Their life support and analysis systems support short and long lasting experiments for a large variety of cells, tissues, animals and plants. On-board reference centrifuges and ground-based controls allow discrimination between effects of microgravity and other Space effects, like radiation. The unique Space environment can also be used for acceleration studies between 0.001- and 2-*g* and to study transient effects occurring during exposure to or removal from microgravity. The concept of multi-user facilities with a generic design of sophisticated support systems and experiment specific hardware in the standard experiment containers opens a large field of new experimentation in many areas of Space Biology.

Acknowledgements

Many thanks to the following colleagues for checking the manuscript for technical details of the respective facilities: Hartmut Helmke (ESA) and Claude Brillouet (ESA) for Biopack, Jean-Michel Bois (ESA) for BIOLAB, Johannes Kreimer (Astrium) for EMCS, Peter Schiller (ESA), Tor-Henning Iversen (Trondheim) and Knut Robert Fossum (Trondheim) for ESE and last not least Hans-Jürg Marthy for final editing of the manuscript.

References

Cogoli, A. (ed.) (1996) Biology under Microgravity Conditions in Spacelab IML-2, J. Biotechnol. 47, 65–403.

De Maziére, A., Gonzalez-Jurado, J., Reijnen, M.J., Narraway, J. and Ubbels, G.A. (1996) Transient effects of microgravity on early embryos of *Xenopus laevis*. Adv. Space Res. 17(617), 219–223.

Iversen, T.H., Fossum, K.R., Svare, H., Johnsson, A., Schiller, P. (2002) Plant growth using EMCS hardware on the ISS. J. Grav. Physio. 9(1), 223–224.

Longdon, N., David, V. (eds.) (1987) *Biorack on Spacelab D1,* ESA SP-1091, ESA Publications Division, ESTEC, Noordwijk, The Netherlands.

Marco, R., Díaz, C., Benguría, A., Mateos, J., de Juan, E. (1999) *Drosophila melanogaster*, a key arthropod model in the study of the evolutionary long term adaptation of multicellular organisms to the space environment. In: Wilson, A. (ed.) *Proceedings of the 2nd Eur. Symp. on the Utilisation of the ISS*, ESA SP-433, pp. 433–440, ESA Publications Division, ESTEC, Noordwijk, The Netherlands.

Mattok, C. (ed.) (1995) *Biorack on Spacelab IML-1,* ESA SP-1162, ESA Publications Division, ESTEC, Noordwijk, The Netherlands.

Perry, M. (ed.) (1999) *Biorack on Spacehab: Experiments on Shuttle to Mir Missions 03, 05 & 06*, ESA SP-1222, ESA Publications Division, ESTEC, Noordwijk, The Netherlands.

Rahmann, H., Anken, R. (2001) Bedeutung der Schwerkraft für die Entwicklung and Orientierung der Tiere. In: Rahmann, H., Hirsch, K.A. (eds.) *Mensch-Leben-Schwerkraf-Kosmos, Perspektiven biowissenschaftlicher Weltraumforschung in Deutschland*, pp. 170–191, Verlag Günter Heimbach, Stuttgart, Germany.

Snyder, R.S. (ed.) (1997) *Second International Microgravity Laboratory (IML-2) Final Report*, NASA Reference Publication 1405, Marshall Space Flight Center, MSFC, Alabama.

Ubbels, G.A., Berendsen, W. and Narraway, J. (1989) Fertilization of frog eggs on a sounding rocket in space. Adv. Space Res. 9(11), 187–197.

Ubbels, G.A. (1997) The role of gravity in the establishment of the dorso–ventral axis in the amphibian embryo (Eggs), Second International Microgravity Laboratory (IML-2)—Final Report. R.S. Snyder, compiler, NASA Reference Publ. 1405, pp. 122–123.

Van Loon, J., Riesselmann, W. (1998) Biopack, a biological research facility. Microgravity News 11 (2): 7–10. ESA Publications Division, ESTEC, Noordwijk, The Netherlands.

Willemsen, H. (1994) Automatic hardware for biological experiments in microgravity. In: *Proceedings of the 2nd Eur. Symp. on Life Sciences Research in Space, Arcachon, France 26 Sept.–1st Oct. 1993*, ESA SP-366, pp. 461–465, ESA Publications Division, ESTEC, Noordwijk, The Netherlands.

List of Main Authors

Ralf H. Anken, PD Dr.
University of Stuttgart-Hohenheim
Zoological Institute
Garbenstrasse 30
D 70593 Stuttgart, Germany
Phone: 0049 (0) 711 459 4020
Fax: 0049 (0) 711 459 3450
E-mail: anken@uni-hohenheim.de

Enno Brinckmann, Dr.
ESA Microgravity Facilities for Columbus Division
ESTEC, Postbus 299
NL 2200 AG Noordwijk, The Netherlands
Phone: 0031 71 565 5014
Fax: 0031 71 565 3111
E-mail: Enno.Brinckmann@esa.int

Augusto Cogoli, Dr. (*Editor of the ASBM series*)
Space Biology Group ETHZ
Biotechnology Space Support Center, BIOTESC
Technoparkstrasse 1
CH-8005 Zurich
Phone: 0041-1-445 1422/1270
Fax: 0041-1-445 1271
E-mail: cogoli@spacebiol.ethz.ch

Christian Dournon, Prof. Dr.
Université Henri Poincaré
EA 3442 "Génétique, Signalisation, Différenciation"
Laboratoire de Biologie expérimentale-Immunologie
Boulevard des Aiguillettes/BP239
F 54506 Vandoeuvre-les-Nancy, France
Phone: 0033 (0) 383 68 42 50; Secretary's office: 42 51
Fax: 0033 (0) 383 90 32 77
E-mail: Christian.Dournon@scbiol.uhp-nancy.fr

Eberhard Horn, Prof. Dr.
Gravitational Physiology,
Department of Neurobiology
Univerisity of Ulm
Albert-Einstein-Allee 11
D 89081 Ulm, Germany
Phone: 0049 (0) 731 50 24565
Fax: 0049 (0) 731 50 24562
E-mail: eberhard.horn@biologie.uni-ulm.de

Akemi Izumi-Kurotani, Dr.
Space Utilization Research Center,
ISAS/Institute of Space and Astronautical Science
3-1-1 Yoshinodai, Sagamihara, Kanagawa 229-8510, Japan
Phone/Fax: 0081 42 759 8360
E-mail: kuro@pub.isas.ac.jp

Kenichi Ijiri, Prof. Dr.
Radioisotope Center, University of Tokyo
2-11-16 Yayoi, Bunkyo-ku, Tokyo 113-0032, Japan
Phone: 0081-3-5841-3054
Fax: 0081-3-5841-3049
E-mail: ijiri@ric.u-tokyo.ac.jp

Roberto Marco, Prof. M. D., D. Sc.
Departmento de Bioquimica de la UAM
Instituto de Investigaciones Biomedicas "Alberto Sols" CSIC-UAM
Facultad de Medicina de la Universidad Autonoma de Madrid
C/Arzobispo Morcillo, 4
E 28029 Madrid, Spain
Phone: 0034-91-3975409
Fax: 0034-91-5854587 & 34-91-3975353
E-mail: roberto.marco@uam.es

Hans-Jürg Marthy, Dr. (*Editor of Volume 9 of the ASBM series*)
CNRS (UMR 7628), UPMC
Observatoire Océanologique
F 66650 Banyuls sur mer, France
Phone: 0033 (0) 468 88 73 16
Fax: 0033 (0) 468 88 73 98
E-mail: marthy@obs-banyuls.fr

Mary. E. Musgrave, Prof. Dr.
Department of Plant Science
University of Connecticut
1376 Storrs Road, Unit 4067
Storrs, CT 06269
Phone: 001 860-486-2925
Fax: 001 860-486-0682
E-mail: mary.musgrave@uconn.edu

Claudia Ricci, Prof. Dr.
University of Milan
Department of Biology
Via Celoria 26
1 20133 Milano, Italy
Phone: 0039 02 50314803
Fax: 0039 02 50314802
E-mail: claudia.ricci@unimi.it

April E. Ronca, Dr.
Life Sciences Division
Gravitational Research Branch
NASA Ames Research Center (MS-261-3)
Moffett Field, CA 94035-1000
Phone: 650-604-2644
Fax: 650-604-0046
E-mail: April.E.Ronca@nasa.gov